Results and Problems in Cell Differentiation

A Series of Topical Volumes in Developmental Biology

20

Editors

W. Hennig, L. Nover, and U. Scheer

Results and Problems in Cell Differentiation

L. Nover (Ed.)

Plant Promoters and Transcription Factors

With 46 Figures and 13 Tables

Springer-Verlag

Berlin Heidelberg New York
London Paris Tokyo
Hong Kong Barcelona
Budapest

Professor Dr. Lutz Nover
Dept. of Molecular Cell Biology
University of Frankfurt Biocenter
Marie-Curie-Straße 9
D-60439 Frankfurt

ISBN 3-540-57288-0 Springer-Verlag Berlin Heidelberg New York
ISBN 0-387-57288-0 Springer-Verlag New York Berlin Heidelberg

Library of Congress Cataloging-in-Publication Data
Plant promoters and transcription factors / L. Nover.
p. cm. -- (Results and problems in cell differentiation ; 20)
Includes bibliographical references.
ISBN 3-540-57288-0 (Berlin : alk. paper). -- ISBN 0-387-57288-0 (New York : alk. paper)
1. Plant genetic regulation. 2. Plant cell differentiation. 3. Promoters (Genetics)
4. Genetic transcription. I. Nover, Lutz. II. Series.
QH607.R4 vol. 20 [QK981.4] 581.87'3223--dc20 93-29266 CIP

Typesetting: Best-set Typesetter Ltd, Hong Kong
31/3145- 5 4 3 2 1 – Printed an acid-free paper

Contents

A: Plant Transcription Systems

1 Structural Organization and Regulation of Transcription by RNA Polymerase I of Plant Nuclear Ribosomal RNA Genes

(With 8 Figures)

VERA HEMLEBEN AND ULRIKE ZENTGRAF

2 RNAPII: A Specific Target for the Cell Cycle Kinase Complex

(With 2 Figures and 1 Table)

LÁSZLÓ BAKÓ, SIRPA NUOTIO, DÉNES DUDITS, JEFF SCHELL, AND CSABA KONCZ

3 Plastid Differentiation: Organelle Promoters and Transcription Factors

(With 3 Figures and 2 Tables)

GERHARD LINK

4 AT-Rich Elements (ATREs) in the Promoter Regions of Nodulin and Other Higher Plant Genes: a Novel Class of *Cis*-Acting Regulatory Element?

(With 1 Figure and 2 Tables)

BRIAN G. FORDE

5 In Vitro Transcription of Plant Nuclear Genes

(With 1 Figure)

PATRICK SCHWEIZER

B: Promoters Regulated by Stress and Developmental Signals

6 Heat Stress Promoters and Transcription Factors

(With 10 Figures and 4 Tables)

KLAUS-DIETER SCHARF, TILO MATERNA, ECKARDT TREUTER, AND LUTZ NOVER

7 Regulatory Elements Governing Pathogenesis-Related (PR) Gene Expression

(With 2 Tables)

IMRE E. SOMSSICH

8 Analysis of Tissue-Specific Elements in the CaMV 35S Promoter

(With 5 Figures)

ERIC LAM

9 Analysis of Ocs-Element Enhancer Sequences and Their Binding Factors

(With 1 Figure)

KARAM B. SINGH, BEI ZHANG, SOMA B. NARASIMHULU, AND RHONDA C. FOLEY

10 Regulation of α-Zein Gene Expression During Maize Endosperm Development

(With 5 Figures)

MILO J. AUKERMAN AND ROBERT J. SCHMIDT

11 Control of Floral Organ Identity by Homeotic MADS-Box Transcription Factors

(With 5 Figures and 1 Table)

BRENDAN DAVIES AND ZSUZSANNA SCHWARZ-SOMMER

12 The *GL1* Gene and the Trichome Developmental Pathway in *Arabidopsis thaliana*

(With 5 Figures and 1 Table)

JOHN C. LARKIN, DAVID G. OPPENHEIMER, AND M. DAVID MARKS

A: Plant Transcription Systems

1 Structural Organization and Regulation of Transcription by RNA Polymerase I of Plant Nuclear Ribosomal RNA Genes

Vera Hemleben and Ulrike Zentgraf

1 Introduction

The RNA components of the plant cytoplasmic ribosomes consist of the 17/18S rRNA (ribosomal RNA) in the 40S ribosome subunit and the 5S, 5.8S, and 25S/26S rRNA in the 60S ribosome subunit. The corresponding genes for the 18S, 5.8S and 25S rRNA, encoded by the nuclear genome, are composed in transcription units which are located as rDNA (ribosomal DNA) repeats in the NOR (nucleolus organizing region) of the chromosome. As in higher animals, the genes for the 5S rRNA are localized separately at other regions in the genome (Hemleben and Grierson 1978). Coordinated regulation of the expression of the different components of the ribosomes can be expected since three RNA polymerases are involved to provide concomitantly the rRNA components and the mRNA for the ribosomal proteins (Sommerville 1986): RNA polymerase I (pol I) is responsible for the 18S–25S rRNA transcription, RNA polymerase III (pol III) produces the 5S rRNA, and the genes for the ribosomal proteins are transcribed by RNA polymerase II (pol II). The question how these different RNA polymerases are coordinately regulated is still a fascinating problem to solve for eukaryotic cells. Until recently, it was believed that the genes transcribed by pols I, II and III, respectively, utilize completely different sets of initiation factors; however, it has now become evident for animal cells that one common factor, the TATA-binding protein, plays a central role in transcription of all three RNA polymerases (for a review, see White and Jackson 1992).

Biosynthesis and processing of plant ribosomal RNA were extensively studied by pulse-chase in vivo labelling of the cell RNAs and gel electrophoresis of the RNA products (Rogers et al. 1970; Grierson and Loening 1972). Although the size of the rRNA precursor molecules is generally smaller (32S to 35S) than that determined for animals (45S), the processing steps resulting in the mature 18S, 5.8S, and 25S rRNA are assumed to occur principally in a similar manner (Rogers et al. 1970; Perry 1976; Rungger and Crippa 1977). The rate of rRNA synthesis can change enormously

Department of Genetics, Biological Institute, University of Tübingen, Auf der Morgenstelle 28, D-72076 Tübingen, FRG

Results and Problems in Cell Differentiation 20
L. Nover (Ed.)
Plant Promoters and Transcription Factors

depending on the developmental and physiological stage of the cells or the environmental conditions. For example, stimulation of rRNA synthesis is observed after light treatment of dark grown plants (Grierson and Loening 1974; Tobin and Silverthorne 1985) or by treatment with 2,4-D, an artificial auxin (Guilfoyle et al. 1975). These changes in rRNA synthesis might reflect changes in the rate of transcription of a given set of genes, however, an increase in rRNA synthesis might also be possible by activating new genes of the ribosomal RNA multigene family, thus raising their accessibility for transcription (Thompson et al. 1988; Baerson and Kaufman 1990). The most detailed biochemical studies on plant pol I carried out in the 1970s were reviewed by Jendrisak (1980); however, these investigations have to be extended to the molecular level.

In higher plants often a large percentage of the nuclear genome represents ribosomal DNA. Variable numbers (between 1000 to more than 30 000) of ribosomal RNA genes forming multigene families arranged in long tandem arrays were reported for a variety of plant species (Ingle et al. 1975; Hemleben et al. 1988). Only recently, it was found that, in contrast to the general assumption of the occurrence of long tandem arrays, the rDNA repeats of flax, although located at a single locus, are frequently interrupted by non-rDNA sequences (Agarwal et al. 1992). In an excellent three-dimensional electron microscope investigation of ribosomal chromatin the spatial organization of rDNA within the fibrillar centers (FC) of the nucleolus was demonstrated using a combined cytochemical, immunocytochemical, and in situ hybridization approach (Motte et al. 1991).

The rDNA repeats containing the 18S–25S rRNA genes exhibit an enormous length and sequence heterogeneity (for reviews, see Rogers and Bendich 1987; Hemleben et al. 1988). Since the coding regions for the 18S, 5.8S, and 25S rRNAs are highly conserved with respect to length and nucleotide sequences, these length and sequence heterogeneities mostly affect the intergenic spacer (IGS) region (*per definitionem* located between the 25S and 18S rRNA coding sequences; Fig. 1).

Plant species of the same genus already differ in nucleotide sequence between 60 and 95%, depending on the IGS region analyzed, e.g., *Vicia faba*, *V. angustifolia* and *V. hirsuta* (Kato et al. 1990; Ueki et al. 1992; Yakura and Nishikawa 1992) or *Cucurbita maxima* and *C. pepo* (Kelly and Siegel 1989; King et al. 1993). Comparing species of different genera within a plant family considerable sequence divergency is observed, e.g., for radish (*Raphanus sativus*), *Sinapis alba* and *Arabidopsis thaliana* (Delcasso-Tremousaygue et al. 1988; Rathgeber and Capesius 1990; Gruendler et al. 1991), *Lycopersicon esculentum* and *Solanum tuberosum* (Schmidt-Puchta et al. 1989; Perry and Palukatis 1990; Borisjuk and Hemleben 1993), or *Triticum aestivum* and *Zea mays* (Toloczyki and Feix 1986; Barker et al. 1988). Nearly no sequence similarity of the IGS (except for short motifs) among representatives of different plant families was found. Therefore, hybridization studies for restriction fragment length polymorphism (RFLP) analysis of closely related plants are possible using rDNA spacer fragments

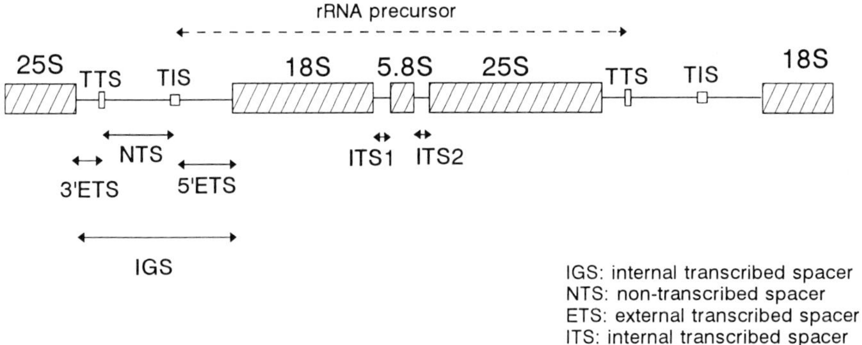

Fig. 1. Schematic drawing of two ribosomal RNA genes arranged in tandem. The 18S, 5.8S and 25S rRNA coding regions internally separated by the internal transcribed spacers, ITS1 and 2, are *boxed*. The intergenic spacer (*IGS*) is divided into the 3′ and 5′ external transcribed spacer separated by the non-transcribed spacer region. Transcription initiation and termination of the rRNA precursor are indicated by *arrows*

(Torres et al. 1989; Torres and Hemleben 1991); more distantly related plants can be only distinguished by RFLPs hybridizing with rDNA probes coding for the mature rRNAs (Hemleben et al. 1992).

The length of an rDNA repeating unit varies from approx. 8 kbp (e.g., *Raphanus sativus*; Delseny et al. 1983) to 14–17 kbp (*Trillium*, *Paris*; Martini et al. 1982; Yakura et al. 1983). Length heterogeneity of the repeats can occur already within an individuum or in different representatives of a species. The high variability in length and sequence of the IGS in plant rDNA offers the opportunity to differentiate even between varieties or cultivars of a species by RFLP analysis as already applied on wheat or barley populations and wild species (Appels and Dvorak 1982; Shagai-Maroof et al. 1984).

In the last decade nucleotide sequence data for the IGS of various higher plants from different plant families were collected and can be analyzed by computer analysis. These data together with recent studies on in vivo and in vitro transcription helped to elucidate the functional role of IGS sequences in the regulation of pol I transcription. Binding assays with IGS elements and crude nuclear or whole-cell extracts provided some evidence for the protein factors involved. In this chapter we will concentrate on the structural and functional organization of the 18S–25S rRNA genes and the role of rDNA methylation and chromatin structure; moreover, we will describe what is emerging from the literature and our own studies on *trans*-acting factors and *cis*-acting elements involved in the regulation of pol I transcription in plants.

2 Structural Organization of the rRNA Genes Transcribed by RNA Polymerase I

The large 18–25S rRNA precursor transcribed by RNA polymerase I, starting at the transcription initiation site(s) (TIS), is encoded by the 5′external transcribed spacer (5′ETS) followed by the 18S rRNA coding region, the 5.8S rDNA sequences – 5′ and 3′ flanked by the internal transcribed spacer 1 and 2 (ITS 1 and 2) – and the 25S rRNA coding region with 3′ETS located downstream (Fig. 1). Non-transcribed spacer (NTS) regions of different length were determined for several plant species (Barker et al. 1988; Delcasso-Tremousaygue et al. 1988; Gerstner et al. 1988; Schiebel et al. 1989; Vincentz and Flavell 1989; Zentgraf et al. 1990).

Various regulatory elements suggested to be involved in pol I transcription are located within the IGS region (Fig. 1). Functional elements are the pol I promoter with the TIS (+1) which is at least in dicots often preceded by an AT-rich region (Zentgraf et al. 1990), the transcription termination site (TTS), putative enhancer sequences binding to regulatory proteins, and processing sites for the rRNA precursor. Replication origins are also assumed to be located within the rDNA IGS as shown for pea rDNA (Hernandez et al. 1988).

A general feature of the IGS is the often complex organization of repeated elements (Fig. 2). Principally, each region of the IGS can be affected by amplification processes and appear as a repeated element within the spacer. In some plants, sequences with similarity to a consensus TIS are duplicated (Gerstner et al. 1988; Kelly and Siegel 1989; Gruendler et al. 1991). Repeated elements characterized by a putative transcriptional terminator site (TTS) for pol I were found in mung bean (*Vigna radiata*; Schiebel et al. 1989); enhancer-like repeated elements are located upstream of the TIS in wheat (*Triticum aestivum*; Flavell et al. 1986; Barker et al. 1988), maize (*Zea mays*; Toloczyki and Feix 1986; Schmitz et al. 1989) and probably occur also in the rDNA of other plants (see Fig. 2). Obviously, even part of the coding regions can be amplified as shown for the cucumber (*Cucumis sativus*) IGS where the 3′ end of the 25S rDNA and 3′ flanking sequences, including the TTS, were duplicated (Ganal et al. 1988; Zentgraf et al. 1990). Mechanisms resulting in the amplification of sequences are considered to be mostly an unequal crossover during recombination occurring within a multigene family (Dover 1986; Rogers et al. 1986; Ganal et al. 1988).

Different numbers of repeated elements in the IGS result in heterogeneous repeat lengths frequently observed in various plants (data compiled in Rogers and Bendich 1987; Hemleben et al. 1988). Remarkably, in plant species with conserved rDNA repeat lengths, difference in length in one region of the IGS is compensated by a different number of subrepeats in the other region; e.g. this phenomenon was observed in a comparison of tomato and potato IGS, both with a repeat length of approx. 9 kbp (Fig. 2; Schmidt-

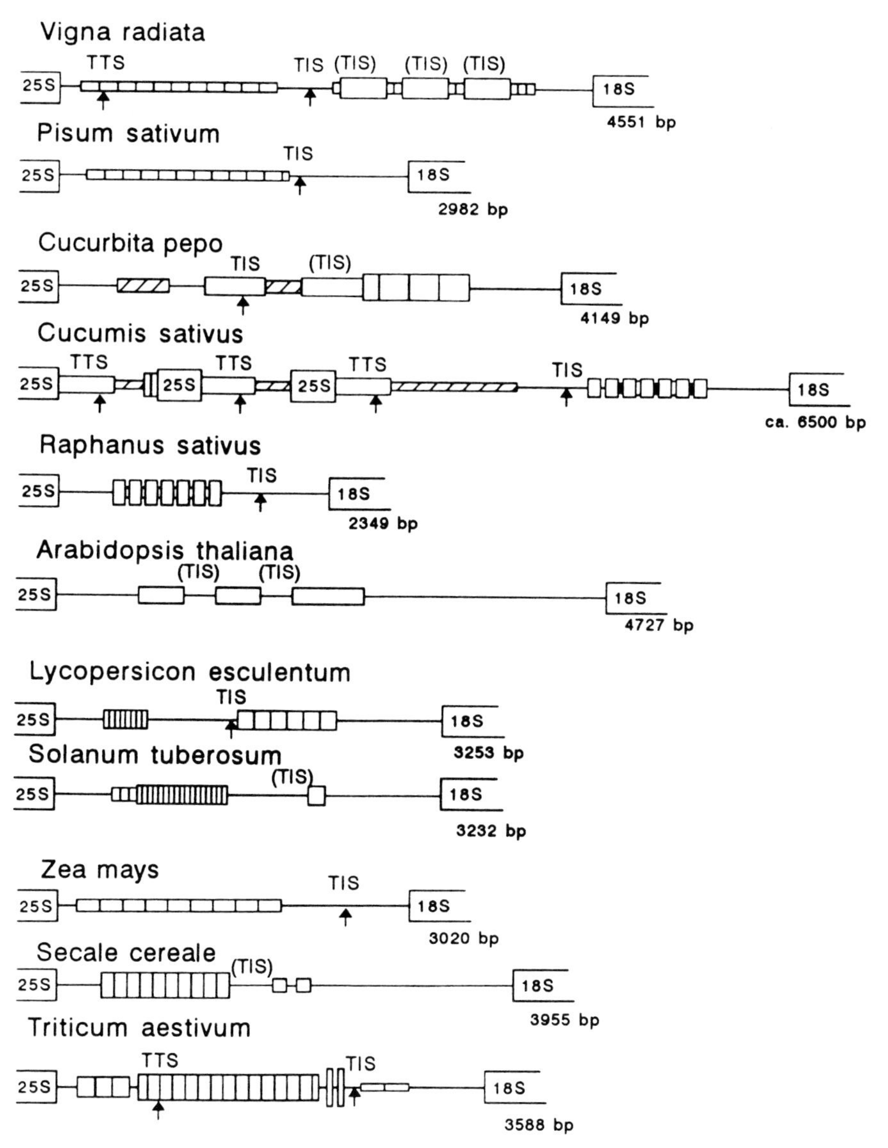

Fig. 2. Schematic drawing of the intergenic spacer (*IGS*) region of a representative of the rDNA repeats of several higher plants. The transcription initiation site (*TIS*) and the transcription termination site (*TTS*) are marked by an *arrow* if they are determined by S1 or mung bean nuclease or by primer extension mapping; shown in *parentheses* are TIS determined by a computer search of the consensus sequence. The size of the IGS is indicated by the numbers of base pairs. References: *Vigna radiata* (Gerstner et al. 1988; Schiebel et al. 1989), *Pisum sativum* (Kato et al. 1990), *Cucurbita pepo* (King et al. 1993), *Cucumis sativus* (Ganal et al. 1988; Zentgraf et al. 1990), *Raphanus sativus* (Delcasso-Tremousaygue et al. 1988), *Arabidopsis thaliana* (Gruendler et al. 1991), *Lycopersicon esculentum* (Schmidt-Puchta et al. 1989; Perry and Palukatis 1990), *Solanum tuberosum* (Borisjuk and Hemleben 1993), *Zea mays* (Toloczyki and Feix 1986), *Secale cereale* (Appels et al. 1986), *Triticum aestivum* (Barker et al. 1988; Vincentz and Flavell 1989)

Puchta et al. 1989; Perry and Palukatis 1990; Boriskjuk and Hemleben 1993).

Whereas the IGS can be highly heterogeneous in length, the size of the 18–25S RNA coding regions, including the ITS 1 and 2, appears relatively constant, mostly approx. 6 kbp. The lengths of the ITS 1 and 2 sequences flanking the 5.8S rRNA coding region were determined in various plant species to be approx. 180 to 250 bp (Torres et al. 1990). Sequence diversity is already observed between related species, however, there seem to be some constraints on these sequences, which are able to form secondary structures, probably with functional relevance for processing. Of particular interest is the observation that the GC content of ITS 1 and 2, compared with another, although variable among species of a given genus, is relatively equal. GC balance, therefore, appears to be a general feature of ITS 1 and 2 (Torres et al. 1990). Possibly, the 5′ ETS singular region reflects a similar behaviour since in the region preceding the 18S rRNA coding sequences, the GC content resembles the percentage of G+C found in ITS 1 and 2 as shown for cucumber and zucchini (King et al. 1993). In contrast, the ETS repeats 5′upstream of the singular ETS region of both plants (see Fig. 2) are characterized by a rather high and similar T+G content, suggesting a functional role of these subrepeats (see Sect. 5.2).

3 Promoter Structure

For several plant rDNAs, the TIS of the rRNA precursor was recognized by S1 or mung bean nuclease or primer extension mapping (Toloczyki and Feix 1986; Delcasso-Tremousaygue et al. 1988; Gerstner et al. 1988; Vincentz and Flavell 1989; Kato et al. 1990; Zentgraf et al. 1990). The sequences directly surrounding the TIS are rather conserved; therefore, this region can now be detected in the IGS sequence by computer search. However, the promoter region located 5′ upstream is only conserved among closely related species (Zentgraf et al. 1990; Gruendler et al. 1991; King et al. 1993); in fact, computer analysis of this region shows approx. 60 to 70% similarity among species of different families, but short common motifs occur which appear to be necessary for specific protein binding (Echeverria et al. 1992; Jackson and Flavell 1992; Nakajima et al. 1992; Zentgraf and Hemleben 1992).

In a newly developed in vitro transcription assay with a crude whole-cell extract from embryonic axes of *Vicia faba* and rDNA fragments of the putative promoter region already determined by S1 nuclease mapping (Kato et al. 1990), correct initiation of pol I transcription was demonstrated. Some longer transcripts were observed starting further upstream of the consensus TIS (Yamashita et al. 1993).

For rRNA genes transcribed by pol I in animal cells a common picture was obtained (reviewed by Sollner-Webb and Mougey 1991): transcription of pol I is mediated by *trans*-acting protein factors binding at a core promoter (CP) preceding the TIS [approx. nucleotides (nts) −45 to +5] and reacting with an upstream control element (UCE) or upstream binding element (UBE; approx. −160 to −110; Sollner-Webb and Tower 1986). Although it was assumed for a long time that the transcription apparatus of pol I is species-specific (Grummt et al. 1982; Miesfeld and Arnheim 1984) and sequence divergency is observed in the promoter region, recent studies showed that the factors binding to UCE can substitute for each other in footprinting assays between human and *Xenopus* (Bell et al. 1989; Pikarrd et al. 1989). For plant cells it was of interest whether *trans*-acting proteins involved in the regulation of rRNA transcription form specific complexes with *cis* regions of the putative promoter (see Sect. 5.2).

As already mentioned, in different plants the transcription initiation site (TIS) previously determined by sequence comparisons with a consensus sequence, and later confirmed by functional analysis by S1 or mung bean nuclease mapping and in vitro transcription assays (Yamashita et al. 1993), is rather similar. Interestingly, some dicots contain several TIS or TIS-like sequences within the IGS (see Fig. 2). For *Cucurbita pepo* it was shown that actually the first (5′ uptream) TIS is prominently used for transcription as investigated by mung bean nuclease mapping experiments (K. King and V. Hemleben unpubl. results). For the duplicated TIS in *Arabidopsis* rDNA the situation is not yet clear; Gruendler et al. (1991) assumed that the two upstream located TIS were "spacer promoters", however, this assumption is made only in analogy to animal rDNA for which the presence of spacer promoters was predominantly described (DeWinter and Moss 1987; Tautz et al. 1987).

Many IGS of dicotyledonous plants investigated so far possess a longer AT-rich sequence element in front of the pol I promoter (Zentgraf et al. 1990). Remarkably, in this region, sequence elements are located with similarity to ARS sequences (autonomous replicating sequences) characterized for years (Broad et al. 1983), which might indicate that replication origins are present in every rDNA repeat. Functional replication origins in the neighborhood of ARS-like sequences (consensus sequence: -A/TTTTATA/GTTTA/T-) have been demonstrated by studies on pea rDNA (Hernandez et al. 1988; Van't Hof and Lamm 1992).

4 Methylation and Chromatin Structure of rDNA

Methylation of DNA in plants at CpG and CpCpG or CpNpG motifs has been repeatedly considered to play a role in the regulation of transcription of nuclear ribosomal RNA genes (Hemleben et al. 1982; Leweke and Hemleben 1982; Flavell et al. 1983, 1988; Scott et al. 1984). In plants with a relatively large number of rDNA repeats a high percentage of the rDNA appears completely methylated at those sites and, in addition, these repeats are probably packaged in a highly condensed and transcriptionally inactive heterochromatic stage. Most dramatically, this phenomenon was observed in various additional lines of wheat carrying a NOR-bearing chromosome of *Aegilops umbellulata*; here, the rDNA repeats in the NOR of the wheat genome were highly methylated and the nucleoli were much smaller than in the parent wheat line lacking the *Aegilops* chromosome (Martini et al. 1982); nucleolar dominance of the rDNA of *Aegilops* and suppression of the wheat rRNA genes were explained by a higher number of subrepeats with enhancer functions in the IGS of *Aegilops* (Flavell et al. 1986), but it might also be influenced by other factors. Moreover, Flavell et al. (1983, 1988) could show that in aneuploid lines of wheat, which include different doses of chromosomes carrying the NORs and which therefore vary in the number rRNA genes (between 3700 and 14000), the proportion of DNA not digestable with the methylation-sensitive restriction enzyme HpaII increased with an increasing amount of rDNA.

In studies on wheat and pea (*Pisum sativum*) rDNA a correlation between DNA methylation and DNase I hypersensitivity as an indication for a differential chromatin structure along the rDNA stretches was proved (Thompson et al. 1988). Interestingly, a relationship between light stimulation of transcription and use of different rDNA length repeat classes could be shown (Kaufman et al. 1987). Length heterogeneity of the rDNA repeats of pea (*Pisum sativum* var. Alaska) originates from a different number approx. 180-bp subrepeats upstream of the TIS (Fig. 2). The longer rDNA repeats (L-variants 9 kbp in length) comprising 74% of rDNA are located on chromosome 7, the shorter repeats (S-variants 8.6 kbp in length) are localized on chromosome 4 (Ellis et al. 1984; Polans et al. 1986). L-variants are probably preferentially transcribed in the dark and in light-grown seedlings indicated by specific, hypersensitive DNase I sites near the promoter region; after light treatment the shorter repeats started to be transcribed and became sensitive to DNase I in this region, correlating with the occurrence of a demethylated HpaII site localized 31 bp downstream of the TIS (Piller et al. 1990).

Furthermore, detailed DNA methylation studies were carried out with members of the Cucurbitaceae generally equipped with a large number of ribosomal DNA (Ingle et al. 1975; Hemleben et al. 1988). *Cucurbita pepo* (zucchini) contains approx. 10000 rDNA repeats per somatic cell; at least 70% of the repeats is completely methylated which was demonstrated after digestion of total DNA with the respective methylation-sensitive restriction endonucleases (Fig. 3). A fraction of approx. 3–5% exhibited only one accessible restriction site per repeat irrespective of the number of restriction sites actually present in the repeats and the methylation-sensitive enzymes used. Another small percentage of the repeats appeared completely unmethylated; these are proposed to represent the actively transcribed portion of rDNA (R.A. Torres and V. Hemleben, in prep.). A similar behaviour of the rDNA repeats was observed for cucumber (*Cucumis sativus*) with an

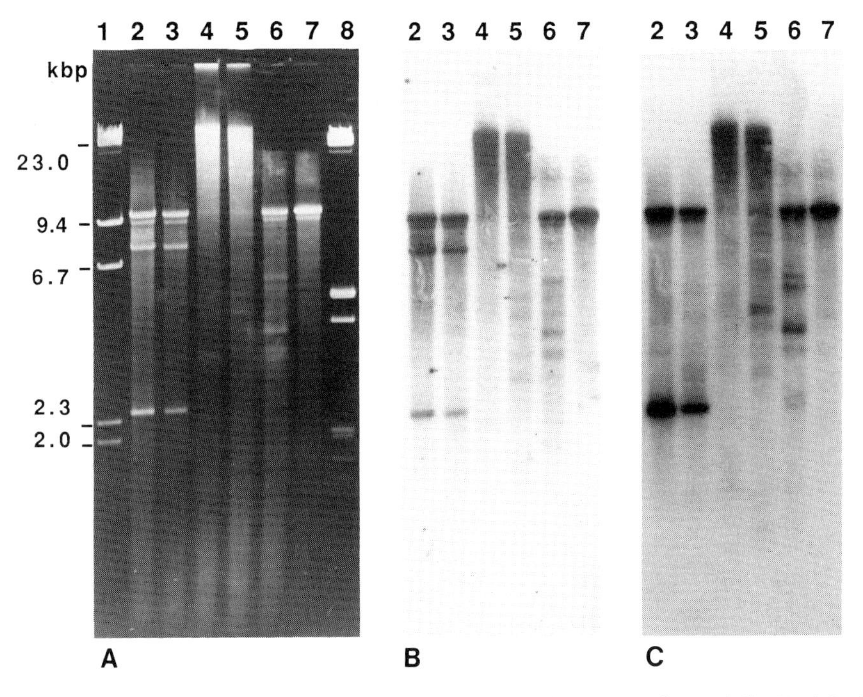

Fig. 3. Methylation analysis of the rDNA of *Cucurbita pepo* (zucchini). Enriched nuclear rDNA was restricted with restriction endonucleases in single or double digests. The DNA was separated on a 1% agarose gel (**A**), Southern-blotted and hybridized to ^{32}P-labelled 18S and 25S rRNA (**B**) or to a ^{32}P-labelled 18S rDNA fragment (**C**). *Lane 1* lambda DNA, HindIII digested; *lane 8* lambda DNA, EcoRI digested; *lanes 2–7* zucchini rDNA digested with HindIII (*lane 2*), HindIII/HpaII (*lane 3*), HpaII (*lane 4*), CfoI (*lane 5*), CfoI/EcoRV (*lane 6*); EcoRV (*lane 7*). Note that a few distinct bands are visible cutting the DNA with the methylation-sensitive restriction enzymes HpaII (CCGG) and CfoI (GCGC); most of the rDNA remained undigested with HpaII and CfoI

even higher amount of rRNA genes (Ganal et al. 1988; Hemleben et al. 1988).

Previous reports that a specific demethylated HpaII site in front of the TIS is a general feature of all actively transcribed repeats (von Kalm et al. 1986; Flavell et al. 1988; Hemleben et al. 1988) are probably not valid for all plants since such a site is not always present in this region. However, studies on *C. pepo* rDNA showed that a specific fraction of rDNA repeats exhibits one randomly distributed demethylated restriction enzyme site irrespective of the methylation-sensitive enzyme tested. Thus, if an HpaII site is present in the promoter region, such a site will be randomly accessible for this restriction enzyme, possibly indicating that these repeats are prepared to be transcribed.

5 Functional Role of Repeated Elements Within the Intergenic Spacer (IGS)

The phenomenon of nucleolar dominance in partial hybrids has already provided some evidence for the functional role of repeated elements in the IGS of plant ribosomal DNA. These observations were further supported by investigations on *Triticale* partial hybrids and by in vitro transcription assays. Moreover, protein binding studies with rDNA fragments of the IGS and nuclear proteins have elucidated the functional role of rDNA subrepeats.

5.1 Transcription-Enhancing Elements Characterized by in Vivo and in Vitro Studies

In vivo, with *Triticum/Aegilops* partial hybrids, it could be demonstrated that only the *Aegilops umbellulata* rDNA repeats, containing a higher number of subrepeats upstream of the TIS supposed to fulfil enhancer functions, were transcriptionally active and formed a prominent nucleolus structure. In contrast, the rRNA genes located on the wheat chromosomes with less subrepeats within the IGS appeared relatively inactive (Flavell et al. 1986). Similar effects could also be observed with wheat-rye hybrids characterized by Appels et al. (1986): microscope observation indicated that the rye nucleolus was largely suppressed in the presence of a wheat NOR locus. In molecular studies it could be shown that the wheat rDNA repeats predominantly contributed to transcriptional activity, whereas the rye rDNA repeats appeared to be repressed (Capesius and Appels 1989). In fact, wheat and rye rDNA subrepeats in the IGS share regions of sequence similarity (Appels and Dvorak 1982; Appels et al. 1986), suggesting a functional role as protein binding sites.

In mung bean (*Vigna radiata*) length heterogenous rDNA repeats are observed differing mostly in the number of complex 5′ ETS subrepeats containing a TIS-like sequence and in the different numbers of 174-bp subrepeats in the 3′ IGS (Gerstner et al. 1988; Schiebel et al. 1989). Only cloned rDNA from longer repeats containing three ETS subrepeats subsequent to the first TIS (see Figs. 2 and 4) was transcribed using in vitro transcription assays with isolated nuclei, whereas with rDNA constructs containing the corresponding region of the shorter rDNA variants with two ETS subrepeats nearly no transcription could be observed (Fig. 4). This result was confirmed by transient expression experiments: the cloned rDNA sequences were introduced into protoplasts of mung bean by electroporation and newly transcribed RNA products of the template were measured. Again, only the rDNA constructs containing more ETS repeats were preferentially transcribed, indicating that subrepeats within the 5′ ETS might have some stimulatory effect on transcription in vivo (K. Unfried and V. Hemleben unpubl. results).

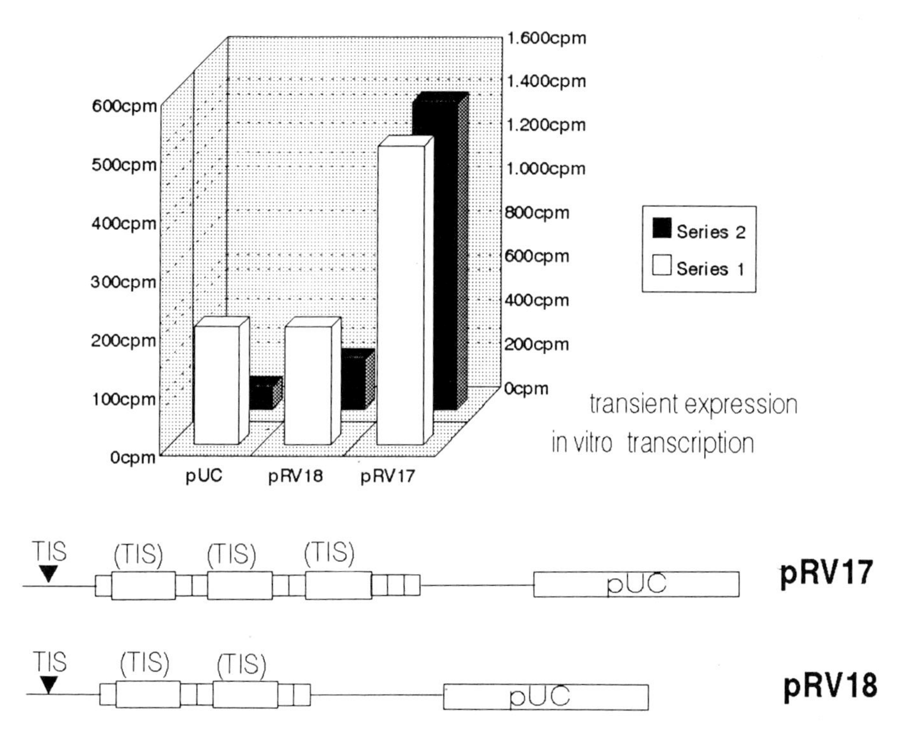

Fig. 4. In vitro transcription and transient expression assays with cloned rDNA fragments of mung bean (*Vigna radiata*). The clone *pRV17*, containing three complex ETS subrepeats, and the clone *pRV18*, containing two ETS subrepeats in addition to the main promoter region (Gerstner et al. 1988) cloned into a pUC vector, were applied to in vitro transcription with isolated nuclei (*series 1*, *white boxes*) or to transient expression assays in protoplasts of mung bean after electroporation (*series 2*, *black boxes*) in the presence of a low concentration of alpha-amanitin (100 µg/ml) which prevented pol II and pol III transcription. The transcription products were blotted onto nitrocellulose and hydridized with ^{32}P-labelled pUC to estimate the amount of newly transcribed RNA after measuring the radioactivity (cpm). A pUC vector was used as control. (K. Unfried and V. Hemleben unpubl.)

The question arose whether subrepeats located upstream *or* downstream of the TIS can fulfil a similar enhancer-like function. For instance, as already mentioned, the length of the potato and tomato IGS is nearly the same (approx. 9 kbp); however, the location and the number of subrepeats in the IGS are different (see Fig. 2). The tomato IGS shows shorter and fewer subrepeats upstream of the TIS, but the difference in length is compensated by another type of subrepeat in the 5′ ETS, where in potato only one of those sequences is present (Borisjuk and Hemleben 1993). Now it will be of interest to see whether a functional compensation of the different repeat types is possible.

5.2 Protein Binding Studies with Promoter Fragments and Repeated Elements Located Upstream or Downstream of the Transcription Initiation Site

The functional significance of a sequence element can be tested by studying DNA-protein interactions, assuming that specific nuclear *trans*-acting proteins are involved in transcriptional regulation of plant rDNA. The first studies reported have been carried out with sequence elements of maize IGS (Schmitz et al. 1989). Specific protein binding was observed at the suggested promoter region and also with repeated elements 5′ upstream of the TIS (see Fig. 2). These findings led to a model showing that sequences in front of the promoter attract proteins similar to the promoter itself, thus having a stimulatory effect on transcription. DNA competition assays on wheat rDNA have shown that subrepeat A, containing fragments located upstream of the TIS, competes with specific promoter fragments for similar proteins (Jackson and Flavell 1992) confirming the enhancer-like function of these repeats as proposed earlier (Flavell et al. 1986).

Echeverria et al. (1992) investigated the capacity of repeated elements upstream of the TIS in radish rDNA (see Fig. 2) to form specific complexes with nuclear proteins, assuming that these subrepeats play an enhancing role

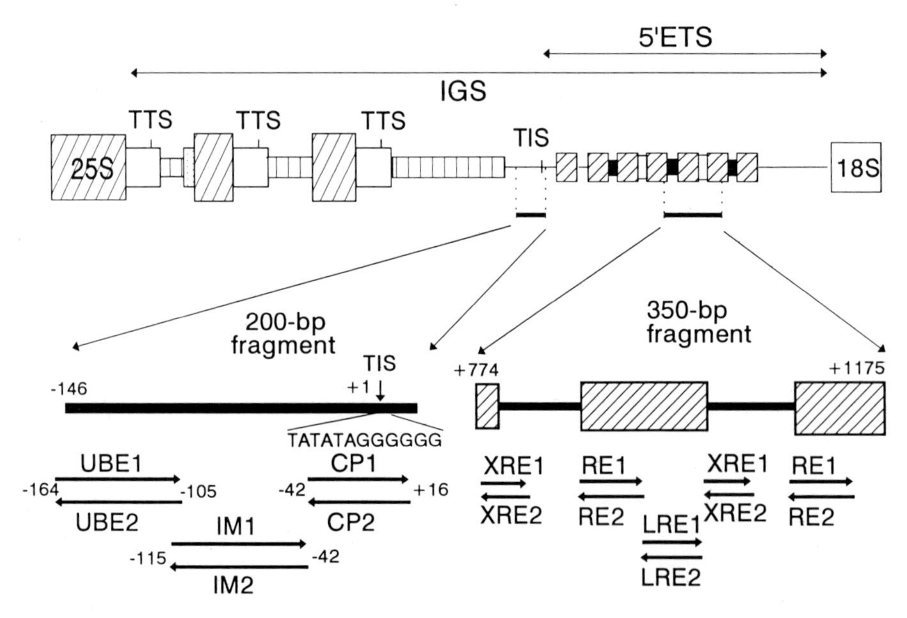

Fig. 5. The intergenic spacer of a rDNA repeat of cucumber. Enlarged are the fragments (or oligonucleotides) used in DNA-protein binding assays: 200-bp promoter fragment containing the TIS and 350-bp fragment containing repeated elements of the 5′ ETS. The *arrows* indicate the oligonucleotides used for protein binding and competition studies in the single- and double-stranded form. *UBE* upstream binding element; *CP* core promoter; *LRE* loop-forming element; *XRE* *Xenopus*-like repeated element; *RE* repeated element

in transcription. Actually, specific protein binding could be demonstrated with the dA/dT-rich regions between the subrepeats and with the dA/dT-rich sequences in the promoter region as proved by competition gel mobility shift and footprinting assays. A functional role of this DNA-protein interaction is as yet unknown.

More detailed information on nuclear proteins specifically recognizing functional rDNA spacer elements is now available for cucumber (*Cucumis sativus*) rDNA (Zentgraf and Hemleben 1992). Preliminary studies have revealed that sequences surrounding the TIS and repeated elements present in the 5′ ETS (Fig. 5) attract similar proteins as shown by competition assays (Fig. 6). Dissecting the promoter region into smaller fragments, which were then applied to gel mobility shift assays with 0.42 M salt nuclear protein extracts from cucumber cotyledons, revealed that an upstream binding element (UBE) and a core promoter (CP) sequence bind the same proteins shown by competition assays. An intermediate sequence (IM) between these regions does not react with any nuclear protein. Remarkably, the DNA-protein interactions obtained with UBE or CP could compete with an LRE ("loop-forming" repeated element, since secondary structures can be formed; Zentgraf et al. 1990) of the 5′ ETS subrepeats. Another part of the ETS subrepeats, called XRE ("*Xenopus*-like" repeated element, containing a sequence motif with high similarity to terminator sequences characterized

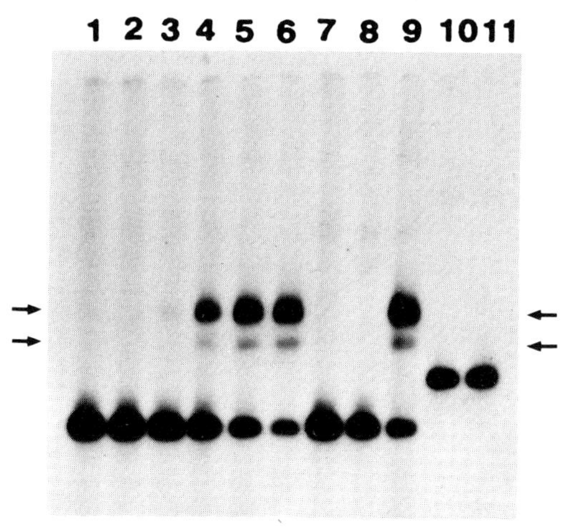

Fig. 6. Gel mobility shift assays with rDNA fragments and a crude nuclear protein extract of cucumber. A ^{32}P-labelled putative promoter fragment spanning from −146 to +41 (see Fig. 5) including the TIS was incubated with increasing amounts of a 0.42 M salt nuclear extract: without nuclear proteins (*lane 1*), 5 µg (*lane 2*), 10 µg (*lane 3*), 20 µg (*lane 4*), 30 µg (*lane 5*), 50 µg (*lane 6*); *lanes 7–9* 200-bp promoter fragment with 50 µg nuclear proteins and a 100-fold excess of unlabelled 200-bp fragment (*lane 7*), with a 100-fold excess of a 350-bp ETS fragment (*lane 8*), and a 100-fold excess of an unspecific 123-bp ladder DNA (*lane 9*). As control a pUC fragment was incubated without (*lane 10*) or with 50 µg nuclear proteins (*lane 11*). The DNA-protein complexes formed are indicated by *arrows*. (Zentgraf and Hemleben 1992)

for *Xenopus* rDNA; Labhart and Reeder 1987; Schiebel et al. 1989), also reacts with nuclear proteins but does not interfere with these promoter complexes. Fragments used in these assays are shown in Fig. 5. The striking fact found in these studies was that for all sequence elements investigated the interactions with nuclear proteins occurred also with DNA fragments in the single-stranded form (Zentgraf and Hemleben 1993).

Possibly, higher order structures within the elements rather than a defined sequence play a role in these interactions, although short sequence motifs may be important (Jackson and Flavell 1992; Zentgraf and Hemleben 1992). For instance, the LRE of cucumber contains a short motif -TGTGGGT- which is present in a corresponding region of the ETS of several higher plants (Gerstner et al. 1988; Schmidt-Puchta et al. 1989).

The proteins reacting with rDNA elements of cucumber were characterized by affinity column chromatography, South-Western blots and after eluting the protein component of a DNA-protein complex in gel mobility shift assays (Zentgraf and Hemleben 1992): three proteins with the molecular masses of 16, 22 and 24 kDa are predominantly involved in the binding at the UBE and CP of the pol I promoter; only the 16-kDa protein binds directly to the LRE of the ETS subrepeats, whereas XRE reacts with a completely different protein of 70 kDa. Again, the function of these proteins can be only speculated.

The 16-kDa protein might be analogous to the p16 protein of rat which is a "HMG-like protein" binding the upstream activating sequences and a segment in the ETS (+352 to +545; Yang-Yen and Rothblum 1988). This correlates very well with the size and the binding of the 16-kDa protein of *Cucumis sativus* to the repeated elements of the ETS, but, in contrast to rat, this protein also shows a clearly detectable affinity to the promoter sequences. Since most proteins characterized for cucumber have a lower molecular mass than those detected in animal systems, smaller proteins in plants, particularly in cucumber, could transmit the contact of other transcription factors and RNA polymerase I to the DNA. South-Western experiments for maize indicated that in contrast to cucumber more proteins with DNA-binding affinity including several smaller ones are involved in complex formation with the RNA polymerase I promoter. This situation could be compared to the species-specific transcription factor SL1 which has no affinity to the DNA itself in humans (only in combination with UBF) but shows DNA binding in rat and mouse (Bell et al. 1990).

Nevertheless, these small proteins might be "HMG-like" proteins. HMG 1 and HMG 2 subgroups (M_r aproximately 25 kDa) are comparable in size and also bind to double- and single-stranded DNA. In general, HMG proteins, including plant HMG proteins as well, are associated with transcriptional active chromatin (Spiker et al. 1983) and might play a role in transcriptional activation processes of RNA polymerase II in plants. Obviously, p16 of rat and the HMG homology boxes of different animal UBF reveal that "HMG-like" proteins seem to be involved in activating RNA polymerase I transcription.

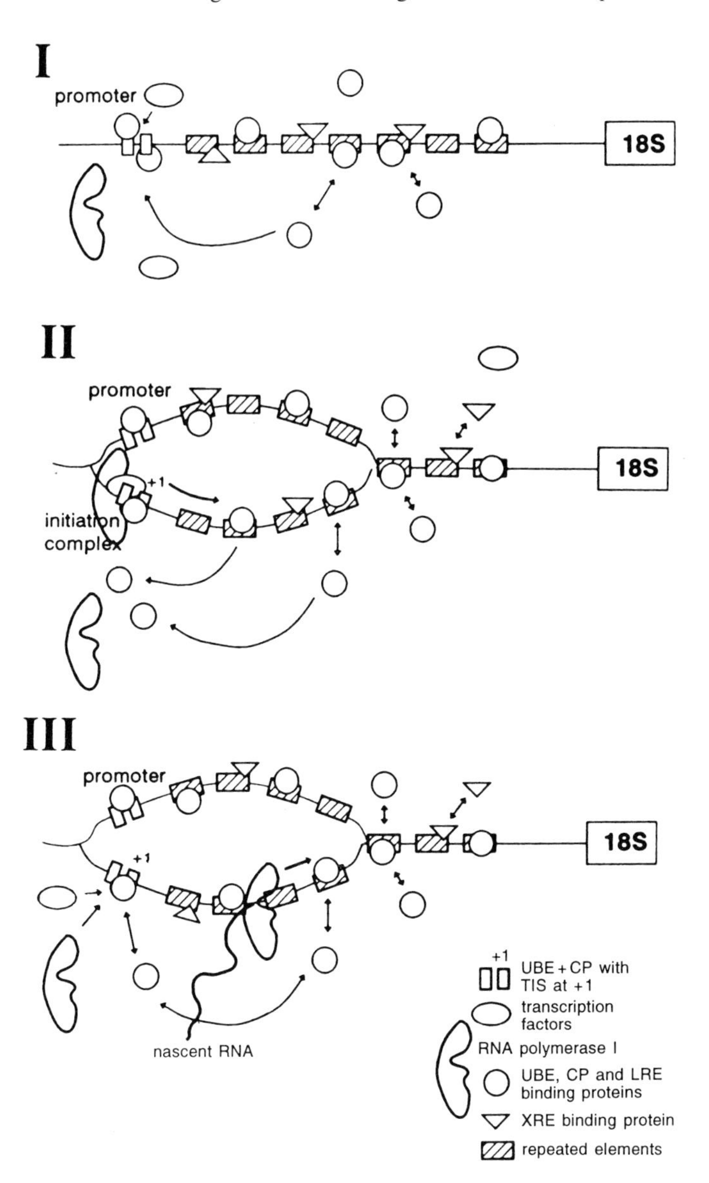

Fig. 7. A model demonstrating how the initiation of RNA polymerase I transcription in cucumber might occur. **I** DNA-binding proteins interact with the promoter and the ETS repeats in a double-stranded stage and are enriched in this nuclear compartment. **II** The complementary strands are separated upstream of UBE, the DNA-binding proteins can bind to the single-stranded DNA form of the respective binding sites, additional transcription factors interact with the promoter region and mediate the contact of RNA polymerase I to the DNA to form the transcription initiation complex. **III** DNA-binding proteins can bind again to the single-stranded DNAs after the RNA polymerase I has passed into the elongation process; they can keep the transcription fork open, thus enhancing the next transcriptional round

The possible role of these DNA-protein interactions in transcriptional regulation has been shown in a model (Fig. 7), illustrating how specific proteins might interact and fulfil a stimulatory function in the transcription process: the complementary DNA strands are separated at an AT-rich region upstream of the promoter. Small proteins previously binding at the ETS subrepeats downstream of the TIS can now bind to the single-stranded UBE and CP region of the promoter, and interact with pol I and possibly with other transcription factors, directing the polymerase to the starting point. During the elongation process, these proteins could keep the DNA template accessible to transcription.

Summarizing the results on in vitro protein binding studies on maize, wheat, radish and cucumber rDNA, it can be concluded that repeated elements located upstream or downstream of the TIS could be involved in stimulating the transcription initiation process. However, more efficient in vivo and in vitro transcription studies are necessary in plant systems.

The questions of how and where termination of transcription occurs and whether termination of pol I transcription is coupled with the initiation process in a "read through enhancement" mechanism, as proposed for animal rDNA (Mitchelson and Moss 1987), are still not completely solved as yet. However, some evidence was obtained by localizing the TTS in the IGS of plant rDNA (see Fig. 2).

6 Termination of RNA Polymerase I Transcription

Transcriptional termination of the rRNA precursor has been determined for radish, mung bean, cucumber, zucchini and wheat by S1 or mung bean nuclease mapping (Delcasso-Tremousaygue et al. 1988; Schiebel et al. 1989; Vincentz and Flavell 1989; Zentgraf et al. 1990; K. King and V. Hemleben unpubl. results). A consensus sequence as observed for the TIS cannot be defined, although for mung bean, a region with 70% similarity to a terminator sequence in *Xenopus* rDNA was found near the termination signal present in earch of the 174-bp repeats in the IGS (Schiebel et al. 1989). Whether this terminator acts concomitantly as enhancer of the transcription initiation process as it was shown for the T3-box (-GACTTGC-box) in *Xenopus* (Labhart and Reeder 1987) awaits further functional analysis. For mung bean a model of a higher order stem-loop structure that would support a "read through enhancement" was proposed (Fig. 8): the presence of inverted sequences within each of the 174-bp subrepeats allows the formation of a complex, higher order structure in the 3′ ETS which brings the transcription termination site into closer proximity of the TIS, thus facilitating initiation of transcription of RNA polymerase I at the subsequent TIS.

Sequences with similarity to the -GACTTGC- box are found in the IGS of other plants, e.g., in the ETS repeats of cucumber (XRE; see Fig. 5),

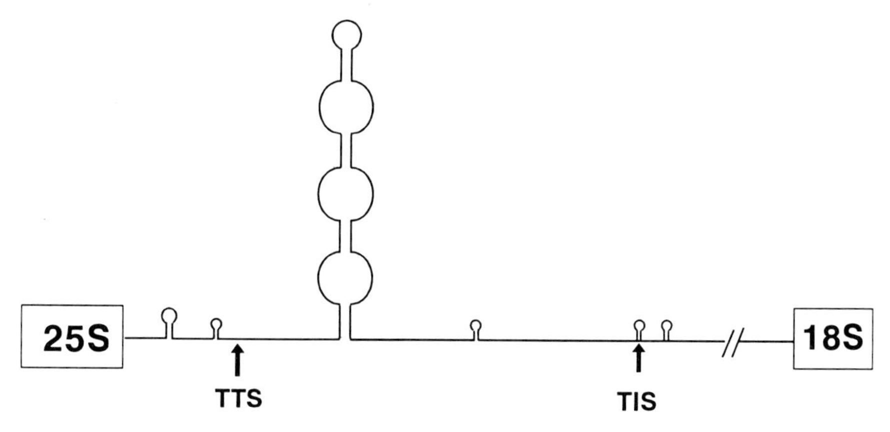

Fig. 8. Possible secondary structure in the intergenic spacer of mung bean (*Vigna radiata*). The occurrence of inverted repeats in each of the 175-bp spacer subrepeats allows the formation of a complex stem-loop structure which brings the TTS into close proximity of the TIS. Small stem loops indicate a processing site and the secondary structures possibly formed near the TTS and TIS, respectively. (Schiebel et al. 1989)

which are suggested to take part in transcriptional regulation (Zentgraf and Hemleben 1992). In other *Vigna* species, however, the 174-bp repeated elements also present in the 3′ IGS region do not contain a perfectly similar -GACTTGC- box, leaving only the possibility that the formation of a stem-loop structure might be of functional relevance for transcriptional termination (Unfried et al. 1991).

Several distinct termination signals were also detected in the 3′ IGS of cucumber which could be correlated with the stem-loop secondary structures probably formed (Zentgraf et al. 1990), since in this region of the IGS duplications of the 3′ end of the 25S rDNA and, subsequently, flanking sequences occur (Ganal et al. 1988). Obviously, in cucumber cells rRNA precursor molecules of different length are synthesized and later processed to the mature rRNAs. Different precursor lengths can be expected also for those plants which exhibit different numbers of 5′ETS subrepeats in their rDNA length variants as shown to occur in mung bean (Gerstner et al. 1988).

7 Conclusions

Detailed analysis of the nucleotide sequence, structure and organization of the ribosomal RNA genes of various higher plants indicated that, in higher plants, a remarkable high variability in length and sequence of the IGS appears. Among representatives of different plant families only the sequences around the TIS show some similarities; in addition, some common motifs occur.

The pol I promoter in higher plants apparently can be divided into a core promoter (CP; between approx. +5 to −64) and an upstream control element (UCE; between approx. −120 to −160), as shown by protein binding studies on cucumber and wheat which suggests that the pol I promoter is similarly organized as in animal cells (see Sollner-Webb and Mougey 1991). The results obtained with protein binding experiments on radish rDNA (Echeverria et al. 1992) demonstrate that specific protein binding is also obtained with a dA/dT-rich region of the promoter. Characterization of the proteins revealed that in cucumber mainly three small proteins, 16, 22 and 24 kDa in size, are involved in binding to the promoter region and to repeats located downstream of the ETS (Zentgraf and Hemleben 1992) and that in maize, several proteins, including some smaller ones, bind to the promoter region and to repeated elements located upstream (Schmitz et al. 1989).

Amplification and recombination processes can affect practically every region of the IGS, resulting in subrepeats which are differently organized; some of them are probably used as transcriptional control elements or as processing signals. The suggestion previously proposed that subrepeats in front of the TIS are used to attract protein factors responsible for the enhancement of transcription could now be extended. Functional analysis by in vivo and in vitro studies gave some evidence that subrepeats upstream *and* downstream of the TIS, respectively, can be involved in transcriptional regulation of pol I. These results led to models suggesting that repeated *cis*-elements can support transcription initiation by interaction with *trans*-factors independent of their localization with respect to the promoter region, therewith following the requirements of enhancer sequences.

More detailed analyses are necessary to prove the importance and functional relevance of short motifs occurring throughout the IGS with similarity to sequences known to be involved in tissue or stimulus specificity of pol II transcription (Weising and Kahl 1990). Transcription of rRNA by pol I can be developmentally or exogenously controlled by up- or down-regulation (during development, hormone control, light stimulation, starvation or heat shock conditions, to mention only some possibilities). Therefore, it can be expected that in some cases coregulation occurs between pol I and pol II (probably also pol III) transcription mediated by common *trans*-acting factors which are concomitantly interacting with *cis*-elements located in the rDNA IGS and in front of pol II-regulated structural genes. However, at this stage it is too early to present a complete picture of the regulation of pol I transcription in plants.

References

Agarwal ML, Aldrich J, Agarwal A, Cullis CA (1992) The flax ribosomal RNA-encoding genes are arranged in tandem at a single locus interspersed by "non-rDNA" sequences. Gene 120:151–156

Appels R, Dvorak J (1982) The wheat ribosomal DNA spacer region: its structure and variation in populations and among species. Theor Appl Genet 63:337–348
Appels R, Moran LB, Gustafson JP (1986) The structrue of DNA from rye (*Secale cereale*) NOR R1 locus and its behaviour in wheat backgrounds. Can J Genet Cytol 28:673–685
Baerson SR, Kaufman LS (1990) Increased rRNA gene activity during a specific window of early pea leaf development. Mol Cell Biol 10:842–845
Barker RF, Harberd NP, Jarvis MG, Flavell RB (1988) Structure and evolution of the intergenic region in a ribosomal DNA repeat unit of wheat. J Mol Biol 201:1–17
Bell SP, Pikaard CS, Reeder RH, Tijan R (1989) Molecular mechanisms governing species specific transcriptiion of ribosomal RNA. Cell 59:489–497
Bell SP, Jantzen H-M, Tjian R (1990) Assembly of alternative multiprotein complexes directs rRNA promotor selectivity. Genes Dev 4:943–954
Borisjuk N, Hemleben V (1993) Nucleotide sequence of the potato rDNA intergenic spacer. Plant Mol Biol 21:381–384
Broad JR, Li Y-Y, Feldman J, Jayaram M, Abraham J, Nasmyth KA, Hicks JB (1983) Localization and sequence analysis of yeast origins of DNA replication. Cold Spring Harbor Symp Quant Biol 47:1165–1173
Capesius I, Appels R (1989) The direct measurement of ribosomal RNA gene activity in wheat-rye hybrids. Genome 32:343–346
Delcasso-Tremousaygue D, Grellet F, Panabieres F, Ananiev ED, Delseny M (1988) Structural and transcriptional characterization of the external spacer of a ribosomal RNA nuclear gene from a higher plant. Eur J Biochem 172:767–776
Delseny M, Cooke R, Penon P (1983) Sequence heterogeneity in radish nuclear ribosomal RNA genes. Plant Sci Lett 30:107–119
DeWinter R, Moss T (1987) A complex array of sequences enhances ribosomal transcription in *Xenopus laevis*. J Mol Biol 196:813–827
Dover GA (1986) Molecular drive in multigene families: how biological novelties arise, spread and are assimilated. Trends Genet 2:159–165
Echeverria M, Delcasso-Tremousaygue D, Delseny M (1992) A nuclear protein fraction binding to dA/dT-rich sequences upstream from the radish rDNA promoter. Plant J 2:211–219
Ellis THN, Davies DR, Castleton JA, Bedford ID (1984) The organization and genetics of rDNA length variants in peas. Chromosoma 91:74–81
Flavell RB, O'Dell M, Thompson WF (1983) Cytosine methylation of ribosomal RNA genes and nucleolus organiser activity in wheat. In: Brandham PE, Bennett MD (eds) Kew Chromosome Conference II. George Allen and Unwin, Sydney, pp 11–17
Flavell RB, O'Dell M, Vincentz M, Sardana R, Barker RF (1986) The differential expression of ribosomal RNA genes. Philos Trans R Soc Lond B 314:385–397
Flavell RB, O'Dell M, Thompson WF (1988) Regulation of cytosine methylation in ribosomal DNA and nucleolus organizer expression in wheat. J Mol Biol 204:523–534
Ganal M, Torres R, Hemleben V (1988) Complex structure of the ribosomal DNA spacer of *Cucumis sativus* (cucumber). Mol Gen Genet 212:548–554
Gerstner J, Schiebel K, von Waldburg G, Hemleben V (1988) Complex organization of the length heterogeneous 5′ external spacer of mung bean (*Vigna radiata*) ribosomal DNA. Genome 30:723–733
Grierson D, Loening UE (1972) Distinct transcription products of ribosomal genes in two different tissues. Nat New Biol 235:80–82
Grierson D, Loening UE (1974) Ribosomal RNA precursors and the synthesis of chloroplast and cytoplasmic ribosomal ribonucleic acids in leaves of *Phaseolus aureus*. Eur J Biochem 44:501–507
Gruendler P, Unfried I, Pascher K, Schweizer D (1991) rDNA intergenic region from *Arabidopsis thaliana*. Structural analysis, intraspecific variation and functional implications. J Mol Biol 221:1209–1222
Grummt I, Roth E, Paule MR (1982) Ribosomal RNA transcription in vitro is species specific. Nature 296:173–174

Guilfoyle TJ, Lin CY, Chen YM, Nagao RT, Key JL (1975) Enhancement of soybean RNA polymerase I by auxin. Proc Natl Acad Sci USA 72:69–72

Hemleben V, Grierson D (1978) Evidence that in higher plants the 25S and 18S genes are not interespersed with genes for 5S rRNA. Chromosoma 65:353–358

Hemleben V, Leweke B, Roth A, Stadler J (1982) Organization of highly repetitive satellite DNA from two Cucurbitaceae species (*Cucumis melo* and *Cucumis sativus*). Nucleic Acids Res 10, 631–644

Hemleben V, Ganal M, Gerstner J, Schiebel K, Torres RA (1988) Organization and length heterogeneity of plant ribosomal RNA genes. In: Kahl G (ed) Architecture of eukaryotic genes. VHC, Weinheim, pp 371–383

Hemleben V, Zentgraf U, King K, Borisjuk N, Schweizer G (1992) Middle repetitive and highly repetitive sequences detect polymorphisms in plants. In: Kahl G, Appelhans H, Kömpf J, Driesel AJ (eds) DNA-polymorphisms in eukaryotic genomes. BTF 10, Adv. Mol Gen 5. Huethig, Heidelberg, pp 157–170

Hernandez P, Bjerknes CA, Lamm SS, Van't Hof (1988) Proximity of an ARS consensus sequence to a replication origin of pea (*Pisum sativum*). Plant Mol Biol 10:413–422

Ingle J, Timmis JN, Sinclair J (1975) The relationship between satellite DNA, ribosomal RNA gene redundancy, and genome size in plants. Plant Physiol 55:496–501

Jackson SD, Flavell RB (1992) Protein-binding to reiterated motifs within the wheat rRNA gene promoter and upstream repeats. Plant Mol Biol 20:911–919

Jendrisak J (1980) Purification, structures and functions of the nuclear RNA polymerases from higher plants. In: Leaver C (ed) Genome organization and expression in plants. NATO Adv Study Inst Ser: Series A, Life Sciences. Plenum Press, New York, pp 77–92

Kato A, Nakajima T, Yamashita J, Yakura K, Tanifuji S (1990) The structure of the large spacer region of the rDNA in *Vicia faba* and *Pisum sativum*. Plant Mol Biol 14:983–993

Kaufman LSJ, Watson JC, Thompson WF (1987) Light regulated changes in DNase I hypersensitive sites in the rDNA genes of *Pisum sativum*. Proc Natl Acad Sci USA 84:1550–1554

Kelly RJ, Siegel A (1989) The *Cucurbita maxima* intergenic spacer has a complex structure. Gene 80:239–248

King K, Torres RA, Zentgraf U, Hemleben V (1993) Molecular evolution of the intergenic spacer in the nuclear ribosomal RNA genes of Cucurbitaceae. J Mol Evol 36:144–152

Labhart P, Reeder RH (1987) A 12-base pair sequence is an essential element of the ribosomal gene terminator in *Xenopus laevis*. Mol Cell Biol 7:1900–1905

Leweke B, Hemleben V (1982) Organization of rDNA in chromatin: plants. In: Busch H, Rothblum L (eds) The cell nucleus: rDNA. Vol XI, part B. Academic Press, New York, pp 225–253

Martini G, O'Dell M, Flavell RB (1982) Partial inactivation of wheat nucleolus organisers by the nucleolus organiser chromosomes from *Aegilops umbellulata*. Chromosoma 84:687–700

Miesfeld R, Arnheim N (1984) Species-specific rDNA transcription is due to promoter-specific binding factors. Mol Cell Biol 4:221–227

Mitchelson K, Moss T (1987) The enhancement of ribosomal transcription by recycling of RNA polymerase I. Nucleic Acids Res 15:9577–9596

Motte PM, Loppes R, Menager M, Deltour R (1991) Three-dimensional electron microscopy of ribosomal chromatin in two higher plants: a cytochemical, immunocytochemical, and in situ hybridization approach. J Histochem Cytochem 39:1495–1506

Nakajima T, Suzuki A, Tanifuji S, Kato A (1992) Characterization of nucleotide sequences that interact with a nuclear protein fraction in rRNA gene of *Vicia faba*. Plant Mol Biol 20:939–949

Perry KL, Palukatis P (1990) Transcription of tomato ribosomal DNA and the organization of the intergenic spacer. Mol Gen Genet 221:102–112

Perry RP (1976) Processing of RNA. Annu Rev Biochem 45:605–629
Pikaard CS, McStay B, Schultz MC, Bell SP, Reeder RH (1989) The *Xenopus* ribosomal gene enhancers bind an essential polymerase I transcription factor, xUBF. Genes Dev 3:1779–1788
Piller KJ, Baerson SR, Polans NO, Kaufman LS (1990) Structural analysis of the short length ribosomal DNA variant from *Pisum sativum* L. cv. Alaska. Nucleic Acids Res 18:3135–3145
Polans NO, Weeden HF, Thompson WF (1986) Distribution, inheritance and linkage relationship of ribosomal DNA spacer length variants in pea. Theor Appl Genet 72:289–295
Rathgeber J, Capesius I (1990) Nucleotide sequence of the intergenic spacer and the 18S ribosomal RNA gene from mustard (*Sinapis alba*). Nucleic Acids Res 18:1288
Rogers MM, Loening UE, Fraser RSS (1970) Ribosomal RNA precursors in plants. J Mol Biol 49:681–692
Rogers SO, Bendich AJ (1987) Ribosomal RNA genes in plants: variability in copy number and in the intergenic spacer. Plant Mol Biol 9:509–520
Rogers SO, Honda S, Bendich AJ (1986) Variation in the ribosomal RNA genes among individuals of *Vicia faba*. Plant Mol Biol 6:339–345
Rungger D, Crippa M (1977) The primary ribosomal DNA transcript in eukaryotes. Proc Biopys Mol Biol 31:247–269
Schiebel K, von Waldburg G, Gerstner J, Hemleben V (1989) Termination of transcription of ribosomal RNA genes of mung bean occurs within a 175 bp repetitive element of the spacer region. Mol Gen Genet 218:302–307
Schmidt-Puchta W, Günther I, Sänger HL (1989) Nucleotide sequence of the intergenic spacer (IGS) of the tomato ribosomal DNA. Plant Mol Biol 13:251–253
Schmitz ML, Maier UG, Brown JWS, Feix G (1989) Specific binding of nuclear proteins to the promoter region of a maize nuclear rRNA gene unit. J Biol Chem 264:1467–1472
Scott NS, Kavanagh TA, Timmis JN (1984) Methylation of rRNA genes in some higher plants. Plant Sci Lett 35:213–217
Shagai-Maroof MA, Soliman KM, Jorgensen RA, Allard RW (1984) Ribosomal DNA spacer length polymorphisms in barley: Mendelian inheritance, chromosomal location, and population dynamics. Proc Natl Acad Sci USA 81:8014–8018
Sollner-Webb B, Mougey EB (1991) News from the nucleolus: rRNA gene expression. Trends Biochem Sci 16:58–62
Sollner-Webb B, Tower J (1986) Transcription of cloned eukaryotic ribosomal RNA genes. Annu Rev Biochem 55:801–830
Sommerville J (1986) Nucleolar structure and ribosome biosynthesis. Trends Biochem Sci 11:438–442
Spiker S, Murray MG, Thompson WF (1983) DNase I sensitivity of transcriptionally active genes in intact nuclei and isolated chromatin of plants. Proc Natl Acad Sci USA 80:815–819
Tautz D, Tautz C, Webb D, Dover GA (1987) Evolutionary divergence of promoters and spacers in the rDNA family of four *Drosophila* species. J Mol Biol 195:525–542
Thompson WF, Flavell RB (1988) DNase I sensitivity of ribosomal RNA genes in chromatin and nucleolar dominance in wheat. J Mol Biol 204:535–548
Thompson WF, Flavell RB, Watson JC, Kaufman LS (1988) Chromatin structure and expression of plant ribosomal RNA genes. In: Kahl G (ed) Architecture of eukaryotic genes. VCH, Weinheim, pp 385–396
Tobin EM, Silverthorne (1985) Light regulation of gene expression in higher plants. Annu Rev Plant Physiol 36:569–593
Toloczyki C, Feix G (1986) Occurrence of 9 homologous repeat units in the external spacer region of a nuclear maize rRNA gene unit. Nucleic Acids Res 14:4969–4986
Torres RA, Hemleben V (1991) Use of ribosomal DNA spacer probes to distinguish cultivars of *Cucurbita pepo* L. and other Cucurbitaceae. Euphytica 53:11–17

Torres RA, Zentgraf U, Hemleben V (1989) Species and genus specificity of the intergenic spacer (IGS) in the ribosomal RNA genes of Cucurbitaceae. Z Naturforsch 44c:1029–1034

Torres RA, Ganal M, Hemleben V (1990) GC balance in the internal transcribed spacers ITS 1 and ITS 2 of nuclear ribosomal RNA genes. J Mol Evol 30:170–181

Ueki M, Uchizawa E, Yakura K (1992) The nucleotide sequence of the rDNA intergenic spacer region in a wild species of the genus *Vicia*, *V. angustifolia*. Plant Mol Biol 18:175–178

Unfried K, Schiebel K, Hemleben V (1991) Subrepeats of rDNA intergenic spacer present as prominent independent satellite DNA in *Vigna radiata* but not in *Vigna angularis*. Gene 99:63–68

Van't Hof J, Lamm SS (1992) Site of initiation of replication of the ribosomal genes of pea (*Pisum sativum*) detected by two-dimensional gel electrophoresis. Plant Mol Biol 20:377–382

Vincentz M, Flavell RB (1989) Mapping of ribosomal RNA transcripts in wheat. Plant Cell 1:579–589

Von Kalm L, Vize PD, Smyth DR (1986) An under methylated region in the spacer of ribosomal RNA genes of *Lilium henryi*. Plant Mol Biol 6:33–39

Weising K, Kahl G (1991) Towards an understanding of plant gene regulation: the action of nuclear factors. Z Naturforsch 46c:1–11

White RJ, Jackson SP (1992) The TATA-binding protein: a central role in transcription by RNA polymerase I, II and III. Trends Genet 8:284–288

Yakura K, Nishikawa K (1992) The nucleotide sequence of the rDNA spacer region between the 25S and 18S rRNA genes in a species of the genus *Vicia*, *V. hirsuta*. Plant Mol Biol 19:537–539

Yakura K, Kato A, Tanifuji S (1983) Structural organization of ribosomal DNA in four *Trillium* species and *Paris verticillata*. Plant Cell Physiol 24:1231–1240

Yamashita J, Nakajima T, Tanifuij S, Kato A (1993) Accurate transcription initiation of *Vicia faba* rDNA in a whole cell extract from embryonic axes. Plant J 3:187–190

Yang-Yen H-F, Rothblum L (1988) Purification and characterization of a high-mobility-group-like DNA-binding protein that stimulates rRNA sythesis in vitro. Mol Cell Biol 8:3406–3414

Zentgraf U, Hemleben V (1992) Complex formation of nuclear proteins with the RNA polymerase I promoter and repeated elements in the external transcribed spacer of *Cucumis sativus* ribosomal DNA. Nucleic Acids Res 20:3685–3691

Zentgraf U, Hemleben V (1993) Nuclear Proteins interact with RNA polymerase I promoter and repeated elements of the 5′ external transcribed spacer of the rDNA of cucumber in a single-stranded stage. Plant Mol Biol 22:1153–1156

Zentgraf U, Ganal M, Hemleben V (1990) Length heterogeneity of the rRNA precursor in cucumber (*Cucumis sativus*). Plant Mol Biol 15:465–474

2 RNAPII: A Specific Target for the Cell Cycle Kinase Complex

László Bakó[1,2], Sirpa Nuotio[2], Dénes Dudits[1], Jeff Schell[2], and Csaba Koncz[1,2]

1 Introduction

Transcription in plants, as in other eukaryotes, is catalyzed by three RNA polymerases (RNAPs). Catalytically active forms of RNAPs were first isolated by Roeder and Rutter (1969) and designated as RNAPI(A), II(B) and III(C). RNAPI transcribes rRNA genes, RNAPII synthesizes the precursors of mRNAs and RNAPIII is involved in the transcription of 5S RNA and tRNA genes. In contrast to prokaryotes in which a single RNA polymerase, consisting of $\beta\beta'\alpha_2$ subunits and associated σ factors (Yura and Ishihama 1979; Helmann and Chamberlin 1988), is sufficient for promoter recognition, the assembly of transcriptionally active initiation complexes in eukaryotes requires specific interactions of RNAPs with multiple transcription factors (TFs) and promoter-specific activator proteins. Studies of the regulation of transcription were started by characterization of the subunit composition of RNAPs.

2 Structure and Function of Eukaryotic RNA Polymerases

A comparison of subunit composition of RNAPs purified from lower eukaryotes, plants, insects, amphibians and mammals was initially obtained by biochemical and immunological methods (for review, see Lewis and Burgess 1982; Guilfoyle 1983; Sentenac 1985). Isolation of yeast cDNAs encoding individual RNAP subunits (Young and Davis 1983; Riva et al. 1986) opened the way to genetic approaches, such as site-specific mutagenesis and epitope tagging of RNAP subunits (for review, see Sentenac 1985; Sawadogo and Sentenac 1990; Young 1991). Phosphorylation, inhibitor binding and chemical cross-linking studies contributed to a comprehensive picture of the structural and functional organization of eukaryotic RNAPs. The yeast system is used as a reference for comparison because most data on RNAP genes and proteins are derived from studies with this organism

[1] Institute of Plant Physiology, Biological Research Center of the Hungarian Academy of Sciences, H-6701 Szeged, Temesvári krt 62, P.O. Box 521, Hungary
[2] Max-Planck Institut für Züchtungsforschung, Carl-von-Linné-Weg 10, D-50829 Köln, FRG

Results and Problems in Cell Differentiation 20
L. Nover (Ed.)
Plant Promoters and Transcription Factors

(Table 1; Guilfoyle and Dietrich 1987; Geiduschek and Tocchini-Valentini 1988; Woychik and Young 1990; Young 1991).

Essential domains involved in DNA binding and RNA synthesis are carried by the two largest subunits of RNAP enzymes which are related in size and sequence in RNAPI, II and III, and show homology to the ββ′ subunits of eubacterial and archaebacterial RNA polymerases (Allison et al. 1985; Sweetser et al. 1987; Mémet et al. 1988; Pühler et al. 1989). The B44.5 subunit of RNAPII(B), and the common subunits AC40 and AC19 of RNAPI(A) and RNAPIII(C) enzymes in yeast are homologous to the α subunit of eubacterial RNA polymerases. Yeast RNAPs contain five common subunits: ABC27, ABC23, ABC14.5, ABC10α and ABC10β. In plants and other eukaryotes three common subunits are shared by all RNAPs.

Thus, RNAPI, II and III of wheat contain identical subunits of 20, 17.8 and 17 kDa, whereas common subunits of 25, 19 and 17.5 kDa are present in RNAPs of cauliflower and other dicotyledonous plants (Guilfoyle et al. 1984; Guilfoyle and Dietrich 1987). In addition, an identical subunit of 38 kDa is present in RNAPI and III, and another subunit of 25 kDa is shared by RNAPII and III in wheat (Table 1).

RNAPI in *S. cerevisiae* contains 13 subunits, whereas in *S. pombe*, plants and mammals it consists of only 6–7 subunits. In *S. pombe* the "core" RNAPI subunits A190(β′), A135(β), AC40(α), and AC19(α) form a complex with the "common" subunits ABC27, ABC23 and ABC14.5 (Sentenac 1985; Woychik et al. 1990). This simple structural organization of RNAPI reflects a conserved feature of transcription factors that regulate the expression of rRNA genes of higher eukaryotes in a similar fashion (Bell et al. 1990; Reeder 1990; Comai et al. 1992).

The ability of RNAPII to recognize TATA-box containing promoters is used as a didactic argument to differentiate between RNAPI and RNAPII. RNAPIII may be considered an "intermediate" form of RNAPs because it initiates transcription on promoters that either contain or lack a TATA element. A functional similarity between RNAPII and III is indicated by the observation that related transcription factors, Oct-1 and Oct-2, regulate the transcription of snRNA and mRNA genes (Tanaka et al. 1992). Nucleotide exchanges in the TATA sequence or alteration of the spacing between the TATA box and upstream activator regions convert the RNAPIII promoters into promoters recognized by RNAPII (for review, see Gabrielsen and Sentenac 1991). In contrast, upstream control elements are not exchangable between RNAPII and RNAPI promoters (Schreck et al. 1989). Transcription factors that interact with the TATA box and upstream activator factors thus direct selective binding of appropriate RNA polymerases to diverse promoters.

Genes encoding most RNAPIII subunits in yeast have been cloned, and all of them were found to be unique and essential for viability. Properties of RNAPIII subunits and their role in transcription have been reviewed recently (Gabrielsen and Sentenac 1991). Due to its pivotal role in transcription

Table 1. Subunit composition of RNA polymerase enzymes purified from yeast and wheat germ (Lewis and Burgess 1982; Guilfoyle et al. 1984; Sentenac 1985; Woychik et al. 1990; Carles et al. 1991; Young 1991)

Yeast			**Wheat germ**			
PNAPI(A)	PNAPII(B)	RNAPIII(C)	RNAPI	RNAPII	RNAPIII	
190	***220***	***160***	***200***	***220***	***150***	β′
135	***150***	***128***	***125***	***140***	***130***	β
					94	
		82				
					55	
		53				
49						
	44.5 (α)					
43						
				42		
40 (α)		**40 (α)**		40		
			38		**38**	
		37				
34.5						
		34				
	32					
		31				
					30	
					28	
27	***27***	***27***		27		
		25		**25**	**25**	
				24	24.5	
23	***23***	***23***				
				21	21.5	
			20	***20***	***20***	
19 (α)		**19 (α)**			19.5	
			17.8	***17.8***	***17.8***	
			17	***17***	***17***	
	17			16.3		
				16		
14.5	***14.5***	***14.5***				
14				14		
	13*					
12.2						
_10_*	_10_*	_10_*				

* Subunits represented by two unrelated proteins of identical molecular mass. α and β mark subunits homologous to prokaryotic and archaebacterial RNAP proteins. Using the nomenclature introduced by Sentenac (1985), the RNAP subunits are designated according to their molecular mass. Subunits of RNAPI(A) are marked by A, subunits of RNAPII(B) and RNAPIII(C) are labelled by B and C, respectively. Common subunits of the RNAP enzymes are designated by the combination of A, B and C letter codes (i.e., AC40 or ABC23). In the nomenclature of Young (1991) the RNAP subunits are designated according to their mobility on SDS-PAGE. Thus, RNAPI(A) subunits are labelled as RPA1, RPA2, etc., whereas RNAPII(B) subunits are designated as RPB1, RPB2, etc. Genes corresponding to these subunits are marked accordingly, such as *RPB1*, *RPB2*, etc. (Young 1991). To avoid confusion, both nomenclatures are used simultaneously in the text. Homologous (i.e., common) subunits of RNAP enzymes are indicated by italic boldface numbers, whereas subunits common between either RNAPI(A) and RNAPII(B) or RNAPII(B) and RNAPIII(C) are labelled by boldface numbers.

of protein coding genes, RNAPII is one of the best-studied enzymes in eukaryotes (Sentenac 1985; Sawadogo and Sentenac 1990; Woychik and Young 1990; Young 1991). Twelve RNAPII subunits and corresponding genes have been identified in yeast (see Table 1; Carles et al. 1991; Young 1991). The subunit organization of RNAPII enzymes isolated from yeast, plants, insects and mammals is very similar. Although RNAPII is thought to perform a more complex function than RNAPIII, genetic studies indicate that mutations in several RNAPII subunit genes (*RPB7*, *RPB4* and *RPB9*) do not cause lethality, whereas the loss of any RNAPIII subunit function is lethal in yeast. The RPB4(B32) and RPB7(B17) subunits of RNAPII form a subcomplex which can dissociate from the enzyme. *RPB4* mutants are temperature-sensitive and lack the RPB7(B17) subunit. RPB4 is homologous to the σ^{70} subunit of prokaryotic RNA polymerases, and together with RPB7 stimulates the formation of preinitiation complexes. Subunit RPB9(B13a), a zinc-binding protein, is only essential for growth at temperature extremes. Gene *RPB11* encodes a second 13-kDa subunit which is required for viability in yeast. Common subunits ABC10α and β are essential zinc-binding proteins, whereas RPB5(ABC27), RPB6(ABC23) and RPB8(ABC14.5) have been suggested to determine nuclear localization and DNA binding (Woychik and Young 1990). RBP6(ABC23) is phosphorylated, and may thus function as a common regulatory protein for all three RNAPs. RBP5(ABC27) is a basic protein, two copies of which are present per enzyme molecule. RNAPII also contains two RBP3(B44.5) subunits with sequence homology to the α-subunit of prokaryotic RNAP enzymes. In prokaryotes the α-subunit regulates the assembly of ββ′ subunits. Mutations affecting either the α, β or β′ subunit all result in defective assembly. Mutations abolishing the assembly of RPB1(B220), RPB2(B150) and RPB3(B45) subunits were similarly localized in the RNAPII subunit genes *RPB1*, *RPB2* and *RPB3* in yeast (Kolodziej et al. 1990; Woychik et al. 1990; Carles et al. 1991; Young 1991). The largest subunit of RNAPII carries a unique C-terminal domain that is absent from homologous subunits of RNAPI and III enzymes, and may regulate RNAPII activity and/or specificity.

3 Functional Domains of the Largest Subunit of RNAPII

In RNAPII enzymes isolated from diverse organisms, three forms of the largest RPB1 subunit, IIa (220 kDa), IIb (180 kDa) and IIo (240 kDa), were detected. Immunological and peptide mapping studies revealed that these RPB1 forms are closely related (Dahmus 1983; Kim and Dahmus 1986). Biochemical analysis indicated that the IIb protein is an artifact that results from protease digestion of the C-terminal domain of RPB1 forms IIa and IIo during enzyme purification (for review, see Lewis and Burgess 1982; Guilfoyle 1983). The IIo form of RBP1 was shown to result from phosphorylation of the C-terminal domain of form IIa (Dahmus 1981). Elegant biochemical and genetic studies revealed that the RPB1(β′) and RPB2(β) subunits are involved in binding of DNA, nucleotides, and amatoxin inhibitors of RNA chain elongation and pyrophosphate exchange reactions (for review, see Sentenac 1985; Riva et al. 1987).

To genetically identify the amatoxin binding RNAPII subunit, stable, conditionally lethal and thermosensitive mutants conferring α-amanitin resistance were isolated in mouse and *Drosophila* (Ingles 1978; Greenleaf et al. 1979). Following genetic mapping of α-amanitin resistance mutations, the $RPII^{C4}$ locus encoding the largest subunit of RNAPII was cloned from *Drosophila* by P-element-mediated gene tagging (Searles et al. 1982; Greenleaf 1983; Jokerst et al. 1989). *RPB1* and *RPB2* genes were subsequently isolated from yeast (Young and Davies 1983) and used for site-specific mutagenesis of the largest subunit genes of RNAPII (Ingles et al. 1984). The *Drosophila* and yeast genes served as probes for cloning the largest subunit genes of RNAPII from *Trypanosoma*, *Crithidia*, *Plasmodium*, *S. pombe*, *Caenorhabditis*, mouse, hamster, man, as well as from higher plants, such as soybean and *Arabidopsis* (Cho et al. 1985; Ahearn et al. 1987; Allison et al. 1988; Bird and Riddle 1989; Evers et al. 1989a,b; Li et al. 1989; Smith et al. 1989; Dietrich et al. 1990; Nawrath et al. 1990; Azuma et al. 1991).

Nucleotide sequence comparison of the largest subunit genes, *RPB1* and *RBP2*, revealed a phylogenetic conservation of 8–9 domains with homology to the β′ and β subunits of prokaryotic RNAPs, and to the largest subunits of RNAPI and III (for review, see Sawadogo and Sentenac 1990; Young 1991). The function of these domains has been deduced by confrontation of genetic and nucleotide sequence data (Allison et al. 1985; Himmelfarb et al. 1987; Nonet et al. 1987a; Sweetser et al. 1987; Jokerst et al. 1989; Nawrath et al. 1990; Scafe et al. 1990). The N-terminal domain of the largest subunit of RNAPII contains a conserved zinc-binding motif $CX_2CX_6CX_2HX_{11}HX_{12}CVCX_2C$. Analogous zinc-binding motifs, essential for interaction between the two largest subunits, also occur in the C-terminal domain I of the RPB2 subunit, and in the largest subunits of RNAPI and III (Yano and Nomura 1991; Young 1991). Several second-site mutations reverting the defects of the RPB1 subunit have been mapped in the C-terminal domain I, close to the nucleotide binding domain H of the RPB2 subunit. A number of mutations correcting the defects of the RPB2 subunit have been located in the domain H of RPB1 (Martin et al. 1990), and in the RPB6 common subunit which participates in DNA binding together with the two largest subunits (Mortin 1990; Archambault et al. 1992). A two-helix motif, known to contact the major groove of DNA, is located in a second N-terminal conserved domain of the RPB1 subunit (Ahearn et al. 1987; Jokerst et al. 1989; Nawrath et al. 1990).

About 120 amino acid residues identical in all eukaryotes define the central region of the largest subunit of RNAPII. A mutation causing α-amanitin resistance in mouse is localized in this region, and results in an asparagine to aspartate substitution in a conserved motif, VGQQ*N*VEG

(Bartolomei and Corden 1987). In yeast the Asn residue is replaced by serine, possibly explaining the insensitivity of yeast RNAPII to amatoxins. The largest subunits of plant RNAPII enzymes also contain a VGQQ*N*VEG motif. Nevertheless, plants can tolerate high concentrations of α-amanitin, possibly because they are capable of metabolically inactivating the toxin (Pitto et al. 1985). However, it cannot be excluded that the lower amatoxin sensitivity of plant RNAPIIs (see Sentenac 1985) results from a different amino acid exchange because the positions of other Ama^R mutations are not yet known. Certain Ama^R mutations cause pleiotropic developmental defects in a gene dosage-dependent fashion. In *Drosophila* the *C4* Ama^R mutation confers an ultrabithorax-like phenotype (*Ub1* locus, Greenleaf et al. 1980), whereas another Ama^R mutation in rat myoblasts prevents muscle differentiation (Crerar et al. 1983). These pleiotropic effects of Ama^R mutations correlate with a decreased rate in RNA elongation by α-amanitin-resistant RNAPII enzymes (Coulter and Greenleaf 1985; Shermoen and O'Farell 1991).

Electron crystallography of a yeast RNAPII lacking the RPB4/RPB7 subcomplex revealed a 25-Å-wide groove formed by the RPB1 and RPB2 subunits capable of accomodating the template DNA (Darst et al. 1991). The three-dimensional crystallographic image supports the genetic data, indicating that functional complementation between structurally altered RPB1 and RPB2 subunits is possible by conformational correction. The overall structure of yeast RNAPII is very similar to that of *E. coli* RNA polymerase, with the exception of a finger-like structure that protrudes from the molecule in the vicinity of the 25-Å groove. This structure corresponds to the hydrophylic C-terminal domain (CTD) of the largest subunit of RNAPII. This domain is composed of YSPTSPS heptameric repeats which are absent from the prokaryotic and archaebacterial RNAPs, as well as from the largest subunits of eukaryotic RNAPI and III enzymes. The number of heptapeptide repeats increases in correlation with the genomic complexity, but the degree of divergence from the consensus YSPTSPS motif varies among the CTD repeats in diverse organisms. In the CTD of *Plasmodium* the consensus heptamer is YSPTSPK and occurs nine times, whereas the number of consensus/variant YSPTSPS repeats is 18/8 in *S. cerevisiae*, 24/5 in *S. pombe*, 11/30 in *Caenorhabditis*, 15/26 in *Arabidopsis*, 2/42 in *Drosophila* and 21/31 in mouse and hamster (for review, see Allison et al. 1985, 1988; Corden et al. 1985; Bird and Riddle 1989; Nawrath et al. 1990; Azuma et al. 1991).

4 Regulation of RNAPII Activity by Phosphorylation of the C-Terminal Domain (CTD) of the Largest Subunit

The repeated pattern of proline residues in the CTD results in consecutive helical β-turns which add up to a tail-like, flexible secondary structure

(Matsushima et al. 1990). This structure is strongly antigenic and particularly sensitive to protease digestion during enzyme purification and when synthesized as part of CTD-fusion proteins in *E. coli* (Christmann and Dahmus 1981; Guilfoyle et al. 1984; Kim and Dahmus 1986; Lee and Greenleaf 1989; Peterson et al. 1992). Removal of the CTD by protease treatments converts the IIa and IIo forms of the largest subunit to form IIb (Dahmus 1983; Guilfoyle et al. 1984; Corden et al. 1985; Guilfoyle and Dietrich 1987). RNAPII enzymes purified from mouse and HeLa cells contain predominantly the IIo form (240 kDa) which can be converted to form IIa by phosphatase treatment (Cadena and Dahmus 1987). Casein kinase II phosphorylates the largest subunit at a single Ser residue of the C-terminal end in vitro (Dahmus 1981), but does not appear to be responsible for the phosphorylation of the CTD in vivo (Lu et al. 1991, 1992). Protease treatment of the in vitro or in vivo ^{32}P-labelled IIo form results in quantitative removal of the label, indicating that the CTD within the IIa form is a unique target for phosphorylation.

The effect of CTD phosphorylation was tested in various in vitro transcription systems (for review, see Sawadogo and Sentenac 1990). Initial studies suggested that all three forms of RNAPII can accurately initiate transcription from the adenovirus core Ad-2 MLP promoter. In these in vitro transcription assays RNAPII forms IIA and IIB were found to be more active than RNAP IIO, suggesting that the CTD or its phosphorylation may not be required for the initiation of transcription (Sentenac 1985; Kim and Dahmus 1989). In contrast, other data demonstrated that CTD-specific antibodies, as well as synthetic CTD heptapeptide repeats, can inhibit the initiation of transcription by RNAPII, but do not affect the transcript elongation (Dahmus and Kedinger 1983; Moyle et al. 1989; Thompson et al. 1989). These experiments also revealed that the CTD is required for transcription initiation from promoters lacking the TATA box, whereas the CTD-less RNA polymerase IIB can accurately initiate transcription from TATA-box containing promoters in vitro. However, when injected into *Xenopus* oocytes, CTD-specific antibodies inhibited the transcription from both the TATA-containing human histone H2b promoter and the TATA-less promoter of U1 small nuclear RNA gene (Thompson et al. 1989). These data showed that the CTD is absolutely required for the stabilization of the initiation complex in the absence of the TATA box, whereas general transcription factors can mediate the binding of RNAPII to TATA-box containing promoters even in the absence of the CTD.

UV cross-linking studies demonstrated that elongating transcripts are associated with the largest subunit of RNAP IIO carrying a phosphorylated CTD (Bartholomew et al. 1986; Cadena and Dahmus 1987). Reconstitution of transcription initiation and elongation complexes using purified transcription factors revealed, on the other hand, that the unphosphorylated form of RNA polymerase (IIA) is required for promoter binding and the formation of stable preinitiation complexes. During the transition from initiation to elongation, inactive RNAP IIA is specifically phosphorylated at the CTD to yield an active enzyme, RNAP IIO (Laybourn and Dahmus 1989; Payne et al. 1989; Lu et al. 1991, 1992; Chesnut et al. 1992). Due to the apparent contradiction between in vitro and in vivo transcription data, the role of CTD in the regulation of transcription has been extensively discussed (for review, see Corden 1990; Sawadogo and Sentenac 1990; Chao and Young 1991; Young 1991).

5 Genetic Analysis of Interactions Between RNAPII and Transcription Regulatory Proteins

Genetic analysis of CTD mutants in yeast, *Drosophila* and mouse provided the first indication that the CTD is essential for the regulation of RNAPII transcription in vivo (Nonet et al. 1987b; Allison et al. 1988; Bartolomei et al. 1988; Zehring et al. 1988). The maintenance of at least 11 CTD heptapeptide repeats is required for viability in yeast. Mutants with 11–12 repeats are conditionally viable, but display heat and cold sensitivity and inositol auxotrophy (Nonet et al. 1987b). In *Drosophila* about 50% of the repeats is essential in vivo (Zehring et al. 1988). In mouse a reduction in the number of repeats to 29–31 reduces the cell growth and division rate and 25 or fewer repeats cause cell lethality (Bartolomei et al. 1988). Replacement of yeast RNAPII CTD with the first 26 repeats of hamster CTD did not affect viability, whereas an exchange for the less conserved *Drosophila* CTD caused lethality in yeast (Allison et al. 1988). Repeats differing in two or three amino acids from the consensus heptapeptide could not substitute for the conserved heptapeptide repeats in mouse cells (Bartolomei et al. 1988).

These observations indicated that both the position and amino acid composition of heptapeptide repeats can effect the CTD function in vivo. Combinations of yeast *INO1* and *GAL10* upstream activator sequences (UAS) with a "core" TATA-box promoter-*lacZ* reporter gene construct revealed a reduction in *INO1*, and *GAL10* mRNAs in a yeast mutant, *rpb1Δ104* that carried only 11 CTD repeats. The deficiency of transcriptional activation by the *GAL10* UAS was shown to result from a defective interaction between the truncated form of RNAPII CTD and the *GAL10*-specific transcription activator protein, GAL4 (Allison and Ingles 1989; Scafe et al. 1990). Using GCN4 and RNAPII affininity columns, Brandl and Struhl (1989) demonstrated that GCN4 and RNAPII can interact directly in vitro. This interaction required the basic DNA-binding domain of GCN4. Studies demonstrating that GAL4 can mediate transcriptional activation in *Drosophila* and mammalian cells underscore the importance of these results and suggest that analogous interactions between the C-terminal repeats of RNA polymerase II and promoter-specific activator proteins generally occur in all eukaryotes (Fischer et al. 1988; Kakidani and Ptashne 1988; Webster et al. 1988).

Isolation of second-site mutations suppressing the defects of CTD deletions in yeast led to the identification of intragenic and extragenic revertants. Intragenic revertants carried either point mutations in the conserved domains of the largest subunits of RNAPII, or duplications of the shortened CTDs. A dominant extragenic suppressor correcting the effect of CTD truncations was mapped to the yeast gene *SRB2*, which encodes a transcription factor that can apparently replace the function of the CTD (Nonet and Young 1989; Koleske et al. 1992). Data indicating that acidic activators, the CTD of RNAPII and transcription factors located at the TATA box can interact with each other opened the way to novel genetic approaches. Deletions removing the TATA box and/or UAS activator binding sites have been exploited for isolation of second-site revertants in yeast. These second-site mutations reverted either the inducibility of transcription by a UAS-specific activator, or reconstructed the basic level of uninduced transcription of diverse genes (i.e., *His3*, *His4*, *Suc2*, etc.). In a second system mutations in genes *SWI1–6*, which are positive regulators of the mating-type endonuclease gene *HO*, were exploited to isolate revertants expressing *HO*.

Isolation of suppressors reverting the defects (i.e., auxotrophy and cell cycle arrest in G1 phase) caused by a deletion of *HIS4* UAS elements yielded four classes of *sit* mutations. *sit1* and *sit2* proved to be mutations in the largest subunit genes of RNAPII. SIT3 is a transcription factor which probably interacts with the GCN4 and TATA-binding proteins. SIT4 is a protein phosphatase required for cell cycle-dependent regulation of gene *SWI4*. Since *sit1–sit4* and *sit2–sit4* double mutants were inviable, it was likely that SIT4 would be required for the dephosphorylation of the largest subunit of RNAPII (Arndt et al. 1989). Combinations of *sit4* alleles with *cdc28* and *bcy1* mutations caused lethality. This indicated that the SIT4 function is connected to regulatory pathways including the CDC28 cell cycle kinase and cAMP-dependent protein kinases harboring a BCY1 regulatory subunit (Hoekstra et al. 1991; Fernandez-Sabaria et al. 1992; Johnston and Lowndes 1992; see below).

CTD truncations of RNAPII and mutations in the *swi1*, *swi2*, and *swi3* genes were also found to prevent the GAL4-mediated activation of genes *GAL1* and *GAL10*, and to negatively affect the UAS-mediated transcriptional control of a large set of genes. In addition to SWI1, SWI2 and SWI3, UAS-dependent transcription from the *HO*, *GAL1*, *INO1*, *SUC2* and other promoters requires further factors, SNF5 and SNF6, which form a functional complex with the SWI1, SWI2 and SWI3 proteins (Laurent and Carlson 1992; Peterson and Herskowitz 1992; Winston and Carlson 1992). These data show that the CTD of RNAPII has a complementary and synergistic function with a transcription factor complex (including proteins SWI1, SWI2, SWI3, SNF5 and SNF6) which "tethers" the UAS-specific activators to other elements of the TATA-box bound initiation complex. Such complementary functions may explain why RNA polymerases lacking a CTD can respond to activation by certain UAS-specific transcription factors (e.g.,

Sp1 in mammals), and why CTD is required for transcription of another set of genes, e.g., those lacking a TATA box (Zhou et al. 1992). In fact, a mutation in the *SIN1* gene was found to suppress the effects of CTD truncation, as well as of *swi1*, *swi2*, and *swi3* gene mutations. *SIN1* encodes a HMG1-type protein which is a chromatin-associated general repressor (Kruger and Herskowitz 1991; Peterson et al. 1991). As the CTD truncations, *swi* and *sit* mutations block the cell cycle progression at the G1 phase.

6 Promoter Recognition and Transcription Initiation by RNAPII: A Phylogenetically Conserved Mechanism

Genetic and biochemical studies in recent years have produced a general concept of transcription initiation (for review, see Gabrielsen and Sentenac 1991; Greenblatt 1991a,b; Roeder 1991; Guarante and Bermingham-McDonogh 1992; Sharp 1992; White and Jackson 1992a,b; Rigby 1993). A brief discussion of this concept helps to understand the function of RNAPII CTD in the regulation of transcription.

Studies of the preinitiation complex (PIC) assembly on class II promoters, such as the adenovirus core promoter (Ad-2 MLP), showed that an ordered binding of RNAPII and several general transcription factors (TFs) to the TATA box and to downstream initiation sequences precedes the initiation of RNA synthesis. Transition of PIC to an active initiation complex (i.e., open complex) requires ATP, indicating the involvement of protein kinase(s) and/or ATPase(s). The rate-limiting step of PIC assembly involves TFs and activator proteins (Buratowski et al. 1991; Wang et al. 1992a). The most extensively characterized factors are TFIIA, -B, -D, -E, -F, -G/J, -H, -I, -S, and -X from HeLa cells which are related to factors α, β, γ, δ, ε and τ from rat liver, and to factors a, b, c, d, e and g from yeast (for review, see Roeder 1991).

The PIC assembly begins with binding of the TFIID to the TATA element and continues with subsequent binding of TFIIA and TFIIB to class II promoters. TFIIA binds upstream of TFIID, whereas the binding site of TFIIB is located 3′ downstream of the TATA box. Genetic analyses of yeast mutations suppressing defects of the TATA box, GAL4 activation domain and RNAPII CTD truncations (see above) support biochemical data, indicating that TFIID is a complex of a TATA-binding protein (TBP) and multiple TBP-associated factors (TAFs). TAFs are thought to act as coactivators mediating the interactions between the TBP, the CTD of RNAPII, TFIIB and acidic activator proteins (Kelleher et al. 1990; Pugh and Tijan 1990; Dynlacht et al. 1991; Flanagan et al. 1991; Meisterernst et al. 1991; Lai et al. 1992; Laurent and Carlson 1992; Takada et al. 1992; Wang et al. 1992b; see below). TATA-binding proteins encoded by HeLa, rat, yeast, *Drosophila* and plant *TBP/TFIIDτ* genes were found to be interchangeable in vitro. Comparison of TBP sequences revealed a phylogenetically conserved C-terminal domain which was shown to bind to the consensus TATA sequence (for review, see Gasch et al. 1990; Cormack et al. 1991; Gill and Tijan 1991; Greenblatt 1991b; Haaß and Feix 1992).

Measurement of RNA polymerase I, II and III activities in a TBP-deficient yeast mutant, *spt15*, led to the discovery that the TATA-binding protein is required for specific promoter recognition by all three eukaryotic RNA polymerases (Cormack and Struhl 1992; Schultz et al. 1992). Biochemical and genetic data so far indicate that general transcription factors (SL1, TFIID and TFIIIB/BRF1, see below) mediating RNAPI, II and III transcription in yeast, *Drosophila* and mammals contain only a single type of TATA-binding protein which is a functional equivalent of TBPs identified in plants. Studies of the crystal structure of *Arabidopsis* TBP protein correlate with the genetic data, and show association of two TBP molecules in a "concave" DNA-fold in which the basic repeat sits astride the DNA. Mutations destroying the DNA-binding map to this region, whereas others influencing RNA polymerase or species specificity cause alterations in the convex surface of the TBP dimer (Nikolov et al. 1992). Lysine residues on the convex surface of the TBF basic repeat are involved in the interaction with TFIIA, which stabilizes the TBP-TATA complex during PIC formation on TATA-box containing promoters (Lee et al. 1992; Nikolov et al. 1992).

Interaction between TBP and TFIIA is not essential for in vitro PIC assembly on core promoters. However, TFIIA is absolutely required for PIC formation when TBP is associated with TAFs in the TFIID complex, and for activator-dependent transcription (for review, see Roeder 1991; Conaway et al. 1992; Sayre et al. 1992a,b; Zhou et al. 1992). TFIIA, consisting of three proteins in HeLa, relieves the effect of inhibitors, such as Nc1, Nc2 and Dr1, which abolish class II gene transcription by binding to the TBP (Roeder 1991; Inostroza et al. 1992). Whereas TFIIA is required for basal transcription, a number of TAF proteins which associate with the TATA-binding protein in the TFIID complex may compete with TFIIA or enhance the interaction of the TFIID-TFIIA complex with gene-specific activators. Recently, using an epitope-labelled TBP, a holo-TFIID was purified from HeLa cells. This holo-TFIID supports transcriptional stimulation by different activation domains provided by transcription activators Sp1, E1a, Zta and GAL4-AH in the presence or absence of the TATA box. It cannot be excluded that on diverse promoters this holo-TFIID is associated with additional TAFs (i.e., transcription factors acting by protein-protein interaction; for discussion, see Pugh and Tijan 1990; Sharp 1992; Zhou et al. 1992). Different regulatory factors may therefore mediate diverse PIC assembly pathways. Thus, PIC assembly on promoters lacking the TATA box is not promoted by TFIIA, but requires TFII-I. TFII-I specifically binds to initiator sequences and acts cooperatively with TFIID (Roeder 1991).

The TFIIA-D-I complex on TATA-box containing promoters and the TFIID-I complex on TATA-less promoters interact with TFIIB. TFIIB-

TFIID interaction enhances the binding of RNAPII to the preinitiation complex. TFIIB has been cloned from yeast (gene *SUA7*) and HeLa cells, and identified as a protein that can associate with RNAPII in solution (Ha et al. 1991; Malik et al. 1991; Tschochner et al. 1992). TFIIB not only forms a bridge between TFIID and RNAPII, but also regulates the start-site selection by interaction with proteins located at the initiator sequences. TFIIB mutations were identified in yeast as suppressors of aberrant initiation (Pinto et al. 1992). Isolation of suppressors correcting a promoter A-block mutation in a tRNA gene and a mutation in the conserved basic domain of the TATA-binding protein led to the discovery of factor BRF, a TFIIB homolog, which is a subunit of the RNAPIII general transcription factor TFIIIB in yeast (Buratowski and Zhou 1992; López-De-León et al. 1992). TFIIB can specifically bind to the acidic activating region of sequence-specific transcription factors, such as GAL4 and HSV-1 VP16. By interacting with TFIIB, which can recruit RNAPII to promoters carrying TFIIA-D-I and TFII-D-I, acidic activators can dramatically increase the rate of PIC formation (Lin and Green 1991; see below).

6.1 The Role of TBP, and TFIIB Homologs in the Formation of Preinitiation Complexes with RNAPI, II and III

Early events leading to the entry of RNA polymerase into the preinitiation complex are very similar for RNAPI, II and III transcribed genes. PIC formation on RNAPI promoters requires two types of transcription factors: a species-specific promoter selectivity factor, SL1, and an upstream binding factor, UBF (for review, see Reeder 1990; Sollner-Webb and Mougey 1991). SL1 does not bind to DNA alone, but forms PIC by interacting with UBF and RNAPI. SL1 consists of the TATA-binding protein TBP, and three associated factors (TAFs). It is yet unknown whether any of the SL1-specific TAFs would be similar to TFIIB or BRF. General properties of UBFs are similar to those of sequence-specific activator proteins of RNAPII promoters (Bell et al. 1990; Jantzen et al. 1990, 1992). Activation of the transcription by the UBF-SL1-RNAPI complex is analogous to the process by which the TFIID-B-I complex and various sequence-specific activators mediate the formation of PIC with RNAPII on TATA-less promoters (Comai et al. 1992).

RNAPIII binds to PICs formed on TATA-less and TATA-containing promoters (Geiduschek and Tocchini-Valentini 1988; Mitchell and Tijan 1989; Reeder 1990; Gabrielsen and Sentenac 1991). TFIII general transcription factors recognize conserved sequence boxes A, B and C in RNAPIII transcribed promoters. In the promoter of 5S RNA genes the binding of TFIIIA (a protein with nine zinc fingers; see Theunissen et al. 1992) to box C is required for subsequent binding of TFIIIC and TFIIIB. In tRNA promoters TFIIIC binds to the A and B boxes and recruits TFIIIB alone. TFIIIA and TFIIIC function as enhancer and proximal element-binding factors which assemble with and position TFIIIB upstream of the transcription start site (Braun et al. 1992; Bartholomew et al. 1993). Transcription initiation on TATA-containing RNAPIII promoters requires TFIIIB and sequence-specific upstream activators (e.g., Oct-1). TFIIIB is a functional analogue of SL1 and TFIID complexes. In association with the TATA-binding protein TBP, TFIIIB carries two subunits, B′/BRF and B″ in yeast, and TAF-172 and TAF-L components in HeLa cells. In yeast the basic domain of TATA-binding protein interacts with the TFIIB-like TFIIIB′/BRF1 subunit and facilitates the assembly with the B″ subunit during binding to TFIIIC (Colbert and Hahn 1992; Kassavetis et al. 1992). Through the B″ subunit TFIIIB interacts with the C34 subunit of RNAPIII which thus enters into the TFIII(A)-C-B preinitiation complex (Bartholomew et al. 1993). The TAF-172 subunit of TFIIIB in HeLa cells functions similarly to BRF. By association with the TBP protein, TAF-172 prevents the TBP from interacting with TATA-box sequences of RNAPII promoters (Lobo et al. 1992; Taggart et al. 1992; White and Jackson 1992a,b; Rigby 1993). Thus, TFIIB, BRF, and probably an analogous subunit of SL1, together with other associated TAFs, can regulate the promoter and RNA polymerase specificity of the TATA-binding factor TBP.

6.2 Transcription Initiation: The Role of TFIIE, F, J/G, H, S and X

The association of RNA polymerase II with promoter-bound TFII(A)DB(I) complexes is aided by RNAPII-associated proteins (RAPs; for review, see Greenblatt 1991a). RAP30 and RAP70 associate to form a tetrameric ($\alpha_2\beta_2$) general transcription factor, TFIIF, which binds to RNAPII through the RAP30 subunit. TFIIF can abolish the aspecific DNA binding activity of RNAPII, and together with TBP and TFIIB, target the RNAPII to the TATA box in vitro (Flores et al. 1991; Finkelstein et al. 1992; Killeen and Greenblatt 1992; Killeen et al. 1992). A second class of RAP factors, TFIIE, binds to PIC together with TFIIF and RNAPII. TFIIE is also a heterotetrameric factor carrying two α (57 kDa) and two β (34 kDa) subunits. β is a modulator of the activity of the RNAPII-binding α-subunit. After binding of the TFII(A)DB(I)-RNAPII-TFIIE/F complex, TFIIE and -F are positioned downstream of the transcription initiation site and induce a dramatic conformational change in the PIC by clearing the start position (Buratowski et al. 1989, 1991; Ohkuma et al. 1990). A further RNAPII associated protein, RAP38 (also known as elongation factor TFIIS), acts as antiattenuator during elongation by preventing the pausing of RNAPII (Izban and Luse 1992). TFIIF, E and S probably remain associated with the elongation complex, and together with TFIIX, increase the rate of mRNA synthesis (Bengal et al. 1991). To convert the initiation complex to an active elongation complex, the binding of two additional factors, TFIIH and J/G, is required. These factors catalyze the phosphorylation of the C-terminal domain of RNA polymerase II which is concomitant with a striking increase in ATP hydrolysis (see below; Sumimoto et al. 1990; Lu et al. 1991, 1992). This general pathway of PIC formation and transcription elongation appears to be phylogenetically conserved, although certain differences exist between the properties of the general transcription factors in diverse organisms (e.g., TFIIE and yeast factor a; for review, see Buratowski et al. 1989; Conaway et al. 1992; Sayre et al. 1992a,b).

6.3 Regulatory Interplay Between TFIID, TFIIB, Sequence-Specific Activators and the C-Terminal Domain of RNA Polymerase II

The assembly of PIC is accelerated by interactions between upstream sequence binding factors, TFIID, TFIIB and the CTD of RNAPII (Greenblatt 1991b; Roeder 1991; Guarante and Bermingham-McDonogh 1992). TFIIB and TBP/TFIID have been shown to specifically interact with activation domains of diverse sequence-specific transcriptional factors (Stringer et al. 1990; Lin and Green 1991; Nikolov et al. 1992). Certain transcriptional activators, such as the USF factor of the Ad-2 MLP promoter, Sp1, the plant zipper TGA1a and yeast GAL4 and GCN4, can function in heterologous transcription systems in vitro and in vivo (Horikoshi et al. 1988a; Kakidani and Ptashne 1988; Webster et al. 1988; Mitchell and Tijan 1989; Katagiri et al. 1990; Meisterernst et al. 1990; Ptashne and Gann 1990). However, in contrast to the TFIID complex, the TBP protein alone fails to promote activator-dependent transcription in vitro. An extensive search for "coactivators, adaptors and mediators" which can "tether or squelch" the effect of transcription activators on the TBP-TFIIB-RNAPII-TFIIE/F complex led to the identification of TAFs and other activator or suppressor proteins (Berger et al. 1990; Kelleher et al. 1990; Pugh and Tijan 1990; Smale et al. 1990; Dynlacht et al. 1991). USA is an interesting representative of such a class of factors. USA can repress basal transcription mediated by general transcription factors, but stimulates transcription in response to activators Sp1, USF, GAL4 or GCN4, particularly on chromatin substrates. USA appears to carry negative (NC1, NC2) and positive (PC1) elements which are probably able to bind TBP (Meisterernst et al. 1991; Roeder 1991).

Genetic and biochemical data demonstrate that a number of transcription factors which bind to TFIID/TBP and TFIIB can also interact with the CTD domain of the largest subunit of RNAPII (Allison and Ingles 1989; Brandl and Struhl 1989; Stringer et al. 1990; Liao et al. 1991; Peterson et al. 1991; Koleske et al. 1992). Thus, the CTD may be considered a major mediator/coactivator of gene-specific transcription factors and of the TATA-binding protein (Liao et al. 1991). The consensus CTD heptameric peptide repeat, when used as an affinity matrix, can quantitatively remove TFIID or TBP from nuclear extracts, indicating that a tight interaction exists between the CTD and TBP. This interaction is regulated by phosphorylation. The phosphorylated form of CTD cannot bind to TBP, which supports the observation that binding to the preinitiation complex requires an unphosphorylated form (IIa) of the largest subunit of RNAPII (Lu et al. 1991; Chesnut et al. 1992; Conaway et al. 1992; Usheva et al. 1992). It is thus possible that, by titrating out TBP, the CTD of RNAPII can regulate transcription by RNA polymerase I and III.

The interaction between CTD and TBP may be abolished by viral oncoproteins, such as the E1A protein of adenovirus, which can specifically bind to the TBP (Horikoshi et al. 1991). E1A has been shown to activate transcription through the action of several transcription factors, such as ATF-2 or E2F (Horikoshi et al. 1988b). Genetic experiments indicate that E1A can directly interact with ATF-2 by forming a bridge to the TATA-binding factor (Lee et al. 1991). In addition to viral genes, the synthesis of E1A protein in adenovirus-infected cells *trans*-activates the transcription of several cellular genes including a heat-shock gene, *hsp70*. Interestingly, the activator-dependent transcription from the Ad-2 MLP and *hsp70* promoters is not affected by the deletion of the carboxyl-terminal domain of the largest subunit of RNAPII in vitro (Buratowski and Sharp 1990; Zehring and Greenleaf 1990; see below).

The interaction between CTD, TFIID, TFIIB and transcription activators may serve as a "short-cut" in the PIC assembly. Once elongation is initiated, the TFIID-B complex remains bound to the TATA-box region thus providing a potential entry site for RNAPII in recurrent transcription cycles (Rice et al. 1991; Roeder 1991). Multiple rounds of transcription on class II promoters are not promoted by general transcription factors, but require a reinitiation factor (RTF; Szentirmay and Sawadogo 1991) and/or transcription activators in vitro. Activator proteins, such as Oct-2, ATF, GAL4 or the EBV Zta protein, promote PIC assembly by interaction with TFIID, TFIIB and possibly CTD, and are continuously required at the promoter for recurrent initiation (for review, see Liebermann and Berk 1991; Arnosti et al. 1993). The effect of remote activators is mediated to proximal activators, and further to CTD, TFIID and TFIIB, by protein-protein interactions, resulting in selective *trans*-activation of various promoters by SV40 T, herpes simplex virus VP16 and adenovirus E1A oncoproteins (Johnson and McKnight 1989; Lai et al. 1992; Seipel et al. 1992; Tanaka et al. 1992).

Transcription, translation, nuclear translocation, DNA- and protein-binding activity of *trans*-acting factors is regulated by phosphorylation. GAL4, which can interact with the CTD, TFIID and TFIIB, is for example active when phosphorylated (Sadowski et al. 1991). The CTD of RNAPII, GAL4, Oct-1, E2F and many other transcription factors is phosphorylated during transcription initiation and at defined stages of the cell cycle (Hunter and Karin 1992).

7 Cell Cycle Regulation of Transcription and DNA Replication

Transcription in dividing cells is overruled by a cell cycle control which, by many interlocked signalling pathways and feedback mechanisms, regulates DNA synthesis, cell growth and division. All these mechanisms directly or indirectly affect the activity, assembly and stability of cyclin-dependent protein kinase complexes which determine the timing of cell cycle progression.

It is thought that in yeast the "cell cycle clock" contains a stable Thr/Ser-specific protein kinase, p34, which is the product of the *CDC28* and *cdc2* genes of *S. cerevisiae* and *S. pombe*, respectively. The p34 kinase is activated by binding to cyclins. The transcription, translation and stability of cyclins are regulated throughout the cell cycle. At the "Start", in late G1, the p34 kinase forms different complexes with "G1-specific" cyclins (CLN1–3) and activates DNA synthesis and replication of the microtubule organizing centre (MTOC). The "Start" is one of the checkpoints controlling cell proliferation by external and internal stimuli that influence the activity of both cyclins and CDC2/CDC28 kinases. After passing the "Start", B-type cyclins accumulate during the S-phase and bind to the CDC2/CDC28 kinase. The cyclin B-p34 complexes, also called pre-MPFs (i.e., mitosis promoting factors), are inactivated by Wee1/Mik1 kinase-mediated phosphorylation. A protein phosphatase, CDC25, converts pre-MPF in fission yeast to an active MPF which induces entry into mitosis at the G2/M boundary. Entry into mitosis is a second checkpoint at which the cell cycle is subjected to a complex physiological control involving feedback mechanisms sensing DNA replication and repair. MPF together with an array of interacting protein kinases and phosphatases is included in the activation of major mitotic events, such as chromosome condensation, destruction of the nuclear envelope and assembly of the mitotic spindle. The exit from mitosis is a third potential checkpoint for physiological control, such as regulation of MPF-activated pathways of protein degradation. Degradation of polyubiquitinated cyclins leads to inactivation of the MPF and release of CDC2 kinase for reinitiation of the cell cycle (for reviews, see Hartwell and Weinert 1989; Murray and Kirschner 1989; Draetta 1990; Nurse 1990; Enoch and Nurse 1991; Hoekstra et al. 1991; Reed 1991; Murray 1992).

Transcription, and the synthesis, repair, and mitotic and meiotic transmission of DNA, represent major downstream processes controlled by the cell cycle. Mutations and inhibitors affecting these processes are known to regulate the activity of CDC2-cyclin complexes at all three checkpoints of the cell cycle (for review and references, see Murray 1992). A simple model explaining coordinate control of transcription and DNA synthesis was derived from the genetic analysis of "Start" in yeast (Andrews and Herskowitz 1990). As described above, deletions of RNAPII CTD and suppressor mutations compensating the effect of CTD truncations cause a cell cycle arrest in G1 phase. Functions sensing the intactness and phosphorylation of RNAPII CTD could thus be involved in the regulation of DNA synthesis. To compensate for the truncation of the CTD, a mutation destroying the function of the SIT4 phosphatase is required in *S. cerevisiae*. This *sit4* mutation results in a deficiency of SWI4/SWI6 transcription factors, which in turn abolishes the expression of genes involved in DNA synthesis and decreases the level of G1/S-specific CLN cyclins. Due to the lack of cyclins, the formation of CDC28/CDC2-CLN cyclin complexes would be inhibited, preventing the phosphorylation and subsequent nuclear transport of SWI6 (and other transcription factors), and the G1/S phase transition in yeast (Lowndes et al. 1991, 1992; Nasmyth and Dirick 1991; Ogas et al. 1991; Fernandez-Sabaria et al. 1992; Johnston and Lowndes 1992; Merrill et al. 1992). The synthesis of CLN cyclins is coordinately regulated by multiple signalling pathways, the major elements of which are heterotrimeric G-proteins, membrane receptors, protein kinases (i.e., TPKs, STE11, STE7, FUS3, KSS1), and transcription factors (i.e., STE12, SWI1–6). In response to starvation or mating-type pheromones these pathways in yeast negatively regulate the levels of CLN cyclins, e.g., by FAR1, FUS3 and SWI6 functions, and cause a G1 cell cycle arrest. An important modulator of this cAMP-dependent pathway is BCY1, the regulatory subunit of A-type protein kinases. At low cAMP concentrations BCY1 inhibits the A kinases and cause a G1 cell cycle arrest, whereas in response to high cAMP levels BCY1 activates A kinases and cell division. The lethality of *sit4/bcy1* double mutants indicates the complementarity of cell cycle control and cAMP-dependent signalling pathways (Sprague 1991; Lew et al. 1992). How these pathways recognize the truncation and phosphorylation of the RNAPII CTD remains an intriguing question.

The situation in plants and other eukaryotes is somewhat more complex. Mammals encode at least nine different cyclins (A, B1–2, C, D1–3, E and F) which have been identified by complementation of yeast *CLN* cyclin mutations. These cyclins may associate with multiple cyclin-dependent kinases, CDC2/CDK1, CDK2, CDK3, CDK4 and CDK5. However, certain restrictions in cyclin-CDK combinations exist. Cyclin D in association with CDK2, CDK4 and CDK5 is now believed to regulate the "Start" in late G1, cyclin E-CDK2 acts in G1-S transition, whereas cyclin A-CDK2 is required for S-phase progression. In late S and G2 phases, cyclin A-CDC2 is a possible activator of cyclin B1-CDC2 and/or cyclin B2-CDC2, which are the regulators of G2-M phase transition (for review, see Clarke et al. 1992; Pagano et al. 1992a,b; Hunter 1993). Recent data indicate that the regulation of the yeast cell cycle may be even more complex as outlined above, and rather comparable in complexity to that of other eukaryotes (Courchesne et al. 1989; Elion et al. 1991; Surana et al. 1991; Yoon and Champbell 1991; Sorger and Murray 1992; Tsuchiya et al. 1992). Characterization of CDK-type kinase and cyclin genes in plants also supports the emerging view on the phylogenetic conservation of cell cycle control elements (John et al. 1989; Doonan 1991; Feiler and Jacobs 1991; Ferreira et al. 1991; Hata 1991;

Hata et al. 1991; Hirayama et al. 1991; Hirt et al. 1991; Hemerly et al. 1992; Nitschke et al. 1992).

7.1 Oncogenes and Tumor Suppressors: Conserved Functions in the Cell Cycle

G1 arrest of the cell cycle induced by starvation or mating-type pheromones in yeast is suppressed by dominant mutations which either cause an overproduction of CLN cyclins or prevent their ubiquitin-mediated degradation (see Cross and Tinkelenberg 1991; Reed 1991; Tyers et al. 1992). Similarly, in animal tumors chromosome translocations and DNA virus insertions were observed to increase the expression of D- or A-type G1/S cyclins. In mammals the passing of "Start or restriction point" at G1 activates the cell cycle by overcoming the G2/M control observed in *S. cerevisiae*. Overexpression of cyclin D is implicated in parathyroid and lymphoid neoplasias (for review, see Motokura and Arnold 1993). G1 progression in mammals, as in yeast, is controlled by limiting nutrients, and can be stimulated by several peptide growth factors. Some of these growth factors, such as CSF-1, activate the transcription of D-type cyclin genes (Matsushime et al. 1991). Cyclins D1, 2 and 3 form complexes with CDK2, CDK4 and CDK5 kinases, which are located in the nucleus during G1. Cyclin D1 was found in association with the proliferating cell nuclear antigen PCNA (Xiong et al. 1992). PCNA is the regulatory subunit of DNA polymerase δ which, together with DNA polymerase α, functions in DNA replication and repair (Yang et al. 1992). Similarly to cyclin D, cyclin A is located in the nucleus from mid-G1 to G2, until it is replaced by cyclin B. Cyclin A overexpression was observed in primary liver cancers, which in a few cases result from hepatitis B virus integration into cyclin A genes (see Bréchot 1993). Cyclin A is known to function at the G1/S transition and in S-phase when it is associated with CDK2, whereas in G2 phase it is bound to CDC2. Cyclin A plays a major role in viral oncogenesis. It has been identified in complex with the E1A protein of adenovirus type 5, as well as in complexes carrying the CDK2 kinase and an associated protein, p107 (for review, see Moran 1993).

As described above, E1A can form a bridge between transcription factors (i.e., ATF-2/CRE-BP1 cAMP response factors) and the TATA-binding protein TBP, and thus mediate *trans*-activation of diverse cellular genes including *N/c-myc*, *c-myb*, *c-fos*, *c-jun* and *hsp70* (Horikoshi et al. 1988b, 1991; Hiebert et al. 1989; Mudryj et al. 1990; Simon et al. 1988; Lee et al. 1991; Shi et al. 1991). E1A and related DNA virus oncoproteins, such as the SV40 T antigen or the E7 protein of human papillomavirus (HPV), have a second major function in the initiation of G1 cell cycle transition and DNA replication. This function is based on their capability to distrupt protein complexes suppressing the cell cycle in differentiated cells. A target for the E1A protein is the retinoblastoma suppressor protein RB, which in

G1 forms a complex with the transcription factor E2F. Other E1A targets are the cyclin E-CDK2-p107 complex in G1 and the cyclin A-CDK2-p107 complex in S, which also carry an E2F subunit (see Devoto et al. 1992; Nevins 1992). E1A dissociates E2F from RB-E2F and cyclin A/E-CDK2-p107 complexes which results in the activation of E2F by phosphorylation (see Bagchi et al. 1989, 1990; Moran 1993).

The recognition sequence and regulatory spectrum of E2F are similar to those of transcription factors SWI4/SWI6, which in yeast activate the transcription of genes encoding cyclins CLN1 and 2, DNA polymerase α and other enzymes involved in DNA synthesis (Andrews and Herskowitz 1990; Gordon and Campbell 1991; Johnston and Lowndes 1992; Merrill et al. 1992). In mammals E2F belongs to a family of homologous factors, binding sites of which have been identified in the promoter region of genes required for G1-S transition and DNA synthesis (i.e., *Myc*, *myb*, dihydrofolate reductase, thymidilate synthase and ribonucleotide reductase). E2F binding sites are also present in the promoter of cyclin A gene (Bréchot 1993). E1A-mediated dissociation of an E2F factor from cyclin E-CDK2-p107 complexes therefore may induce the transcription of the cyclin A gene which in turn leads to the activation of the cyclin A-CDK2-p107-E2F complex in G1/S (Lees et al. 1992; Pagano et al. 1992b). Phosphorylation of E2F results in the transcription of human DNA polymerase α gene. As in yeast, DNA synthesis in mammals is coordinately regulated by signalling pathways responding to growth factor stimuli (Bagchi et al. 1989; Pearson et al. 1991). Whereas the transcription of DNA polymerase α is activated by E2F and CDK2 kinase in G1-S, the enzyme is inactivated by CDC2-cyclin B-mediated phosphorylation in G2/M (Nasheuer et al. 1991).

The adenovirus E1A oncoprotein binds to the retinoblastoma suppressor RB by splitting the RB-E2F complex. The released E2F in turn activates the cell cycle and DNA replication. Formation of RB-E2F complex in differentiated cells inhibits the cell cycle. Unlike its homologue p107, RB cannot bind cyclins and thus prevents the formation of E2F-cyclin-CDK complexes (Ewen et al. 1991, 1992; Goodrich et al. 1991; Cobrinik et al. 1992; Faha et al. 1992; Hamel et al. 1992). In contrast to RB, p107 may perform a positive regulatory function by recruiting E2F and cyclin E/A to activate the CDK2 kinase (Cao et al. 1992). RB inhibits the E2F-induced expression of *CDC2* kinase, cyclin A, *c-myc* and *c-myb* genes (Dalton 1992; Hiebert et al. 1992; Pagano et al. 1992a). c-Myb counteracts RB by activating the *cdc2* gene in response to growth factor stimuli (Ku et al. 1993). It is intriguing that the transcription of a *cdc2/cdk* gene was also observed to be under growth factor control in higher plants (Hirt et al. 1991). In mammals RB can be inactivated in somatic cells by protein kinases which induce RB phosphorylation in response to growth factor and mitogenic stimuli (Chen et al. 1989). During the cell cycle, RB is phosphorylated in at least three steps (in G1, S and G2/M) probably by the CDK4-cyclin D, CDK2-cyclin E and CDK2/CDC2-cyclin A kinases (Hinds et al. 1992; Matsushime et al. 1992; Shirodkar et al. 1992). During mitosis, RB binds to the catalytic subunit of phosphoprotein phosphatase PP1 and is activated by dephosphorylation (for review, see Hollingsworth et al. 1993).

RB alone, or in complex with E2F and E1A, can act as transcriptional repressor or activator of E2F- and E1A-regulated genes, respectively. Nuclear localization of RB, p107, E2F, E1A, CDK2 complexes and collaborating phosphatases indicate that these proteins can interact with general and site-specific transcription factors and RNA polymerases to coordinately regulate DNA replication and transcription. In contrast to E2F, the RB-E2F complex was shown to inhibit the activation of adenovirus promoters (Weintraub et al. 1992). RB acts as a tissue-specific activator of several genes. Thus, RB expression in lung epithelial cells stimulates, whereas in mouse fibroblasts represses the *c-fos* promoter (Robbins et al. 1990). RB binds to similar sequence elements as the transcription factor Sp1 (RB control elements, RCEs). RB control elements occur in combination with E2F and Myc/Max binding sites in many promoters (Blackwood et al. 1992; see regulation of *hsp70* promoter above).

Overexpression of c-Myc in various tumors can override the RB-caused G1 arrest. The amino-terminal domain of c/N-Myc was observed to interact with RB in vitro, and plays a role in the downregulation of RB activity (Rustgi et al. 1991; Evan and Littlewood 1993). The N-terminal domain of the E1A oncoprotein can interact in an analogous way with RB and with other similar putative tumor suppressor proteins, such as p130 and p300. p300 is a ubiquitous nuclear protein conserved between mammals and is involved in the regulation of genes encoding members of AP transcription factor and Hsp70 heat-shock protein families (see Moran 1993).

Coregulation of transcription and DNA replication is further illustrated by the fact that RB can also interact with the SV40 T antigen. The SV40 T antigen and topoisomerase I can efficiently initiate DNA replication in the presence of a replication factor RF-S. RF-S contains a CDK-cyclin complex and a replication protein, RPA, which is activated by phosphorylation during G1 (D'Urso et al. 1990; Dutta and Stillman 1992; Fotedar and Roberts 1992; see above). The SV40 T antigen also binds to p53, a second class of major tumor suppressor proteins (see Perry and Levine 1993). In animals, p53 acts as a negative regulator of the cell cycle, which in response to DNA damage (i.e., γ-irradiation) causes a G1 arrest (Kastan et al. 1992). When expressed in yeast, p53 similarly induces a G1 arrest which is stimulated by coexpression of human CDC2Hs (Nigro et al. 1992). p53 has been found in association with cell cycle kinase complexes in transformed cells, and to be phosphorylated in vitro by CDK2/CDC2-cyclin A and CDC2-cyclin B complexes (Bischoff et al. 1990; Milner et al. 1990). CDC2/CDK2 has been found in association with a complex consisting of SV40 T antigen, cyclin A and RF-A single-stranded DNA-binding protein in replication forks, which provided an indication for an interaction between CDC2/CDK2-cyclin A and p53 in vivo (see above; Dutta et al. 1991). Although the role of p53 in cell cycle progression is not completely deciphered, it is likely that p53 is phosphorylated, as RB, during late S and G2. In contrast to RB, p53 is inactivated by dephosphorylation during mitosis.

Similarly to RB, p53 is also a transcription factor, which binds to specific sequences located upstream of rRNA genes and *c-fos* (Farmer et al. 1992; Zambetti et al. 1992). p53 may thus be implicated in the regulation of rRNA transcription. Moreover, p53 was found to bind to the *hsp70* promoter and to inhibit heat-shock gene expression by interacting with the CCAAT factor, a component of the differentiation switch (for review, see Umek et al. 1991; Rorth and Montell 1992; Agoff et al. 1993). p53 also binds to sequences adjacent to the SV40 replication origin and inhibits DNA replication through competition with the SV40 T antigen (Bargonetti et al. 1991, 1992). Conversely, the SV40 T antigen was observed to inhibit p53 in *trans*-activation or repression of diverse promoters (Mietz et al. 1992). As tumor suppressor, p53 performs a redundant function with RB, while as a transcription factor it counteracts RB.

7.2 Cell Cycle Regulation of RNAPII: CTD and CDC Kinases

From the data described above it is apparent that the transcriptional activity of RNA polymerases is regulated by many diverse pathways and cellular functions. Is there a major mechanism modulating the activity of the RNA polymerase enzymes themselves? So far one can logically argue both ways. No, because evidently the transcription of RNAP genes, the translation, secondary modification and assembly of RNAP subunits and their interaction with transcription factors provide endless combinations for regulation. Yes, because many connections between cell cycle regulation, transcription factors, *trans*-activating oncoproteins and tumor suppressors point to the role of RNA polymerase II CTD as a major mediator between transcription and diverse regulatory pathways. In the light of the complexity of these regulatory interactions it is not surprising that there is considerable confusion in explaining the role of the CTD (see Corden 1990; Sawadogo and Sentenac 1990; Woychik and Young 1990; Young 1991).

Once assembled, the activity of RNAPII is primarily regulated by phosphorylation of the C-terminal YSPTSPS repeated domains (CTD) of the largest subunit. As reviewed above, the CTD plays a major role in interactions with the TATA-binding factor TBP, transcription activators and repressors, and is also involved in the regulation of the "Start" during the cell cycle. These functions of the CTD are reminiscent of those of viral oncoproteins. Phosphorylation of the CTD provides a twofold regulation. CTD phosphorylation of "free" RNAPII inactivates the enzyme and prevents its binding to promoters. Once RNAPII is stably bound to preinitiation complexes in the chromatin, however, the phosphorylation of CTD is required for the activation of RNAPII and initiation of transcription. Data described above show that phosphorylation of the CTD in initiation complexes is concomitant with the phosphorylation of transcription activators and repressors which interact with the CTD. Consequently, protein kinases phosphorylating the CTD are integral components of transcriptionally active chromatin. It is logical to assume that multiple protein kinases involved in the regulation of the cell cycle, transcription, DNA replication and repair control the phosphorylation of the CTD.

In fact, studies of CTD-specific protein kinases indicate that this assumption is correct. Thus far, several classes of CTD kinases have been isolated from yeast (Lee and Greenleaf 1989), *Aspergillus* (Stone and Reinberg 1992), plants (Guilfoyle 1989), mouse (Cisek and Corden 1989, 1991; Zhang and Corden 1991a,b), rat liver (Serizawa et al. 1992) and human cells (Dvir et al. 1992; Lu et al. 1992; Peterson et al. 1992; Gottlieb and Jackson 1993; Payne and Dahmus 1993). Because of the double effect of CTD phosphorylation, these kinases can be classified into two families: (a) template- or DNA-bound kinases which cannot phosphorylate histone H1 but are able to phosphorylate the RNAPII when bound to PIC or DNA, and (b) cell cycle kinases which display histone H1 kinase activity in vitro and can efficiently phosphorylate the RNAPII CTD in solution.

Template-dependent CTD kinases belong to two families. The first family includes general transcription factors, such as TFIIH from HeLa cells, δ factor from rat liver, and factor b from yeast which catalyze ATP hydrolysis and phosphorylation of the RNAPII CTD during the initiation of transcription (Arias et al. 1991; Feaver et al. 1991; Conaway et al. 1992; Flores et al. 1992; Lu et al. 1992; Serizawa et al. 1992; Payne and Dahmus 1993). TFIIH factors from HeLa cells were shown to contain two to five subunits (92/95, 62, 43, 40 and 35/33 kDa), whereas the most purified mouse δ factor carries eight subunits (94, 85, 68, 46, 43, 40, 38 and 35 kDa), and the yeast factor b consists of three subunits (85, 75 and 50 kDa). All three factors catalyze the ATP-dependent phosphorylation of the CTD after binding to PIC. TFIIE is a major stimulator of the TFIIH CTD kinase activity and is required for the entry of TFIIH into the preinitiation complex. TFIIJ binding to PIC further increases the activity of TFIIH. TFIIH alone or in complex with general transcription factors TFIIADBEF poorly phosphorylates the CTD in solution. The CTD kinase activity of TFIIH is greatly stimulated by DNA templates with TATA or initiator elements. The DNA-dependent ATPase activity of TFIIH-like factors accounts for a helicase function which was thought to be performed by TFIIE or F previously (Greenblatt 1991a). Properties of the less characterized *Aspergillus* KI and HeLa CTDK2 kinases resemble those of the TFIIH family. KI has been shown to phosphorylate Ser at the fifth position of YSPTS*PS repeats, thus other members of the TFIIH family are also expected to display a similar substrate specificity (Stone and Reinberg 1992). TFIIH and related kinases are required for PIC formation and fulfil a CTD kinase function required in both differentiated and dividing cells.

The second family of DNA-dependent kinases is represented by the human Ku antigen-associated DNA-PK CTD kinase (Dvir et al. 1992; Peterson et al. 1992; Gottlieb and Jackson 1993). DNA-PK is a well-characterized protein kinase of 300/350 kDa. DNA-PK-Ku phosphorylates the CTD at both Ser the Thr residues but only when RNAPII is bound to DNA or in the preinitiation complex. Moreover, DNA-PK specifically phosphorylates the Ku antigen, p53, the SV40 T antigen, and the transcription factors Sp1, Oct1, Oct2, c-Fos and c-Myc in DNA-bound forms (see Gottlieb and Jackson 1993). Ku is known to bind to double-stranded DNA ends, and is associated with chromatin in a cell cycle-dependent manner. Ku binds to DNA ends aspecifically in cooperative interaction with several transcription factors (see above substrates for DNA-PK) and slides to specific internal sequences which are similar to Sp1, Oct1, Oct2 and Ap1 (Jun, Fos, Myc) recognition sites. DNA-PK-Ku may have multiple functions. By phosphorylation of the CTD, DNA-PK can probably inhibit the aspecific binding of RNAPII to free DNA ends. Phosphorylation of Thr residues in the CTD suggests that DNA-PK may also phosphorylate RNAPII during elongation which may lead to irreversible inactivation of RNAPII in the absence of Thr-specific phosphatases. DNA-PK may thus arrest transcription in response to DNA damage. The p70 subunit of Ku is also referred to as transcription factor PSE1/TREF1/TREF2, which is an activator of the U1 snRNA promoter (Reeves and Sthoeger 1989; Gunderson et al. 1990; Knuth et al. 1990). Therefore, an alternative function for DNA-PK-Ku could be the inhibition of RNAPII activity in response to transcription by RNAPI or RNAPIII.

The second major class of CTD kinases is represented by the cyclin-dependent cell cycle kinases described above. Using CTD heptapeptide repeats as substrates Cisek and Corden (1989) have identified CTD kinase activities in mouse, human, hamster, and yeast cell extracts. Purification of CTD kinase activities from mouse Ehrlich ascites cells resulted in two major enzymes, E1 and E2, which phosphorylate the CTD at serines in the second and fifth positions of the YS*PTS*PS heptamer. E1 has proved to be a G2/M-specific cell cycle kinase complex consisting of a p34 kinase and a p62 cyclin B subunit. E2 carried the p34 CDC2 kinase subunit in combination with p58, a yet unidentified cyclin A-like subunit (Cisek and Corden 1991; Zhang and Corden 1991a,b). Lee and Greenleaf (1989) detected similar CTD kinase activities in yeast, *Drosophila* and HeLa cells, and purified an enzyme from *S. cerevisiae* which consists of 58, 38 and 32 kDa subunits (Lee and Greenleaf 1989). A related CTD kinase, KIII, displaying histone H1 kinase activity, was purified from *Aspergillus*, but did not cross-react with antibodies raised against the conserved PSTAIR motif of CDK2/CDC2 kinases (Stone and Reinberg 1992).

7.3 A Plant CTD Kinase Provides a Link Between Regulation of RNAPII Transcription, DNA Replication and Cell Cycle

Recently, we have initiated an extensive purification and characterization of cell cycle kinase complexes from alfalfa cell cultures, and analyzed their CTD kinase activity using either CTD fusion proteins or RNAPII holoenzyme as substrates. CDC2/CDK2 kinases were identified by antibodies raised against their conserved PSTAIR motif and purified by means

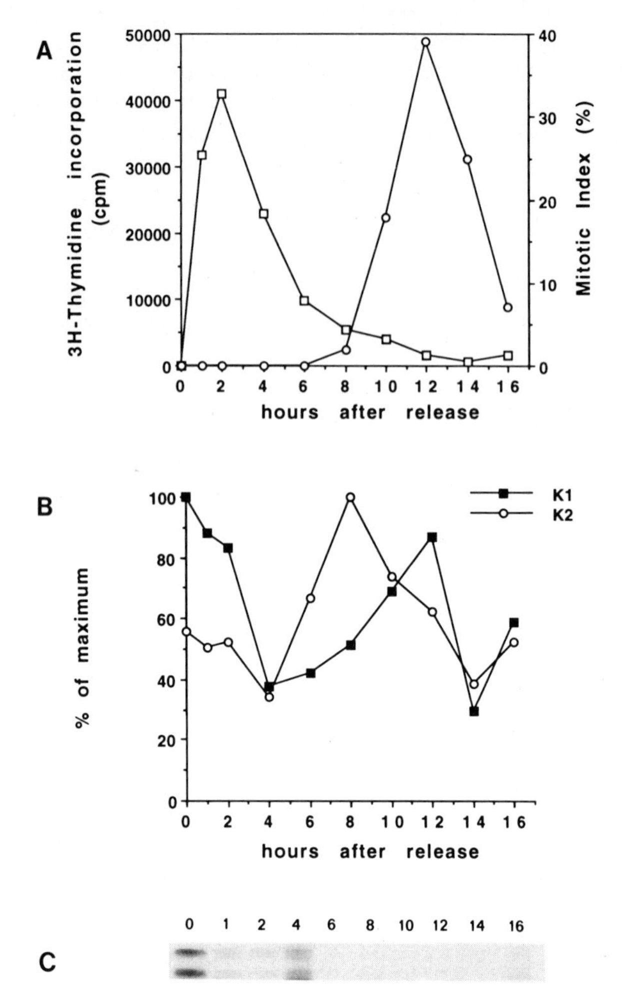

Fig. 1. Oscillation of CTD kinase activity and transition of diverse cell cycle kinase complexes throughout the cell cycle in plants. **A** DNA synthesis and mitotic division in an alfalfa cell suspension following the release of aphidicolin caused G1/S phase arrest. *Squares* show [^{3}H]thymidine incorporation in pulse-labelled cells after removing the DNA synthesis inhibitor; *circles indicate* the mitotic index (%) determined by counting an average of 1000 nuclei. **B** Cell extracts, prepared at defined time intervals after the release of the aphidicolin block, were resolved by DEAE-Sepharose chromatography to *K1* and *K2* cell cycle kinase fractions, which were characterized previously. Equivalent amounts of proteins from K1 and K2 fractions were bound to p13-Sepharose beads, and after extensive washes, the matrix-bound CDC2/CDK2 activities were determined using histone H1, as substrate. **C** Oscillation of CTD kinase activities during the cell cycle. K1 kinase fractions described above were incubated with a CTD fusion protein carrying the CTD domain of *Arabidopsis* RNAPII fused to *Staphylococcus* protein A, and separated by SDS-PAGE. The autoradiography shows typical gel shifts of CTD-fusion proteins which are due to phosphorylation by CTD kinase(s) associated with the K1 enzyme fraction

of their ability to interact with p13, the product of yeast *Suc1* gene (Brizuela et al. 1987; Ducommun et al. 1991). Higher plants, as other eukaryotes, accumulate diverse CDK-cyclin complexes during the cell cycle. Alfalfa cells arrested at G1/S phase by aphidicolin (an inhibitor of DNA polymerase α) accumulate a mixture of CDK kinases, p31 and p32, which bind to p13, phosphorylate histone H1, display CTD kinase activity and occur in different complexes of about 300 kDa (Fig. 1, K1). After releasing the cells from the G1/S block, the CTD kinase activity decreases throughout the S-phase, but a second peak of CTD-kinase activity rises during the onset of G2/M and mitosis. A histone kinase activity appearing in G2/M is associated with complexes of 75 kDa which contain only the p32 kinase subunit, a homologue of yeast CDC28 kinase (Fig. 1, K2; Hirt et al. 1991). This G2/M phase-specific kinase, K2, however, does not phosphorylate the CTD, indicating either that in our experiment some cells have escaped the aphidicolin block and contaminated the G2-M samples, or that the CTD kinase activity is associated with another type of CDC/CDK kinase complex in G2/M. Preliminary data indicate that the p31 kinase subunit of K1 complexes, which mediate CTD phosphorylation, is probably a homologue of animal CDK2 kinases. As CDK2, the p31 subunit of K1 complexes can bind p13 and is recognized by PSTAIR antibodies. In addition to K1 and K2 complexes, PSTAIR-reactive proteins have also been detected in association with DNA, but failed to phosphorylate the RNAPII in solution (Bakó unpubl.).

The overall analogy of substrate specificity, timely distribution and complexity between plant and animal cell cycle kinase complexes prompted us to address the specific question whether plant cell cycle kinases would carry a subunit with E2F binding specificity. A screening for specific binding to the consensus E2F recognition sequence revealed that, similarly to animals, the G1/S specific plant cell cycle kinase fraction, K1, carries an E2F binding activity. Since the same kinase fraction was implicated in specific phosphorylation of the RNAPII CTD, it appeared logical to assay whether RNAPII binding to these cell cycle kinase complexes can regulate the E2F binding activity. As illustrated by Fig. 2, RNAPII binding to the G1/S kinase fraction K1 dramatically increases the E2F binding activity, probably by dissociation of an E2F-like factor from the K1 cell cycle kinase complex(es).

8 Conclusion

The role of RNAPII in interaction with cell cycle kinase complexes is probably multiple. In analogy to viral oncogenes, this process may induce DNA synthesis by E2F and inactivate the free RNAPII pool in G1/S phase. By phosphorylation of RNAPII the cell cycle kinase could also differentially regulate transcription, since phosphorylation abolishes the ability of the CTD to titrate out the TATA-binding protein, which in turn would enhance

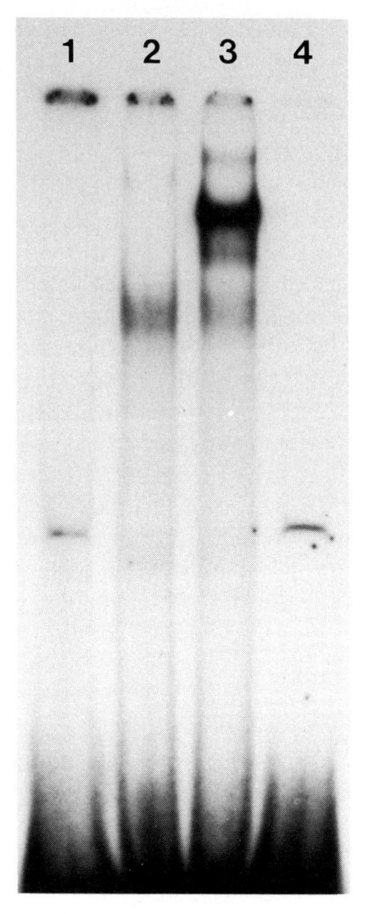

Fig. 2. E2F binding activity is associated with the K1 CTD kinase fraction and inducible by RNAPII. An oligonucleotide carrying the consensus TTTGGCGGGAA E2F binding site was end-labelled by T4 polynucleotide kinase and incubated in the presence of a 1000-fold excess of poly(dI/dC) competitor (*1–4*), with a 300-kDa fraction of purified K1 CTD kinase (*2*), with the K1 CTD kinase and tomato RNAPII holoenzyme (*3*), or with tomato RNAPII alone (*4*). A similar increase in E2F binding and formation of novel oligonucleotide-protein complexes were observed using RNAPII enzymes from alfalfa and wheat. These complexes could be specifically competed with unlabelled E2F oligonucleotides, but not with increasing amounts of poly(dI/dC) (data not shown)

the formation of SL1 and TFIIIB complexes required for RNAPI and RNAPIII transcription. As outlined above, certain genes would be able to escape this regulation, because transcription from their promoter may not require the CTD function. Alternatively, the binding of RNAPII to CDK-cyclin complexes could perform a *trans*-activating function by recruiting to promoters cell cycle kinases which would facilitate recurrent transcription by phosphorylating the RNAPII and interacting with transcription regulatory proteins. In animals a competition of RNAPII with viral oncoproteins, such as adenovirus E1A or SV40 T antigen, may ultimately result in the accumulation of free phosphorylated RNAPII and E2F, which would trigger the replication of viral DNAs and deregulate transcription. The data described above suggest that many additional regulatory pathways exist, the uncovering of which will contribute to further understanding of the connections between transcription, DNA replication and cell cycle control.

References

Agoff N, Hou J, Linzer DIH, Wu B (1993) Regulation of the human hsp70 promoter by p53. Science 259:84–87

Ahearn JM, Bartolomei MS, West M, Cisek LS, Corden JL (1987) Cloning and sequence analysis of the mouse genomic locus encoding the largest subunit of RNA polymerase II. J Biol Chem 262:10695–10705

Allison LA, Ingles CJ (1989) Mutations in RNA polymerase II enhance or suppress mutation in *GAL4*. Proc Natl Acad Sci USA 86:2794–2798

Allison LA, Moyle M, Shales M, Ingles CJ (1985) Extensive homology among the largest subunits of eukaryotic and prokaryotic RNA polymerases. Cell 42:599–619

Allison LA, Wong JK-C, Fitzpatrick VD, Moyle M, Ingles JC (1988) The C-terminal domain of the largest subunit of RNA polymerase of *Saccharomyces cerevisiae*, *Drosophila melanogaster*, and mammals: a conserved structure with an essential funtion. Mol Cell Biol 8:321–329

Andrews BJ, Herskowitz I (1990) Regulation of cell cycle-dependent gene expression in yeast. J Biol Chem 265:14057–14060

Archambault J, Drebot MA, Stone JC, Friesen JD (1992) Isolation and phenotypic analysis of conditional-lethal, linker-insertion mutations in the gene encoding the largest subunit of RNA polymerase II in *Saccharomyces cerevisiae*. Mol Gen Genet 232:408–414

Arias JA, Peterson S, Dynan WS (1991) Promoter-dependent phosphorylation of RNA polymerase II by a template-bound kinase. J Biol Chem 266:8055–8061

Arndt KT, Styles CA, Fink GR (1989) A suppressor of a *HIS4* transcriptional defect encodes a protein with homology to the catalytic subunit of protein phosphatases. Cell 56:527–537

Arnosti DN, Merino A, Reinberg D, Schaffner W (1993) Oct-2 facilitates funtional preinitiation complex assembly and is continuously required at the promoter for multiple rounds of transcription. EMBO J 12:157–166

Azuma Y, Yamagishi M, Ueshima R, Ishihama A (1991) Cloning and sequence determination of the *Schizosaccharomyces pombe rpb1* gene encoding the largest subunit of RNA polymerase II. Nucleic Acids Res 19:461–468

Bagchi S, Raychaudhuri P, Nevins JR (1989) Phosphorylation-dependent activation of the adenovirus-inducible E2F transcription factor in a cell-free system. Proc Natl Acad Sci USA 86:4352–4356

Bagchi S, Raychaudhuri P, Nevins JR (1990) Adenovirus E1A proteins can dissociate heteromeric complexes involving the E2F transcription factor: a novel mechanism for E1A *trans*-activation. Cell 62:659–669

Bargonetti J, Friedman PN, Kern SE, Vogelstein B, Prives C (1991) Wild-type but not mutant p53 immunopurified proteins bind to sequences adjacent to the SV40 origin of replication. Cell 65:1083–1091

Bargonetti J, Raynisdóttir I, Friedman PN, Prives C (1992) Site-specific binding of wild-type p53 to cellular DNA is inhibited by SV40 T antigen and mutant p53. Genes Dev 6:1886–1898

Bartholomew B, Dahmus ME, Meares CF (1986) RNA contacts subunits IIo and IIc in HeLa RNA polymerase II transcription complexes. J Biol Chem 261:14226–14231
Bartholomew B, Durkovich D, Kassavetis GA, Geiduschek EP (1993) Orientation and topography of RNA polymerase III in transcription complexes. Mol Cell Biol 13:942–952
Bartolomei MS, Corden JL (1987) Localization of an α-amanitin resistance mutation in the gene encoding the largest subunit of mouse RNA polymerase II. Mol Cell Biol 7:586–594
Bartolomei MS, Halden NF, Cullen CT, Corden JL (1988) Genetic analysis of the repetitive carboxyl-terminal domain of the largest subunit of mouse RNA polymerase II. Mol Cell Biol 8:330–339
Bell SP, Jantzen H-M, Tijan R (1990) Assembly of alternative multiprotein complexes directs rRNA promoter selectivity. Genes Dev 4:943–954
Bengal E, Flores O, Krauskopf A, Reinberg D, Aloni Y (1991) Role of the mammalian transcription factors IIF, IIS and IIX during elongation by RNA polymerase II. Mol Cell Biol 11:1195–1206
Berger SL, Cress WD, Cress A, Triezenberg SJ, Guarente L (1990) Selective inhibition of activated but not basal transcription by the acidic activation domain of VP16: evidence for transcriptional adaptors. Cell 61:1199–1208
Bird DM, Riddle DL (1989) Molecular cloning and sequencing of *ama-1*, the gene encoding the largest subunit of *Caenorhabditis elegans* RNA polymerase II. Mol Cell Biol 9:4119–4130
Bischoff JR, Friedman PN, Marshak DR, Prives V, Beach D (1990) Human p53 is phosphorylated by p60-cdc2 and cyclin B-cdc2. Proc Natl Acad Sci USA 87: 4766–4770
Blackwood EM, Lüscher B, Eisenman RN (1992) Myc and Max associate in vivo. Genes Dev 6:71–80
Brandl CJ, Struhl K (1989) Yeast GCN4 transcriptional activator protein interacts with RNA polymerase II in vitro. Proc Natl Acad Sci USA 86:2652–2656
Braun BR, Bartolomew B, Kassavetis GA, Geiduschek EP (1992) Topography of transcription factor complexes on the *Saccharomyces cerevisiae* 5S RNA gene. J Mol Biol 228:1063–1077
Bréchot C (1993) Oncogenic activation of cyclin A. Curr Opinion Genet 3:11–18
Brizuela L, Draetta G, Beach D (1987) $p13^{SUC1}$ acts in fission yeast cell division cycle as a component of the $p34^{cdc2}$ protein kinase. EMBO J 6:3507–3514
Buratowski S, Sharp PA (1990) Transcription initiation complexes and upstream activation with RNA polymerase II lacking the C-terminal domain of the largest subunit. Mol Cell Biol 10:5562–5564
Buratowski S, Zhou H (1992) A suppressor of TBP mutations encodes an RNA polymerase III transcription factor with homology to TFIIB. Cell 71:221–230
Buratowski S, Hahn S, Guarante L, Sharp PA (1989) Five intermediate complexes in transcription initiation by RNA polymerase II. Cell 56:549–561
Buratowski S, Sopta M, Greenblatt J, Sharp PA (1991) RNA polymerase II-associated proteins are required for a DNA conformation change in the transcription initiation complex. Proc Natl Acad Sci USA 88:7509–7513
Cadena DL, Dahmus ME (1987) Messenger RNA synthesis in mammalian cells is catalyzed by the phosphorylated form of RNA polymerase II. J Biol Chem 256:3332–3339
Cao L, Faha B, Dembski M, Tsai L-H, Harlow E, Dyson N (1992) Independent binding of the retinoblastoma protein and p107 to the transcription factor E2F. Nature 355:176–179
Carles C, Treich I, Bouet F, Riva M, Sentenac A (1991) Two additional common subunits, ABC10α and ABC10β, are shared by yeast RNA polymerases. J Biol Chem 266:24092–24096
Chao DM, Young RA (1991) Tailored tails and transcription initiation: the carboxyl terminal domain of RNA polymerase II. Gene Expr 1:1–4

Chen P-L, Scully P, Shew J-Y, Wang JYJ, Lee W-H (1989) Phosphorylation of the retinoblastoma gene product is modulated during the cell cycle and cellular differentiation. Cell 58:1193–1198
Chesnut JD, Stephens JH, Dahmus ME (1992) The interaction of RNA polymerase II with the adenovirus-2 major late promoter is precluded by phosphorylation of the C-terminal domain of subunit IIa. J Biol Chem 267:10500–10506
Cho KWY, Khalili K, Zadomeni R, Weinmann R (1985) The gene encoding the large subunit of human RNA polymerase II. J Biol Chem 260:15204–15210
Christmann JL, Dahmus ME (1981) Monoclonal antibody specific for calf thymus RNA polymerase II_o and II_A. J Biol Chem 256:11798–11803
Cisek LJ, Corden JL (1989) Phosphorylation of RNA polymerase by the murine homologue of the cell-cycle control protein cdc2. Nature 339:679–684
Cisek LJ, Corden JL (1991) Purification of protein kinases that phosphorylate the repetitive carboxyl-terminal domain of eukaryotic RNA polymerase II. Methods Enzymol 200:301–325
Clarke PR, Leiss D, Pagano M, Karsenti E (1992) Cyclin A- and cyclin B-dependent protein kinases are regulated by different mechanisms in *Xenopus* egg extracts. EMBO J 11:1751–1761
Cobrinik D, Dowdy SF, Hinds PW, Miitnacht S, Weinberg RA (1992) The retinoblastoma protein and the regulation of cell cycling. Trends Biochem 17:312–315
Colbert T, Hahn S (1992) A yeast TFIIB-related factor involved in RNA polymerase III transcription. Genes Dev 6:1940–1949
Comai L, Tanese N, Tijan R (1992) The TATA-binding protein and associated factors are integral components of the RNA polymerase I transcription factor, SL1. Cell 68:965–976
Conaway JW, Bradsher JN, Conaway RC (1992) Mechanism of assembly of the RNA polymerase II preinitiation complex. J Biol Chem 267:10142–10148
Corden JL (1990) Tails of RNA polymerase II. Trends Biochem 15:383–387
Corden JL, Cadena DL, Ahearn JM, Dahmus ME (1985) A unique structure at the carboxyl terminus of the largest subunit of eukaryotic RNA polymerase II. Proc Natl Acad Sci USA 82:7934–7938
Cormack BP, Struhl K (1992) The TATA-binding protein is required for transcription by all three nuclear RNA polymerases in yeast cells. Cell 69:685–696
Cormack BP, Strubin M, Ponticelli AS, Struhl K (1991) Functional differences between yeast and human TFIID ara localized to the highly conserved region. Cell 65:341–348
Coulter DE, Greenleaf AL (1985) A mutation in the largest subunit of RNA polymerase II alters RNA chain elongation in vitro. J Biol Chem 260:13190–13198
Courchesne WE, Kunisawa R, Thorner J (1989) A putative protein kinase overcomes pheromone-induced arrest of cell cycling in *S. cerevisiae*. Cell 58:1107–1119
Crerar MM, Leather R, David E, Pearson ML (1983) Myogenic differentiation in L6 rat myoblasts: evidence for pleiotropic effects on myogenesis by RNA polymerase II mutations to α-amanitin resistance. Mol Cell Biol 3:946–955
Cross FR, Tinkelenberg AH (1991) A potential positive feed-back loop controlling *CLN1* and *CLN2* gene expression at the start of the yeast cell cycle. Cell 65:875–883
Dahmus ME (1981) Phosphorylation of eukaryotic DNA-dependent RNA polymerases. J Biol Chem 256:3332–3339
Dahmus ME (1983) Structural relationship between the large subunits of calf thymus RNA polymerase II. J Biol Chem 258:3956–3960
Dahmus ME, Kedinger C (1983) Transcription of adenovirus-2 major late promoter inhibited by monoclonal antibody against RNA polymerases II_o and II_A. J Biol Chem 258:2303–2307
Dalton S (1992) Cell cycle regulation of the human *cdc2* gene. EMBO J 11:1797–1804

Darst SE, Edwards AM, Kubalek EW, Kornberg RD (1991) Three-dimensional structure of yeast RNA polymerase II at 16Å resolution. Cell 66:121–128
Devoto SH, Mudryj M, Pines J, Hunter T, Nevins JR (1992) A cyclin A-protein kinase complex possesses sequence-specific DNA binding activity: $p33^{cdk2}$ is a component of the E2F-cyclin A complex. Cell 68:167–176
Dietrich MA, Prenger JP, Guilfoyle TJ (1990) Analysis of the genes encoding the largest subunit of RNA polymerase II in *Arabidopsis* and soybean. Plant Mol Biol 15:207–223
Doonan JH (1991) Cycling plant cells. Plant J 1:129–132
Draetta G (1990) Cell cycle control in eukaryotes: molecular mechanism of cdc2 activation. Trends Biochem 15:378–383
Ducommun B, Brambilla P, Draetta G (1991) Mutations at sites involved in SUC1 binding inactivate Cdc2. Mol Cell Biol 11:6177–6184
D'Urso G, Marracino RL, Marshak DR, Roberts JM (1990) Cell cycle control of DNA replication by a homologue from human cells of the $p34^{cdc2}$ protein kinase. Science 250:786–791
Dutta A, Stillman B (1992) CDC2 family kinases phosphorylate a human cell DNA replication factor, RPA, and activate DNA replication. EMBO J 11:2189–2199
Dutta A, Din S, Brill SJ, Stillman B (1991) Phosphorylation of replication protein A: a role for cdc2 kinase in G1/S regulation. Cold Spring Harbor Symp Quant Biol LVI:315–324
Dvir A, Peterson SR, Knuth MW, Lu H, Dynan WS (1992) Ku autoantigen is the regulatory component of a template-associated protein kinase that phosphorylates RNA polymerase II. Proc Natl Acad Sci USA 89:11920–11924
Dynlacht BD, Hoey T, Tijan R (1991) Isolation of coactivators associated with the TATA-binding protein that mediate transcriptional activation. Cell 66:563–576
Elion A, Brill JA, Fink GR (1991) FUS3 represses CLN1 and CLN2 and in concert with KSS1 promotes signal transduction. Proc Natl Acad Sci USA 88:9392–9396
Enoch T, Nurse P (1991) Coupling M phase and S phase: controls maintaining the dependence of mitosis on chromosome replication. Cell 65:921–923
Evan GI, Littlewood TD (1993) The role of *c-myc* in cell growth. Curr Opinion Genet Dev 3:44–49
Evers R, Hammer A, Köck J, Jess W, Borst P, Mémet S, Cornelissen CA (1989a) *Trypanosoma brucei* contains two RNA polymerase II largest subunit genes with an altered C-terminal domain. Cell 56:585–597
Evers R, Hammer A, Cornelissen WCA (1989b) Unusual C-terminal domain of the largest subunit of RNA polymerase II of *Crithidia fasciculata*. Nucleic Acids Res 17:3403–3413
Ewen ME, Xing Y, Lawrance BJ, Livingston DM (1991) Molecular cloning, chromosomal mapping, and expression of the cDNA for p107, a retinoblastoma gene product-related protein. Cell 66:1155–1164
Ewen ME, Faha B, Harlow E, Livingston DM (1992) Interaction of p107 with cyclin A independent of complex formation with viral oncoproteins. Science 255:85–87
Faha B, Ewen ME, Tsai L-H, Livingston DM, Harlow E (1992) Interaction between human cyclin A and adenovirus E1A-associated p107 protein. Science 255:87–90
Farmer G, Bargonetti J, Zhu H, Friedman P, Prywes R, Prives C (1992) Wild-type p53 activates transcription in vitro. Nature 358:83–86
Feaver WJ, Gileadi O, Li Y, Kornberg RD (1991) CTD kinase associated with yeast RNA polymerase II initiation factor b. Cell 67:1223–1230
Feiler HS, Jacobs TW (1991) Cloning of the pea *cdc2* homologue by efficient immunological screening of PCR products. Plant Mol Biol 17:321–333
Fernandez-Sabaria MJ, Sutton A, Zhong T, Arndt KT (1992) SIT4 protein phosphatase is required for the normal accumulation of *SWI4*, *CLN1*, *CLN2*, and *HCS26* RNAs during late G1. Genes Dev 6:2417–2428
Ferreira PCG, Hemerly AS, Villarroel R, Van Montagu M, Inzé D (1991) The *Arabidopsis* functional homology of the $p34^{cdc2}$ protein kinase. Plant Cell 3:531–540

Finkelstein A, Kostrub KF, Li J, Chavez DP, Wang BQ, Fang SM, Greenblatt J, Burton ZF (1992) A cDNA encoding RAP74, a general initiation factor for transcription by RNA polymerase II. Nature 355:464–467
Fischer J, Giniger E, Maniatis T, Ptashne M (1988) GAL4 activates transcription in *Drosophila*. Nature 332:853–856
Flanagan PM, Kelleher RJ, Sayre MH, Tschochner H, Kornberg RG (1991) A mediator required for activation of RNA polymerase II transcription in vitro. Nature 350:436–438
Flores O, Lu H, Killeen M, Greenblatt J, Burton ZF, Reinberg D (1991) The small subunit of transcription factor IIF recruits RNA polymerase II into the preinitiation complex. Proc Natl Acad Sci USA 88:9999–10003
Flores O, Lu H, Reinberg D (1992) Factors involved in specific transcription by mammalian RNA polymerase II. J Biol Chem 267:2786–2793
Fotedar R, Roberts JM (1992) Cell cycle regulated phosphorylation of RPA-32 occurs within the replication complex. EMBO J 11:2177–2187
Gabrielsen OS, Sentenac A (1991) RNA polymerase III(C) and its transcription factors. Trends Biochem 16:412–416
Gasch A, Hoffmann A, Horikoshi M, Roeder RG, Chua N-H (1990) *Arabidopsis thaliana* contains two genes for TFIID. Nature 346:390–394
Geiduschek EP, Tocchini-Valentini GP (1988) Transcription by RNA polymerase III. Annu Rev Biochem 57:873–914
Gill G, Tijan R (1991) A highly specific conserved domain of TFIID displays species specificity in vivo. Cell 65:333–340
Goodrich DW, Wang NP, Quian Y-W, Lee EY-HP, Lee W-H (1991) The retinoblastoma gene product regulates progression through the G1 phase of the cell cycle. Cell 67:293–302
Gordon CB, Campbell JL (1991) A cell cycle-responsive transcriptional control element and a negative control element in the gene encoding DNA polymerase α in *Saccharomyces cerevisiae*. Proc Natl Acad Sci USA 88:6058–6062
Gottlieb TM, Jackson SP (1993) The DNA-dependent protein kinase: requirement for DNA ends and association with Ku antigen. Cell 72:131–142
Greenblatt J (1991a) RNA polymerase-associated transcription factors. Trends Biochem 16:408–412
Greenblatt J (1991b) Roles of TFIID in transcriptional initiation by RNA polymerase II. Cell 66:1067–1070
Greenleaf AL (1983) Amanitin-resistant RNA polymerase II mutations are in the enzyme's largest subunit. J Biol Chem 258:13403–13406
Greenleaf AL, Borsett LM, Jimachello PF, Coulter DE (1979) α-Amanitin resistant *D. melanogaster* with an altered RNA polymerase II. Cell 18:613–622
Greenleaf AL, Weeks JR, Voelker RA, Ohnishi S, Dickson B (1980) Genetic and biochemical characterization of mutants at an RNA polymerase II locus in *D. melanogaster*. Cell 21:785–792
Guarante L, Bermingham-McDonogh O (1992) Conservation and evolution of transcriptional mechanisms in eukaryotes. Trends Genet 8:27–32
Guilfoyle TJ (1983) DNA-dependent RNA polymerases of plants and lower eukaryotes. In: Samson TJ (ed) Biochemistry and molecular biology of the cell nucleus, vol II. CRC Press, Boca Raton, pp 1–42
Guilfoyle TJ (1989) A protein kinase from wheat germ that phosphorylates the largest subunit of RNA polymerase II. Plant Cell 1:827–836
Guilfoyle TJ, Dietrich MA (1987) Plant RNA polymerases: structures, regulation and genes. In: Bruening G, Harada J, Kosuge T, Hollaender A (eds) Tailoring genes for crop improvement. Plenum Press, New York, pp 87–100
Guilfoyle TJ, Hagen G, Malcolm S (1984) Immunological studies on plant DNA-dependent RNA polymerases with antibodies raised against individual subunits. J Biol Chem 259:640–648
Gunderson Si, Knuth MW, Burgess RR (1990) The U1 snRNA promoter correctly initiates transcription and is activated by PSE1. Genes Dev 4:2048–2060

Ha I, Lane WS, Reinberg D (1991) Cloning of a human gene encoding the general transcription initiation factor IIB. Nature 352:689–695

Haaß MM, Feix G (1992) Two different cDNAs encoding TFIID proteins of maize. FEBS Lett 301:294–297

Hamel PA, Gallie BL, Phillips RA (1992) The retinoblastoma protein and cell cycle regulation. Trends Genet 8:180–185

Hartwell LH, Weinert TA (1989) Check-points: controls that ensure the order of cell cycle events. Science 246:629–634

Hata S (1991) cDNA cloning on a novel cdc2$^+$/CDC28-related protein kinase from rice. FEBS Lett 279:149–152

Hata S, Kouchi H, Suzuka I, Ishii T (1991) Isolation and characterization of cDNA clones for plant cyclins. EMBO J 10:2681–2688

Helmann JD, Chamberlin MJ (1988) Structure and function of bacterial sigma factors. Annu Rev Biochem 57:839–872

Hemerly A, Bergounioux C, Van Montagu M, Inzé D (1992) Genes regulating the plant cell cycle: isolation of a mitotic-like cyclin from *Arabidopsis thaliana*. Proc Natl Acad Sci USA 89:3295–3299

Hiebert SW, Lipp M, Nevins JR (1989) E1A-dependent *trans*-activation of the human *MYC* promoter is mediated by the E2F factor. Proc Natl Acad Sci USA 86:3594–3598

Hiebert SW, Chellappan SP, Horowitz JM, Nevins JR (1992) The interaction of RB with E2F coincides with an inhibition of the transcriptional activity of E2F. Genes Dev 6:177–185

Himmelfarb HJ, Simpson EM, Friesen JD (1987) Isolation and characterization of temperature-sensitive RNA polymerase II mutants of *Saccharomyces cerevisiae*. Mol Cell Biol 7:2155–2164

Hinds PW, Mittnacht S, Dulic V, Arnold A, Reed SI, Weinberg RA (1992) Regulation of retinoblastoma protein functions by ectopic expression of human cyclins. Cell 70:993–1006

Hirayama T, Imajuku Y, Anai T, Matsui M, Oka A (1991) Identification of two cell-cycle-controlling *cdc2* gene homologs in *Arabidopsis thaliana*. Gene 105:159–165

Hirt H, Pay A, Györgyey J, Bakó L, Németh K, Bögre L, Schweyen RJ, Heberle-Bors E, Dudits D (1991) Complementation of a yeast cell cycle mutant by an alfalfa cDNA encoding a protein kinase homologous tp p34^{cdc2}. Proc Natl Acad Sci USA 88:1636–1640

Hoekstra MF, Demaggio AJ, Dhillon N (1991) Genetically identified protein kinases in yeast II: DNA metabolism and meiosis. Trends Genet 7:293–297

Hollingsworth RE, Carmel EH, Lee W-H (1993) Retinoblastoma protein and the cell cycle. Curr Opinion Genet Dev 3:55–62

Horikoshi M, Carey MF, Kakidani H, Roeder RG (1988a) Mechanism of action of a yeast activator: direct effect of GAL4 derivatives on mammalian TFIID-promoter interactions. Cell 54:665–669

Horikoshi M, Hai T, Lin Y-S, Green MR, Roeder RG (1988b) Transcription factor ATF interacts with the TATA factor to facilitate establishment of a preinitiation complex. Cell 54:1033–1042

Korikoshi N, Maguire K, Kralli A, Maldonado E, Reinberg D, Weinmann R (1991) Direct interaction between adenovirus E1a protein and the TATA box binding transcription factors II D. Proc Natl Acad Sci USA 88:5124–5128

Hunter T (1993) Oncogenes and cell proliferation. Curr Opinions Genet Dev 3:1–4

Hunter T, Karin M (1992) The regulation of transcription by phosphorylation. Cell 70:375–387

Ingles CJ (1978) Temperature-sensitive RNA polymerase II mutations in Chinese hamster ovary cells. Proc Natl Acad Sci USA 75:405–409

Ingles JC, Himmelfarb HJ, Shales M, Greenleaf AL, Friesen JD (1984) Identification, molecular cloning, and mutagenesis of *Saccharomyces cerevisiae* RNA polymerase genes. Proc Natl Acad Sci USA 81:2157–2161

Inostroza JA, Mermelstein FH, Ha I, Lane WS, Reinberg D (1992) DR1, a TATA-binding protein-associated phosphoprotein and inhibitor of class II gene transcription. Cell 70:477–489

Izban MG, Luse DS (1992) The RNA polymerase II ternary complex cleaves the nascent transcript in a 3′–>5′ direction in the presence of elongation factor SII. Genes Dev 6:1342–1356

Jantzen M-H, Admon A, Bell SP, Tijan R (1990) Nucleolar transcription factor hUBF contains a DNA-binding motif with homology to HMG proteins. Nature 344:830–836

Jantzen H-M, Chow AM, King DS, Tijan R (1992) Multiple domains of the RNA polymerase I activator hUBF interact with the TATA-binding protein complex hSL1 to mediate transcription. Genes Dev 6:1959–1963

John PCL, Sek FJ, Lee MG (1989) A homolog of the cell cycle control protein $p34^{cdc2}$ participates in the division cycle of *Chlamydomonas* and a similar protein is detectable in higher plants and remote taxa. Plant Cell 1:1185–1193

Johnson PF, McKnight SL (1989) Eukaryotic transcriptional regulatory proteins. Annu Rev Biochem 58:799–839

Johnston LH, Lowndes NF (1992) Cell cycle control of DNA synthesis in budding yeast. Nucleic Acids Res 20:2403–2410

Jokerst RS, Weeks JR, Zehring WA, Greenleaf AL (1989) Analysis of the gene encoding the largest subunit of RNA polymerase II in *Drosophila*. Mol Gen Genet 215:266–275

Kakidani H, Ptashne M (1988) GAL4 activates gene expression in mammalian cells. Cell 52:161–167

Kassavetis GA, Joazeiro CAP, Pisano M, Geiduschek EP, Colbert T, Hahn S, Blanco JA (1992) The role of the TATA-binding protein in the assembly and function of the multisubunit yeast RNA polymerase III transcription factor, TFIIIB. Cell 71:1055–1064

Kastan MB, Zhan Q, El-Deiry WS, Carrier F, Jacks T, Walsh WV, Plunkett BS, Vogelstein B, Fornace AJ (1992) A mammalian cell cycle check-point pathway utilizing p53 and *GADD45* is defective in Ataxia-Telangiectasia. Cell 71: 587–597

Katagiri F, Yamazaki K, Horikoshi M, Roeder RG, Chua N-H (1990) A plant DNA-binding protein increases the number of active preinitiation complexes in a human in vitro transcription system. Genes Dev 4:1899–1909

Kelleher III RJ, Flanagan PM, Kornberg RD (1990) A novel mediator between activator proteins and the RNA polymerase II transcription apparatus. Cell 61:1209–1215

Killeen MT, Greenblatt JF (1992) The general transcription factor RAP30 binds to RNA polymerase II and prevents it from binding nonspecifically to DNA. Mol Cell Biol 12:30–37

Killeen M, Coulombe B, Greenblatt J (1992) Recombinant TBP, transcription factor IIB, and RAP30 are sufficient for promoter recognition by mammalian RNA polymerase II. J Biol Chem 267:9463–9466

Kim W-Y, Dahmus ME (1986) Immunological analysis of mammalian RNA polymerase II subspecies. J Biol Chem 261:1419–1425

Kim W-Y, Dahmus ME (1989) The major late promoter of adenovirus-2 is accurately transcribed by RNA polymerases IIO, IIA and IIB. J Biol Chem 264:3169–3176

Knuth MW, Gunderson SI, Thompson NE, Strasheim LA, Burgess RR (1990) Purification and characterization of proximal sequence element-binding protein 1, a transcription activating protein related to Ku and TREF that binds to proximal sequence element of the human U1 promoter. J Biol Chem 265:17911–17920

Koleske A, Buratowski S, Nonet M, Young RA (1992) A novel transcription factor reveals a functional link between the RNA polymerase II CTD and TFIID. Cell 69:883–894

Kolodziej PA, Woychik N, Liao S-N, Young RA (1990) RNA polymerase II subunit composition, stoichiometry, and phosphorylation. Mol Cell Biol 10: 1915–1920

Kruger W, Herskowitz I (1991) A negative regulator of *HO* transcription, SIN1 (SPT2), is a nonspecific DNA-binding protein related to HMG1. Mol Cell Biol 11:4135–4146

Ku D-H, Wen S-C, Engelhard A, Nicolaides NC, Lipson KE, Marino TA, Calabretta B (1993) *c-myb* transactivates *cdc2* expression via Myb binding sites in the 5′ flanking region of the human *cdc2* gene. J Biol Chem 268:2255–2259

Lai L-S, Cleary MA, Herr W (1992) A single amino acid exchange transfers VP16-induced positive control from the Oct-1 to the Oct-2 homeo domain. Genes Dev 6:2058–2065

Laurent BC, Carlson M (1992) Yeast SNF2/SWI2, SNF5, and SNF6 proteins function coordinately with the gene-specific transcriptional activators GAL4 and Bicoid. Genes Dev 6:1707–1715

Laybourn PJ, Dahmus ME (1989) Transcription dependent structural changes in the C-terminal domain of mammalian RNA polymerase subunit IIa/o. J Biol Chem 264:6693–6698

Lee DK, Dejong J, Hashimoto S, Horikoshi M, Roeder RG (1992) TFIIA induces conformational changes in TFIID via interactions with the basic repeat. Mol Cell Biol 12:5189–5196

Lee JM, Greenleaf AL (1989) A protein kinase that phosphorylates the C-terminal repeat domain of the largest subunit of RNA polymerase II. Proc Natl Acad Sci USA 86:3624–2628

Lee WS, Kao C, Bryant GO, Liu X, Berk AJ (1991) Adenovirus E1A activation domain binds the basic repeat in the TATA box transcription factor. Cell 67:365–376

Lees E, Faha B, Dulic V, Reed SI, Harlow E (1992) Cyclin E/cdk2 and cyclin A/cdk2 kinases associate with p107 and E2F in a temporally distinct manner. Genes Dev 6:1874–1885

Lew DL, Marini NJ, Reed SI (1992) Different cyclins control the timing of cell cycle commitment in mother and daughter cells of the budding yeast *S. cerevisiae.* Cell 69:317–327

Lewis MK, Burgess RR (1982) Eukaryotic RNA polymerases. In: Boyer PD (ed) The enzymes, vol XV. Academic Press, New York, pp 109–153

Li W-O, Bzik DJ, Gu H, Tanaka M, Fox BA, Inselburg J (1989) An enlarged largest subunit of *Plasmodium falciparum* RNA polymerase II defines conserved and variable RNA polymerase domains. Nucleic Acids Res 17:9621–9636

Liao S-M, Taylor ICA, Kingston RE, Young RA (1991) RNA polymerase II carboxy-terminal domain contributes to the response to multiple acidic activators in vitro. Genes Dev 5:2431–2440

Liebermann PM, Berk AJ (1991) The Zta *trans*-activator protein stabilizes TFIID association with promoter DNA by direct protein-protein interaction. Genes Dev 5:2441–2454

Lin Y-S, Green MR (1991) Mechanism of action of an acidic transcription activator in vitro. Cell 64:971–981

Lobo SM, Tanaka M, Sullivan ML, Hernandez N (1992) A TBP complex essential for transcription from TATA-less but not TATA-containing RNA polymerase III promoters is part of the TFIIIB fraction. Cell 71:1029–1040

López-De-León A, Librizzi M, Puglia K, Willis IM (1992) *PCF4* encodes an RNA polymerase III transcription factor with homology to TFIIB. Cell 71:211–220

Lowndes NF, Johnston AL, Johnston LH (1991) Coordination of expression of DNA synthesis genes in budding yeast by a cell-cycle regulated *trans*-factor. Nature 350:247–250

Lowndes NF, Johnston AL, Breeden L, Johnston LH (1992) SWI6 protein is required for transcription of the periodically expressed DNA synthesis genes in budding yeast. Nature 357:505–508

Lu H, Flores O, Weinmann R, Reinberg D (1991) The nonphosphorylated form of RNA polymerase II preferentially associates with the preinitiation complex. Proc Natl Acad Sci USA 88:10004–10008

Lu H, Zawel L, Fisher L, Egly J-M, Reinberg D (1992) Human general transcription factor IIH phyosphorylates the C-terminal domain of RNA polymerase II. Nature 358:641–645

Malik S, Hisatake K, Sumimoto H, Horikoshi M, Roeder RG (1991) Sequence of general transcription factor TFIIB and relationship to other initiation factors. Proc Natl Acad Sci USA 88:9553–9557
Martin C, Okamura S, Yound RA (1990) Genetic exploration of interactive domains in RNA polymerase II subunits. Mol Cell Biol 10:1908–1914
Matsushima N, Creutz CE, Kretsinger RH (1990) Polyproline, β-turn helices. Novel secondary structures proposed for the tandem repeats within rhodopsin, synatophysin, synexin, gliadin, RNA polymerase II, hordein and glutein. Proteins 7:125–155
Matsushime H, Roussel MF, Ashmun RA, Sherr CJ (1991) Colony-stimulating factor 1 regulates novel cyclins during the G1 phase of the cell cycle. Cell 65:701–713
Matshushime H, Ewen ME, Strom DK, Kato J-Y, Hanks SK, Roussel MF, Sherr CJ (1992) Identification and properties of an atypical catalytic subunit ($p34^{PSK-J3}$/cdk4) for mammalian D type G1 cyclins. Cell 71:323–334
Meisterernst M, Roeder RG (1991) Family of proteins that interact with TFIID and regulate promoter activity. Cell 67:557–567
Meisterernst M, Horikoshi M, Roeder RG (1990) Recombinant yeast TFIID, a general transcription factor, mediates activation by the gene-specific factor USF in a chromatin assembly assay. Proc Natl Acad Sci USA 87:9153–9157
Meisterernst M, Roy AL, Lieu HM, Roeder RG (1991) Activation of class II gene transcription by regulatory factors is potentiated by a novel activity. Cell 66: 981–993
Mémet S, Saurin W, Sentenac A (1988) RNA polymerases B and C are more closely related to each other that to RNA polymerase A. J Biol Chem 263:10048–10051
Merrill GF, Morgan BA, Lowndes NF, Johnston LH (1992) DNA synthesis control in yeast: an evolutionary conserved mechanism for regulating DNA synthesis genes? BioEssays 14:823–830
Mietz JA, Unger T, Huibregtse JM, Howley PM (1992) The transcriptional transactivation function of wild-type p53 is inhibited by SV40 large T-antigen and by HPV-16 E6 oncoprotein. EMBO J 11:5013–5020
Milner J, Cook A, Mason J (1990) p53 is associated with $p34^{cdc2}$ in transformed cells. EMBO J 9:2885–2889
Mitchell PJ, Tijan R (1989) Transcriptional regulation in mammalian cells by sequence-specific DNA-binding proteins. Science 245:371–378
Moran E (1993) DNA tumor virus transforming proteins and the cell cycle. Curr Opinion Genet Dev 3:63–70
Mortin MA (1990) Use of second-site suppressor mutations in *Drosophila* to identify components of the transcriptional machinery. Proc Natl Acad Sci USA 87:4864–4868
Motokura T, Arnold A (1993) Cyclin D and oncogenesis. Curr Opinion Genet Dev 3:5–10
Moyle M, Lee JS, Anderson WF, Ingles JC (1989) The C-terminal domain of the largest subunit of RNA polymerase II and transcription initiation. Mol Cell Biol 9:5750–5753
Mudryj M, Hiebert SW, Nevins JR (1990) A role for the adenovirus inducible E2F transcription factor in a proliferation dependent signal transduction pathway. EMBO J 9:2179–2184
Murray AW (1992) Creative blocks: cell-cycle checkpoints and feedback controls. Nature 359:599–604
Murray AW, Kirschner MW (1989) Dominoes and clocks: the union of two views of the cell cycle. Science 246:614–621
Nasheuer H-P, Moore A, Wahl AF, Wang TS-F (1991) Cell cycle-dependent phosphorylation of human DNA polymerase α. J Biol Chem 266:7893–7903
Nasmyth K, Dirick L (1991) The role of *SWI4* and *SWI6* in the activity of G1 cyclins in yeast. Cell 66:995–1013

Nawrath C, Schell J, Koncz C (1990) Homologous domains of the largest subunit of eukaryotic RNA polymerase II are conserved in plants. Mol Gen Genet 223:65–75

Nevins JR (1992) E2F: a link between the Rb tumor suppressor protein and viral oncoproteins. Science 258:424–429

Nigro JM, Sikorski R, Reed SI, Vogelstein B (1992) Human p53 and *CDC2Hs* genes combine to inhibit the proliferation of *Saccharomyces cerevisiae*. Mol Cell Biol 12:1357–1365

Nikolov DB, Hu S-H, Lin J, Gasch A, Hoffmann A, Horikoshi M, Chua N-H, Roeder RG, Burley SK (1992) Crystal structure of TFIID TATA-box binding protein. Nature 360:40–46

Nitschke K, Fleig U, Schell J, Palme K (1992) Complementation of the cs *dis2*-11 cell cycle mutant of *Schizosaccharomyces pombe* by a protein phosphatase from *Arabidopsis thaliana*. EMBO J 11:1327–1333

Nonet ML, Young RA (1989) Intragenic and extragenic suppressors of mutations in the heptapeptide repeat domain of *Saccharomyces cerevisiae* RNA polymerase II. Genetics 123:715–724

Nonet M, Scafe C, Sexton J, Young R (1987a) Eucaryotic RNA polymerase conditional mutant that rapidly ceases mRNA synthesis. Mol Cell Biol 7:1602–1611

Nonet M, Sweetser D, Young RA (1987b) Functional redundancy and structural polymorphism in the large subunit of RNA polymerase II. Cell 50:909–915

Nurse P (1990) Universal control mechanism regulating onset of M-phase. Nature 344:503–508

Ogas J, Andrews BJ, Herskowitz I (1991) Transcriptional activation of *CLN1*, *CLN2*, and a putative new G1 cyclin (*HCS26*) by SWI4, a positive regulator of G1-specific transcription. Cell 66:1015–1026

Ohkuma Y, Sumimoto H, Horikosho M, Roeder RG (1990) Factors involved in specific transcription by mammalian RNA polymerase II: purification and characterization of general transcription factor TFIIE. Proc Natl Acad Sci USA 87:9163–9167

Pagano M, Pepperkok R, Verde F, Ansorge W, Draetta G (1992a) Cyclin A is required at two points in the human cell cycle. EMBO J 11:961–971

Pagano M, Draetta G, Jansen-Dürr P (1992b) Association of cdk2 kinase with the transcription factor E2F during S phase. Science 255:1144–1147

Payne JM, Dahmus ME (1993) Partial purification and characterization of two distinct protein kinases that differentially phosphorylate the carboxyl-terminal domain of RNA polymerase subunit IIa. J Biol Chem 268:80–87

Payne JM, Laybourn PJ, Dahmus ME (1989) The transition of RNA polymerase II from initiation to elongation is associated with phosphorylation of the carboxy-terminal domain of subunit IIa. J Biol Chem 264:19621–19629

Pearson BE, Nasheuer H-P, Wang TS-F (1991) Human DNA polymerase α gene: sequences controlling expression in cycling and serum-stimulated cells. Mol Cell Biol 11:2081–2095

Perry ME, Levine AJ (1993) Tumor-suppressor p53 and the cell cycle. Curr Opinion Genet Dev 3:50–54

Peterson CL, Herskowitz I (1992) Characterization of the yeast *SWI1*, *SWI2*, and *SWI3* genes, which encode a global activator of transcription. Cell 68:573–583

Peterson CL, Kruger W, Herskowitz I (1991) A functional interaction between the C-terminal domain of RNA polymerase II and the negative regulator SIN1. Cell 64:1135–1143

Peterson SR, Dvir A, Anderson CW, Dynan WS (1992) DNA binding provides a signal for phosphorylation of the RNA polymerase II heptapeptide repeats. Genes Dev 6:426–438

Pinto I, Ware DE, Hampsey M (1992) The yeast *SUA7* gene encodes a homolog of human transcription factor TFIIB and is required for normal start site selection in vivo. Cell 68:977–988

Pitto L, Schiavo L, Terzi M (1985) α-Amanitin resistance is developmentally regulated in carrot. Proc Natl Acad Sci USA 82:2799–2803

Ptashne M, Gann AAF (1990) Activators and targets. Nature 346:329–331

Pugh BF, Tijan R (1990) Mechanism of transcriptional activation by Sp1: evidence for coactivators. Cell 61:1187–1197

Pühler G, Leffers H, Gropp F, Palm P, Klenk H-P, Lottspeich F, Garrett RA, Zillig W (1989) Archaebacterial DNA-dependent RNA polymerases testify to the evolution of the eukaryotic nuclear genome. Proc Natl Acad Sci USA 86:4569–4573

Reed SI (1991) G1-specific cyclins: in search of an S-phase-promoting factor. Trends Genet 7:95–99

Reeder RH (1990) rRNA synthesis in the nucleolus. Trends Genet 6:390–395

Reeves WH, Sthoeger ZM (1989) Molecular cloning of cDNA encoding the p70 (Ku) lupus autoantigen. J Biol Chem 264:5047–5052

Rice GA, Kane CM, Chamberlin MJ (1991) Footprinting analysis of mammalian RNA polymerase II along its transcript: an alternative view of transcription elongation. Proc Natl Acad Sci USA 88:4245–4249

Rigby PW (1993) Three in one and one in three: it all depends on TBP. Cell 72:7–10

Riva M, Mémet S, Micouin J-Y, Huet J, Treich I, Dassa J, Young R, Buhler J-M, Sentenac A, Fromageot P (1986) Isolation of structural genes for yeast RNA polymerases by immunological screening. Proc Natl Acad Sci USA 83:1554–1558

Riva M, Shäffner AR, Sentenac A, Hartmann GR, Mustaev AA, Zaychikov EF, Grachev MA (1987) Active site labeling of the RNA polymerase A, B and C from yeast. J Biol Chem 262:14377–14380

Robbins PD, Horowitz JM, Mulligan RC (1990) Negative regulation of human *c-fos* expression by the retinoblastoma gene product. Nature 346:668–671

Roeder RG (1991) The complexities of eukaryotic transcription initiation: regulation of preinitiation complex assembly. Trends Biochem 16:402–408

Roeder RG, Rutter WJ (1969) Multiple forms of DNA-dependent DNA polymerases in eukaryotic organisms. Nature 224:234–237

Rorth P, Montell DJ (1992) *Drosophila* C/EBP: a tissue-specific DNA-binding protein required for embryonic development. Genes Dev 6:2299–2311

Rustgi AK, Dyson N, Bernards R (1991) Amino-terminal domains of c-Myc and N-myc proteins mediate binding to the retinoblastoma gene product. Nature 352: 541–544

Sadowski I, Niedbala D, Wood K, Ptashne M (1991) GAL4 is phosphorylated as a consequence of transcriptional activation. Proc Natl Acad Sci USA 88:10510–10514

Sawadogo M, Sentenac A (1990) RNA polymerase B(II) and general transcription factors. Annu Rev Biochem 59:711–754

Sayre MH, Tschochner H, Kornberg RD (1992a) Reconstruction of transcription with five purified initiation factors and RNA polymerase II from *Saccharomyces cerevisiae*. J Biol Chem 267:23376–23382

Sayre MH, Tschochner H, Kornberg RD (1992b) Purification and properties of *Saccharomyces cerevisiae* RNA polymerase II general initiation factor a. J Biol Chem 267:23383–23387

Scafe C, Martin C, Nonet M, Podos S, Okamura S, Young RA (1990) Conditional mutations occur predominantly in highly conserved residues of RNA polymerase II subunits. Mol Cell Biol 10:1270–1275

Schreck R, Carey MF, Grummt I (1989) Transcriptional enhancement by upstream activators is brought about by different molecular mechanisms for class I and II RNA polymerase genes. EMBO J 8:3011–3017

Schultz MC, Reeder RH, Hahn S (1992) Variants of the TATA-binding protein can distinguish subsets of RNA polymerase I, II and III promoters. Cell 69:697–702

Searles L, Jokerst RS, Bingham PM, Voelker RA, Greenleaf AL (1982) Molecular cloning of sequences from a *Drosophila* RNA polymerase II locus by P element transposon tagging. Cell 31:585–592

Seipel K, Georgiev O, Schaffner W (1992) Different activation domains stimulate transcription from remote ("enhancer") and proximal ("promoter") positions. EMBO J 11:4961–4968

Sentenac A (1985) Eukaryotic RNA polymerases. CRC Crit Rev Biochem 18:31–90

Serizawa H, Conaway RC, Conaway JW (1992) A carboxyl-terminal-domain kinase associated with RNA polymerase II transcription factor δ from rat liver. Proc Natl Acad Sci USA 89:7476–7480

Sharp PA (1992) TATA-binding protein is a classless factor. Cell 68:819–821

Shermoen AW, O'Farell PH (1991) Progression of the cell cycle through mitosis leads to abortion of nascent transcripts. Cell 67:303–310

Shi Y, Seto E, Chang L-S, Shenk T (1991) Transcriptional repression by YY1, a human GLI-Krüppel-related protein, and relief of repression by adenovirus E1A protein. Cell 67:377–388

Shirodkar S, Ewen M, DeCaprio JA, Morgan J, Livingston DM; Chittenden T (1992) The transcription factor E2F interacts with the retinoblastoma product and a p107-cyclin A complex in a cell cycle-regulated manner. Cell 68:157–166

Simon CM, Fisch TM, Benecke BJ, Nevins JR, Heintz N (1988) Definition of multiple, functionally distinct TATA elements, one of which is a target in the *hsp70* promoter for E1A regulation. Cell 52:723–729

Smale S, Schmidt MC, Berk AJ, Baltimore D (1990) Transcriptional activation by Sp1 as directed through TATA or initiator: specific requirement for mammalian transcription factor IID. Proc Natl Acad Sci USA 87:4509–4513

Smith JL, Levin JR, Ingles JC, Agabian N (1989) In *Trypanosomes* the homolog of the largest subunit of RNA polymerase II is encoded by two genes and has a highly unusual C-terminal domain structure. Cell 56:815–827

Sollner-Webb B, Mougey EB (1991) News from the nucleolus. Trends Biochem 16:58–62

Sorger PK, Murray AW (1992) S-phase feedback control in budding yeast independent of tyrosine phosphorylation of $p34^{cdc2}$. Nature 355:365–368

Sprague GF Jr (1991) Signal transduction in yeast mating: receptors, transcription factors and the kinase connections. Trends Genet 7:393–397

Stone N, Reinberg D (1992) Protein kinases from *Aspergillus nidulans* that phosphorylate the carboxyl-terminal domain of the largest subunit of RNA polymerase II. J Biol Chem 267:6353–6360

Stringer KF, Ingles J, Greenblatt J (1990) Direct and selective binding of an acidic activation domain to the TATA-box factor TFIID. Nature 345:783–786

Sumimoto H, Ohkuma Y, Yamamoto T, Horikoshi M, Roeder RG (1990) Factors involved in specific transcription by mammalian RNA polymerase II: identification of general transcription factor TFIIG. Proc Natl Acad Sci USA 87:9158–9162

Surana U, Robitsch H, Prince C, Schuster T, Fitch I, Futcher BA, Nasmyth K (1991) The role of CDC28 and cyclins during mitosis in the budding yeast *S. cerevisiae*. Cell 65:145–161

Sweetser D, Nodet M, Young RA (1987) Prokaryotic and eukaryotic RNA polymerases have homologous core subunits. Proc Natl Acad Sci USA 84:1192–1196

Szentirmay MN, Sawadogo M (1991) Transcription factor requirement for multiple rounds of initiation by human RNA polymerase II. Proc Natl Acad Sci USA 88:10691–10695

Taggart AKP, Fischer TS, Pugh BF (1992) The TATA-binding protein and associated factors are components of PolIII transcription factor TFIIIB. Cell 71:1015–1028

Takada R, Nakatani Y, Hoffmann A, Kokubo T, Hasegawa T, Roeder RG, Hirokoshi M (1992) Identification of human TFIID components and direct interaction between a 250-kDa polypeptide and the TATA box-binding protein (TFIIDτ). Proc Natl Acad Sci USA 89:11809–11813

Tanaka M, Lai J-S, Herr W (1992) Promoter-selective activation domains in Oct-1 and Oct-2 direct differential activation of an snRNA and mRNA promoter. Cell 68:755–767

Theunissen O, Rudt F, Guddat U, Mentzel H, Pieier T (1992) RNA and DNA binding zinc fingers in *Xenopus* TFIIIA. Cell 71:679–690

Thompson NE, Steinberg TH, Aronson DB, Burgess RR (1989) Inhibition of in vivo and in vitro transcription by monoclonal antibodies prepared against wheat germ RNA polymerase II that react with the heptapeptide repeat of eukaryotic RNA polymerase II. J Biol Chem 264:11511–11520

Tschochner H, Sayre MH, Flanagan PM, Feaver WJ, Kornberg RD (1992) Yeast RNA polymerase II initiation factor e: isolation and identification as the functional counterpart of human transcription factor IIB. J Biol Chem 89:11292–11296

Tsuchiya E, Uno M, Kiguchi A, Masuoka K, Kanemori Y, Okabe S, Mikayawa T (1992) The *Saccharomyces cerevisiae NPS1* gene, a novel *CDC* gene which encodes a 160 kDa nuclear protein involved in G2 phase control. EMBO J 11:4017–4026

Tyers M, Tokiwa G, Nash R, Futcher B (1992) The Cln3-Cdc28 kinase complex of *S. cerevisiae* is regulated by proteolysis and phosphorylation. EMBO J 11:1773–1784

Umek RM, Friedman AD, McKnight SL (1991) CCAAT-enhancer binding protein: a component of a differentiation switch. Science 251:288–292

Usheva A, Maldonado E, Goldring A, Lu H, Houbavi C, Reinberg D, Aloni Y (1992) Specific interaction between the nonphosphorylated form of RNA polymerase II and the TATA-binding protein. Cell 69:871–881

Wang W, Carey M, Gralla JD (1992a) Polymerase II promoter activation: closed complex formation and ATP-driven start site opening. Science 255:450–453

Wang W, Gralla JD, Carey M (1992b) The acidic activator GAL4-AH can stimulate polymerase II transcription by promoting assembly of a closed complex requiring TFIID and TFIIA. Genes Dev 6:1716–1727

Webster N, Jin JR, Green S, Hollis M, Chambon P (1988) The yeast UAS_G is a transcriptional enhancer in human HeLa cells in the presence of the GAL4 *trans*-activator. Cell 52:169–178

Weintraub SJ, Prater CA, Dean DC (1992) Retinoblastoma protein switches the E2F site from positive to negative element. Nature 358:259–261

White RJ, Jackson SP (1992a) The TATA-binding protein: a central role in transcription by RNA polymerase I, II and III. Trends Genet 8:284–288

White RJ, Jackson SP (1992b) Mechanism of TATA-binding protein recruitment to a TATA-less class III promoter. Cell 71:1041–1053

Winston F, Carlson M (1992) Yeast SNF/SWI transcriptional activators and the SPT/SIN chromatin connection. Trends Genet 8:387–391

Woychik NA, Young RA (1990) RNA polymerase II: subunit structure and function. Trends Biochem 15:347–351

Woychik NA, Liao S-M, Kolodziej PA, Young RA (1990) Subunits shared by eukaryotic nuclear RNA polymerases. Genes Dev 4:313–323

Xiong Y, Zhang H, Beach D (1992) D type cyclins associate with multiple protein kinases and the DNA replication and repair factor PCNA. Cell 71:505–514

Yang C-L, Chang L-S, Zhang P, Ha H, Zhu L, Toomey NL, Lee MYWT (1992) Molecular cloning of the cDNA for the catalytic subunit of human DNA polymerase δ. Nucleic Acids Res 20:735–745

Yano R, Nomura M (1991) Suppressor analysis of temperature-sensitive mutations of the largest subunit of RNA polymerase I in *Saccharomyces cerevisiae*: a suppressor gene encodes the second-largest subunit of RNA polymerase I. Mol Cell Biol 11:754–764

Yoon H-J, Campbell JL (1991) The CDC7 protein of *Saccharomyces cerevisiae* is a phosphoprotein that contains protein kinase activity. Proc Natl Acad Sci USA 88:3574–3578

Young RA (1991) RNA polymerase II. Annu Rev Biochem 60:689–715

Young RA, Davies RW (1983) Yeast RNA polymerase II genes: isolation with antibody probes. Science 222:778–782

Yura T, Ishihama A (1979) Genetics of bacterial RNA polymerases. Annu Rev Genet 13:59–97

Zambetti GP, Bargonetti J, Walker K, Prives C, Levine AJ (1992) Wild-type p53 mediates positive regulation of gene expression through a specific DNA sequence element. Genes Dev 6:1143–1152

Zehring WA, Greenleaf AL (1990) The carboxyl-terminal repeat domain of RNA polymerase II is not required for transcription factor Sp1 to function in vitro. J Biol Chem 265:8351–8353

Zehring WA, Lee JM, Weeks JR, Jokerst RS, Greenleaf AL (1988) The C-terminal repeat domain of RNA polymerase II largest subunit is essential in vivo but is not required for accurate transcription initiation in vitro. Proc Natl Acad Sci USA 85:3698–3702

Zhang J, Corden JL (1991a) Identification of phosphorylation sites in the repetitive carboxyl-terminal domain of the mouse RNA polymerase II largest subunit. J Biol Chem 266:2290–2296

Zhang J, Corden JL (1991b) Phosphorylation causes a conformational change in the carboxyl-terminal domain of the mouse RNA polymerase II largest subunit. J Biol Chem 266:2297–2302

Zhou QZ, Liebermann PM, Boyer TG, Berk AJ (1992) Holo-TFIID supports transcriptional stimulation by diverse activators and from a TATA-less promoter. Genes Dev 6:1964–1974

3 Plastid Differentiation: Organelle Promoters and Transcription Factors

Gerhard Link

1 Introduction

One of the most distinctive features of plants is the existence of plastids, i.e., the typical DNA-containing organelles that usually occur in multiple copies per cell, originate by division, and are capable of differentiation (for review, see Kirk and Tilney-Bassett 1967). Various developmental pathways are known, each leading from the undifferentiated small proplastids of meristematic cells to the specialized end products such as the photosynthetically active chloroplasts, the etioplasts in dark-grown angiosperm seedlings, and several other pigmented (chromoplasts) or unpigmented plastid forms (leucoplasts). Most of these specialized organelles can be interconverted in response to internal or external cues such as hormones and light (for review, see e.g. Thomson and Whatley 1980; Baker and Barber 1984). Despite this flexibility, it appears as a general principle that each plant cell harbors a uniform plastid population of a type which precisely matches the differentiation state of the entire cell. In molecular terms, several aspects of plastid differentiation are notable (Fig. 1).

First, plastids contain their own genome in the form of multiple, double-stranded circular DNA molecules coding for the organellar rRNAs and tRNAs, as well as for mRNAs specifying almost 100 plastid-localized proteins. The latter can be grouped into proteins involved in photosynthesis and CO_2 fixation and those involved in plastid gene expression itself, such as ribosomal proteins, translation factors, and components of the plastid transcription apparatus (for review, see Weil 1987; Bogorad and Vasil 1991a,b; Sugiura 1992).

Second, despite the high coding capacity of the organelle genome, the majority of the plastid proteins are nuclear-encoded and posttranslationally imported from the cytoplasm, including many components of the plastid gene expression apparatus itself. Hence, as is the case for mitochondria, the plastids are genetically semiautonomous, i.e., they depend on, and are controlled by, products of nuclear genes (Rochaix 1992). Vice versa, the plastids seem to control the status of gene expression in the nucleocytoplasmic

University of Bochum, Plant Cell Physiology and Molecular Biology, D-44780 Bochum, FRG

Results and Problems in Cell Differentiation 20
L. Nover (Ed.)
Plant Promoters and Transcription Factors

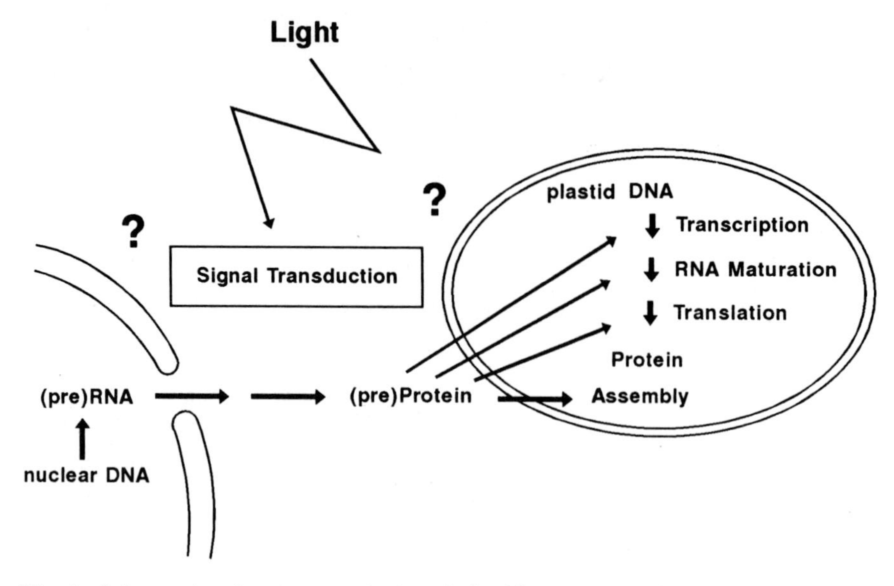

Fig. 1. Scheme showing the complexity of plastid gene expression and its regulation. Indicated are the principal steps in gene expression in both the plastid and nucleocytoplasmic compartments of the plant cell. Light is taken as an example for the various environmental, cellular, and developmental signals that can affect plastid differentiation, although the molecular details of signal transduction remain to be established (*question marks*)

compartment, suggesting a balanced network of regulatory interactions within each plant cell (Taylor 1989; Bogorad 1991; Rapp and Mullet 1991).

Third, the organization of the organellar DNA does not seem to vary in different plastid types of the same plant species. Nevertheless, there are distinct differences in plastid-type-specific gene expression patterns, both at RNA and protein levels. These reflect, and complement, changes in developmentally regulated gene expression in the surrounding nucleo-cytoplasmic cell compartment (for review, see Herrmann et al. 1992).

Chloroplast formation, which represents the probably best-characterized and most important plastid differentiation process, is strictly dependent on light in higher plant systems. A detailed picture is available already of the light-mediated changes in the expression of nuclear genes for chloroplast proteins such as the small subunit of ribulose-1.5-bisphosphate carboxylase/oxygenase (RubisCo) and the light harvesting LHCII proteins of photosystem II (Tobin and Silverthorne 1985; Thompson and White 1991). The photoreceptor systems for the light signals that ultimately modulate the expression of these nuclear genes, i.e., phytochrome and blue-UV photoreceptor(s), have been intensely studied, and the connecting signal transduction pathway(s) is beginning to emerge (Thompson and White 1991).

The aim of this review is to summarize what is known about the control of gene expression inside the organelle, with emphasis on the role of plastid transcription. What might be the principal determinants for the transmission of cell-specific, developmental, and external signals at this level of gene expression? These questions were addressed in several recent reviews (Link 1988, 1991; Mullet 1988; Herrmann et al. 1992). The current focus will be on the role of plastid promoters and transcription factors.

2 Plastid Differentiation

2.1 Organelle Genes

With few exceptions, the plastid DNA (ptDNA) molecules from many different plants were found to be in a size range of 120–160 kbp, depending on the species. The "typical" ptDNA molecule has four characteristic regions, i.e., the two copies of a large, inverted repeat, interspersed by the large and small "unique" or "single-copy" regions. The complete ptDNA sequence was published for the two higher plants, tobacco and rice, as well as for the liverwort, *Marchantia polymorpha*, and extensive sequence data are already available for ptDNA from a number of other species (for review, see Sugiura 1992). This work has established a certain degree of variability in the genome organization of different plants, involving insertions, deletions, and inversions, each of which can affect the structure, function, and mode of expression of certain plastid genes. Moreover, observations consistent with the occasional loss of a gene from (or recruitment to) the plastid genome during evolution were reported (for review, see Palmer 1991). Taken together, these data stress the importance of first analyzing the genomic context of a plastid gene in a new plant species before studying its expression and function. Nevertheless, for higher plants the "tobacco-type" ptDNA organization has thus far proved to be typical (Sugiura 1992; Fig. 2).

2.2 Gene Expression Levels

Similar to the situation found in bacterial cells, most chloroplast genes are part of di- or polycistronic transcription units. Other features of the genomic organization, however, are more reminiscent of that found in nuclear DNA of eukaryotic cells, one example being the existence of split genes. The latter give rise to primary transcripts which undergo (*cis* and *trans*) splicing, in addition to the 5′ and 3′ processing reactions found for transcripts from both split and unsplit genes (Rochaix 1992; Sugiura 1992). Some plastid transcripts require RNA editing in order to become functional (Hoch et al. 1991). Hence, the variability in the structural organization of plastid genes is reflected by a wide array of gene expression mechanisms, which together determine the abundance and functional activity of the final gene products.

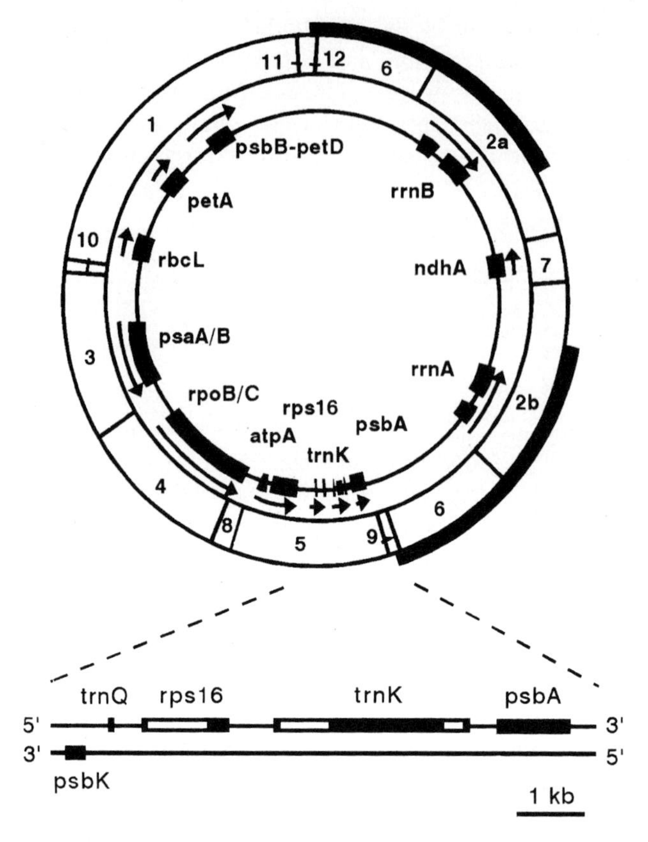

Fig. 2. Physical map of the 158-kbp chloroplast DNA from the crucifer mustard (*Sinapis alba* L.), showing the location and orientation of selected genes. The *outer circles* delimit the PstI restriction fragments (numbered *1–12*) as well as the two copies of the large inverted repeat (*thickened areas on outside*) (Dietrich and Link 1985). The *black fields on the inner circle* mark the position of plastid genes and gene clusters named according to current convention (Hallick 1989). The *arrows* indicate the direction of transcription. *Below* the *psbK – psbA* region is shown on an expanded scale. *Filled boxes* Coding regions, including the maturase-like intron-ORF of the *trnK* gene (Neuhaus and Link 1987); *open boxes*: noncoding intron regions (Link and Langridge 1984; Neuhaus et al. 1989; Neuhaus and Link 1990)

One striking feature of ptDNA organization is the dispersal of genes that encode functionally related proteins. For instance, the genes for photosystem I or II proteins are scattered throughout the entire genome rather than being grouped into closely spaced clusters. As a consequence, genomic regions will often contain genes for a variety of functions in close physical proximity, even within a single transcription unit. On the other hand, this does not imply that closely spaced plastid genes follow the same mode of expression. Often, a considerable variability is observed as a result of differential control at the posttranscriptional and translational level.

For instance, a number of mustard chloroplast genes coding for a variety of products and differing in their expression mode are clustered within a 10-kbp DNA region on the same strand (Fig. 2; Neuhaus and Link 1990). The downstream *psbA* gene (encoding the D1 reaction center protein of photosystem II) is unsplit. The next two genes in 5′ direction, *trnK* for $tRNA^{Lys}$ and *rps16* for ribosomal protein CS16, are both split by long introns. The *trnK* intron contains an open reading frame for a putative regulatory protein that shares sequence similarity with mitochondrial maturases involved in RNA splicing (Neuhaus and Link 1987). The most upstream gene (*trnQ*, coding for $tRNA^{Gln}$) is an unsplit tRNA gene (for nomenclature of plastid genes, see Hallick 1989). The steady-state RNA levels assessed by Northern hybridization are much higher for the *psbA* mRNA than for the *rps16* mRNA and the long *trnK* precursor transcript (Hughes et al. 1987; Neuhaus et al. 1989; Neuhaus and Link 1990). In addition, there are differences with regard to the temporal mode of transcript accumulation during seedling development, with the mature *rps16* transcripts reaching plateau levels early ("constitutive mode"), *trnK* precursor transcripts showing a transient rise and subsequent fall in concentration ("transient mode"), and *psbA* transcripts beginning to accumulate relatively late during development ("late enhanced mode"). The latter continue to accumulate to high levels in light- but not dark-grown seedlings, whereas *rps16* and *trnK* transcripts appear to be relatively unaffected by light/dark regimes (Link 1988, 1991; Neuhaus and Link 1990).

Regulation of plastid gene expression was shown at various levels, ranging from DNA modification, transcription, to RNA maturation, translation, and posttranslational modifications (for review, see Herrmann et al. 1992). There has long been a debate as to which mechanism might be the most relevant for plastid differentiation and function. As more detailed knowledge has accumulated, it is becoming increasingly clear now that plastid gene expression is not regulated by one single dominating mechanism. On the contrary, different genes located on the same ptDNA molecule can be regulated quite differently. This is also true sometimes for one and the same gene in different plant species, or in the same species but at different developmental stages and in different tissues. Moreover, the various gene expression levels are interrelated and, hence, it will be necessary to gain information on all of them in order to appreciate the full potential of plastid gene regulation.

Research on plastid transcription has mainly focused on two aspects: (1) the architecture of plastid promoters and (2) the proteins of the transcription apparatus. Questions relevant to the first aspect are: Which basic sequence elements are essential for protein binding and transcription initiation? Are there additional regulatory *cis*-elements that might account for gene-specific transcription or might be involved, e.g., in stage-specific or photoregulated gene expression? Questions relevant to the second aspect are: What is the basic structural organization of the transcription apparatus? How many different organellar RNA polymerases exist? If there are multiple enzymes, how are they related? In addition to the "core" RNA polymerase(s), are there other (regulatory) *trans*-factors that either bind to DNA directly or act indirectly via protein-protein interactions? Where is the intracellular coding site for the polypeptides comprising the plastid transcription apparatus? As will be discussed in the next chapter, answers to some of these questions are already available or are beginning to emerge, while other aspects will require further, extensive work.

3 Determinants of Transcription Control

3.1 Organelle Promoters

By using sequencing and mapping techniques, it has become possible to locate numerous plastid genes and identify their coding regions. In addition, this work has provided information on the 5′ and 3′ noncoding and flanking regions that might be expected to contain the sequences required for the controlled expression of these genes. The general notion from the earlier part of this work was that sequence elements resembling the "−35" (TTGACA) and "−10" (TATAAT) consensus elements typical for many *E. coli* promoters (Hawley and McClure 1983; Reznikoff et al. 1985) are located 5′ to a number of plastid genes and hence might indicate "prototype" plastid promoters (Kung and Lin 1985; Hanley-Bowdoin and Chua 1987; Weil 1987). More recently, however, it became clear from more extensive sequence comparisons and functional studies that this picture might be incomplete. Indeed, as for prokaryotes (Reznikoff et al. 1985), considerable deviations from the consensus motifs have been noted for various chloroplast genes.

For instance, −35/−10-like motifs are present immediately in front of the mapped RNA 5′ end of a number of mustard chloroplast genes, including the *rrn* operon (Przybyl et al. 1984), *psbA* (Link and Langridge 1984), *trnK* (Neuhaus and Link 1987), and *trnQ* (Neuhaus and Link 1990). In the case of the *rps16* gene, however, a "−10"-like element but no "−35" element was found (Table 1; Neuhaus et al. 1989). A similar situation was reported for the *atpB* gene from *Chlamydomonas* (Klein et al. 1992), which likewise lacks a typical −35 region.

An even more extreme deviation from the "classical" sigma^{70}-type (Reznikoff et al. 1985) promoter organization of many bacterial genes was recently noted for a putative blue light-responsive promoter (LRP.) upstream of the *psbD-psbC* operon in several plant species (Christopher et al. 1992). The latter shows only a slight resemblance to the −35/−10 consensus sequence and, instead, reveals GATA motifs reminiscent of light-responsive plant nuclear genes (Gilmartin et al. 1990). Although a functional analysis of this promoter remains to be done, there are other examples that point in the same direction. For instance, the mustard *psbA* promoter

Table 1. Promoter regions of mustard plastid genes

Promoter	Sequence[h]		
rrnB[a]	GTGGGA**TTGACG**TGAGGGGGTAGGGGTAGC		**TATATT**TCTGGGAGC
trnQ[b]	ATTGAA**TTGACG**AACAACCAATAGCAATAT		**TACTCT**TTTAGTAGT
trnH[c]	TGTTTA**TTGAAA**TCGATGCGTTTTTATTT		**TATTAT**TTTAGACAA
trnK[d]	ATCATA**TTGACA**ACAGTGTATTGACCAAA		**TATAAT**TCTCTCATG
trnK*[d]	AAAGAT**TTGACA**TTATCCTACTCGGCT		**TAGTTT**TTATTCATA
trnG[e]	CTTTGT**TCGACA**AAAGGTCAATTTATA		**TACAAT**AATTGCATT
trnG*[e]	TCTGCT**TTGACA**AATTTTATTAAGTCTGCGC		**TAGAAT**TTTCTCATT
psbA[f]	CATTGG**TTGACA**TGGC**TATATAA**GTCATG		**TATACT**GTTCAATAA
rps16[g]	TCTGTTTAAAAATCCAA**TATTTTT**TAAGAA		**TAAAAT**ATTGAAATG
psbK[b]	TTTCGG**TTGAAA**TAATCAAAAAGAAGTT		**TATTCT**TCTTTTCAT
Consensus sequences			
	−35		−10
E. coli σ^{70} promoter	TTGaca ←	15–21 bp →	TAtaaT 5–7 bp Ṗu
Plastid "σ^{70}-like" promoter	TTGaca ←	11–24 bp →	TAtaaT
Nuclear polII promoter TATA box	TAT$^{A}_{T}$A$^{A}_{T}$A ←	22–27 bp →	PyṖuPyPyPy

[a] Przybyl et al. (1984). [b] Neuhaus and Link (1990). [c] Nickelsen and Link (1990). [d] Neuhaus and Link (1987). [e] K. Liere, J. Nickelsen, H. Neuhaus and G. Link (unpubl.). [f] Link and Langridge (1984). [g] Neuhaus et al. (1989). [h] Sequence motifs that conform to one of the consensus sequences are in boldface and underlined. The transcription start site is marked by a dot.

(Link and Langridge 1984) contains the typical −35/−10 regions and, in addition, an interspersed sequence element, TATATAA, resembling the TATA box of many nuclear genes transcribed by RNA polymerase II (Breathnach and Chambon 1981). It was shown by deletion and point mutation analysis (Link 1984; Eisermann et al. 1990) that this TATA box-like sequence element is indeed required for faithful transcription in a chloroplast in vitro system. A sequence element, TATTTTT, which also conforms to the TATA-box consensus sequence (Breathnach and Chambon 1981), is present in the *rps16* promoter (Neuhaus et al. 1989), which lacks a −35-type element.

The *Chlamydomonas* "−35-less" *atpB* promoter (Klein et al. 1992) was shown to require about 22 bp upstream and 60 bp downstream of the transcription start site for maximum activity. Since even a DNA fragment extending from positions −10 to +55 still showed residual promoter activity, these experiments provide strong arguments that further upstream regions are not required for in vivo activity in this case. The −10-like region contains a palindromic octamer motif, TATAATAT, which is conserved in the promoter sequences of several other *Chlamydomonas* chloroplast genes that also lack the −35 region. Furthermore, the *Chlamydomonas atpB* promoter differs from its higher plant equivalent, as the *atpB* promoters from maize and spinach have been reported to contain typical −35-like elements that are essential for in vitro transcription activity (Bradley and Gatenby 1985; Chen et al. 1990).

The lack of a typical −35-like element and the presence of a TATA box-like motif may be clues to the possible existence of more than one transcription system within the plastid compartment. The notion that an "eukaryotic" transcription system similar to that found in the nucleus might be present in addition to the "prokaryotic" system responsible for transcription from typical −35/−10 promoters is supported by additional observations. For instance, the tobacco *psbA* promoter was shown to be active in a nuclear background in vivo (Cornelissen and Vandewiele 1989). Furthermore, sequence analyses of plastid *trnR* and *trnS* genes from spinach (Gruissem el al. 1986), as well as of *trnS*, *trnQ* (Neuhaus et al. 1989, 1990) and *trnH* from mustard (Nickelsen and Link 1990), have revealed the presence of sequence elements that resemble the internal A- and B-block elements of nuclear tRNA promoters transcribed by RNA polymerase III (Galli et al. 1981). In the case of the spinach genes, a functional analysis involving deletion of 5′ flanking sequences and in vitro transcription indicated that there was no requirement for this region, thus lending support to the notion that indeed the internal boxes might be the functional determinants (Table 2; Gruissem et al. 1986).

On the other hand, even amongst *E. coli* promoters, considerable deviation from the −35/−10 consensus (Reznikoff et al. 1985) was noted, in particular for positively regulated (sigma^{54}-type) promoters (Raibaud and Schwartz 1984; Gralla 1991; Kustu et al. 1991). Bacteriophage T3, T7, and late T4 promoters were all shown to be structurally different from "typical"

Table 2. Internal promoter-like elements in plastid tRNA genes

Gene	Promoter sequence				
trnS[a]	GGAGAGA	TGGCTGAGTGG	ACTAAAGCGTTGGATTGCTAATCCAT-18-CGAG	GGTTCGAATCC	CTCTCTTTCCC
trnQ[b]	TGGGGCG	TAGCCAAGCGG	TA AGGCAACGGGTTTTGGTCCCGC-4- CGGA	GGTTCGAATCC	TTCCGTCCCAG
trnK[c]	GGGTTGC	TAACTCAACGG	TA GAGTACTCGGCTTTTAACCGAC-4- TCCG	GGTTCGAGTCC	CGGGCAACCCA
trnH[d]	GGCGGATG	TAGCCAAGTGG	ATTAAGGCAGTGGATTGTGAATTCAC-4- CGCG	GGTTCAATTCC	CGTCGTTCGGC
trnG[e]	GCGGGTA	TAGTTTAGTGG	TAAAACCTTAGC -15-TGCG	GGTTCGATTCC	CGCTACCCGCT
Consensus		T$^{AGC}_{gat}$YNA$^{GT}_{aC}$GG		GGTTC$^{G}_{A}$ANTCC	
Nuclear polIII Consensus		TGGCNNAGTGG A-box	← 31–52 bp →	GGTTCGANNCC B-box	

[a] Neuhaus et al. (1989). [b] Neuhaus and Link (1990). [c] Neuhaus and Link (1987). [d] Nickelsen and Link (1990). [e] K. Liere, J. Nickelsen, H. Neuhaus and G. Link (unpubl.).

bacterial promoters and are transcribed by different (or modified) RNA polymerases (Geiduschek 1991).

Mitochondrial promoters in *Saccharomyces cerevisiae* typically contain a conserved nonanucleotide sequence, ATATAAGTA, which is reminiscent of both the nuclear TATA box and the prokaryotic −10 element (Costanzo and Fox 1990). It was shown that yeast mitochondrial RNA polymerase is capable of transcribing chimeric genes from nuclear TATA box-containing promoters (Marczynski et al. 1989). On the other hand, the mitochondrial enzyme appears structurally related to the T3/T7 type of RNA polymerases (Masters et al. 1987), but not to the bacterial and eukaryotic (nuclear) enzymes, which share a remarkable degree of sequence similarity (Broyles and Moss 1986). In view of these findings it is tempting to suggest the possibility of multiple transcription systems inside the plastid compartment, while it appears premature to designate them as "prokaryotic" versus "eukaryotic".

It is tempting to group plastid promoters according to their strength, based on transcription efficiency in organellar run-on experiments (Deng and Gruissem 1987; Klein and Mullet 1987; Rapp et al. 1992). According to this criterion, the *rrn* and *psbA* promoters are particularly "strong", although it remains unclear how this relates to the presence or absence of certain nucleotides at defined positions. A conserved 14-bp sequence element was noted in *rrn* promoters from several plants, which does not seem to be present in published sequences upstream of plastid genes coding for tRNAs or mRNAs (Baeza et al. 1991). Since this sequence element is involved in protein binding, it may have a crucial role in establishing the strength of the *rrn* promoter.

Thus far, only few mutational analyses have been carried out with plastid promoters, using chloroplast in vitro transcription as functional assay (Link 1984; Bradley and Gatenby 1985; Gruissem and Zurawski 1985a,b; Hanley-Bowdoin and Chua 1987; Westhoff 1985; Gruissem et al. 1986; Sun et al. 1989; Eisermann et al. 1990). Moreover, with the advent of plastid transformation, it has become possible to carry out a similar function analysis in vivo, i.e., to study plastid promoters in their natural context (Blowers et al. 1990; Klein et al. 1992). Clearly, this approach will help to gain a better understanding of the criteria for a "strong" versus "weak" plastid promoter, particularly if it can be used with higher plant systems with similar efficiency as with *Chlamydomonas* (Staub and Maliga 1992).

Promoter strength (as inferred from organellar run-on transcription efficiency) should not necessarily be viewed as a constant, since it is conceivable that one and the same promoter interacts differently with the transcription apparatus in different developmental situations and cell types. For instance, evidence was obtained for differential usage of the *psbA* promoter from mustard by in vitro transcription systems from chloroplasts and etioplasts (Eisermann et al. 1990). Positions of the −35-like region were found to be required in the chloroplast but not in the etioplast system, while the TATA box-like region was essential in both cases. The latter was previously found by deletion mutagenesis to be required for basal transcription (Link 1984), which does not, however, preclude a regulatory role of this region. It is interesting to note that the TATA box-like motif is not strictly conserved in published sequences of *psbA* promoters from some other plant species and, hence, it remains to be seen if differences in this region might be reflected by species-specific differences in *psbA* gene expression in vivo.

While in the case of the mustard *psbA* gene there is an apparent switch between sequence elements of the same promoter (Link 1984; Eisermann et al. 1990), switching between different promoters appears to be another mechanism for transcriptional regulation of plastid gene expression. Based on the analysis of complex transcription units that account for multiple RNAs, evidence was obtained for the existence of "alternative" promoters either upstream or within transcribed regions (Chen et al. 1990; Haley and Bogorad 1990; Neuhaus and Link 1990; Vera et al. 1992). It was shown for the *psbD-psbC* operon from barley that multiple transcripts arise from alternative promoters and different transcript patterns were observed for light- versus dark-grown seedlings (Berends Sexton et al. 1990). These changes were shown to be at least in part the result of differential transcription and, hence, this work provides evidence for a molecular switch involving alternative plastid promoters. An even more complex situation was noted for the two adjacent gene clusters *psbE-psbF-psbL*-ORF40 and *petE*-ORF42 from maize that are transcribed in opposite directions (Haley and Bogorad 1990). Evidence was obtained in this case for alternative upstream promoters that direct synthesis of overlapping RNAs from both clusters. Light/dark changes in the transcript pools generated in vivo suggest selective promoter usage in this system. Moreover, the close physical proximity of promoters on opposite strands, leading to the production of antisense transcripts, raises the possibility of promoter interdependence (Goodrich and McClure 1991) as an important mechanism for regulation of plastid gene expression.

The total number of active chloroplast promoters was determined to be in the order of 30–40 (Woodbury et al. 1989). However, this appears to be a minimum number, since additional promoters can be detected by more sensitive techniques (Vera and Sugiura 1992). Moreover, stage-specific differential usage of alternative promoters has to be taken into account.

Evidence was obtained that a fraction of transcripts from the mustard *psbA* gene region is cotranscribed with sequences of the preceding *trnK* gene (Nickelsen and Link 1991). Although their exact 5′ and 3′ ends were not mapped, the observed read-through *trnK-psbA* transcripts suggest the presence of an alternative promoter within this region. This might explain the existence of a minor D1-related protein (Callahan et al. 1990), which could originate via an alternative gene expression pathway. It has been recently shown that in *Pinus contorta* an additional copy of the *psbA* gene is present upstream of *trnK* and, as a result of transcriptional fusion, *trnK* is under the control of the *psbA* promoter and has acquired *psbA*-like expression characteristics (Link 1988) in this species (Lidholm and Gustafsson 1992).

The *trnK-psbA* region and other plastid transcription units that might be connected by RNA read-through (Nickelsen and Link 1991) appear reminiscent of what has been termed "superoperons" in purple, nonsulfur bacteria such as *Rhodobacter capsulatus* (Wellington et al. 1992). Transcription read-through from pigment biosynthetic operons into downstream operons that encode pigment-binding proteins is viewed as an efficient mechanism that links the expression of genes required for adaptation to changes in environmental conditions. It will be interesting to see if there is a similar role for the complex transcriptional organization of plastid genes.

3.2 Plastid Transcription Factors

In analogy to bacteria, plastid DNA-dependent RNA polymerase might be expected to consist of a multisubunit core associated with a sigma factor that confers specific DNA-binding and transcription initiation to the holoenzyme. Along the same line, additional DNA-binding proteins might be required as activators and/or repressors. At least 25 years of research in chloroplast molecular biology have shown that, in principle, this picture seems to be correct, although the various aspects have still been clarified only in part and many details have turned out to be more complicated than anticipated. Since plastid RNA polymerase has been the subject of recent reviews (Bogorad 1991; Herrmann et al. 1992; Igloi and Kössel 1992; Sugiura 1992), the results are only briefly summarized here.

Upon solubilization and extensive purification, chloroplast enzyme preparations with DNA-dependent RNA polymerase activity are usually found to contain multiple polypeptides that could be bona fide subunits. Protein sequencing (Hu and Bogorad 1990; Hu et al. 1991) has provided evidence that several polypeptides of a highly purified maize enzyme are encoded by plastid genes that share regional sequence similarity with the *E. coli* genes *rpoA*, *rpoB*, and *rpoC* for the α, β, and β′ subunits, respectively. The ptDNA region homologous to *E. coli rpoC* is split into two genes, *rpoC1* and *rpoC2* (for review, see Bogorad 1991; Igloi and Kössel 1992; Sugiura 1992). Hence, based on sequence similarity, four polypeptides of the maize enzyme appear to be functional subunits: the 120-kDa *rpoB* gene product, most likely the β equivalent, the 38-kDa *rpoA* product (α equivalent), as well as the 180-kDa *rpoC2* and 78-kDa *rpoC1* products designated β″ and β′, respectively. Two bands associated with the purified maize enzyme were shown to represent material related to the large subunit of RubisCo (55 kDa) and the a and b subunits of the RubisCo subunit-binding protein (Hemmingsen et al. 1988) (61 kDa), respectively. The identities of two other bands, at 85 and 64 kDa, remain to be clarified (Hu and Bogorad 1990; Hu et al. 1991).

DNA-dependent RNA polymerase preparations from pea (Rajasekhar et al. 1991), spinach (Lerbs et al. 1988), and mustard chloroplasts (T. Pfannschmidt and G. Link, unpubl. data) show a similar, yet even more complex polypeptide pattern compared to the maize enzyme. These differences cannot be attributed solely to variations in the purification procedures, since preparations containing substantially different polypeptide patterns were obtained from pea and mustard chloroplasts using the same procedure (M. Theismann and G. Link, unpubl.). The *rpo* genes from various plant species differ in size (for review, see Igloi and Kössel 1992; Sugiura 1992), which is reflected by the different electrophoretic mobilities of polypeptides. Nevertheless, it is possible that plastid RNA polymerase might exist in a variety of different forms, resulting from various combinations of polypeptides that comprise the enzyme. Some of these polypeptides could be stage- and/or tissue-specific, thus accounting for heterogeneity of enzyme preparations (Fig. 3).

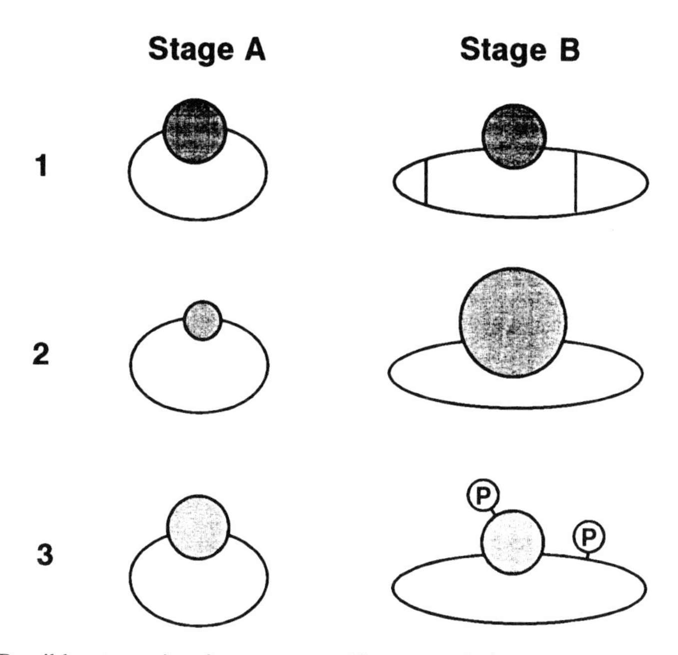

Fig. 3. Possible strategies for stage-specific transcription by plastid RNA polymerase. The scheme assumes the presence of one single class of (multisubunit) core polymerase that interacts with a single specificity factor (*gray circle*). Obviously, the details would become more complicated if multiple core enzymes and factors are involved. Transcription complexes that differ in a stage-specific way can result in either of three situations: *1* The core enzyme at plastid differentiation *stage A* consists of polypeptides that differ in number and/or size from those in *stage B* (indicated by the different sizes and shapes as well as the internal fields generated by the vertical lines); the specificity factor is the same in both stages. *2* The same core but different factors are present in *stages A* and *B*. This is likely to change the conformation of the core (indicated by the different shapes) and the size of the entire holoenzyme complex. *3* Both the core and factor polypeptides are the same in *stages A* and *B*. They differ, however, by reversible protein modification such as phosphorylation-dephosphorylation. While the first two scenarios cannot be excluded, there is evidence that the third one plays a definite role in chloroplast versus etioplast transcription (see text)

A related possibility would be that more than one single class of plastid RNA polymerases might be present in the organelle. As discussed in the previous section, the architecture of plastid promoters can be reconciled with the idea of multiple RNA polymerases, each one specific for a given type of promoter, similar to the situation in the nucleus (Breathnach and Chambon 1981). Evidence for the presence of at least two distinct enzymes was initially presented by Hallick et al. (1976), who isolated from *Euglena gracilis* chloroplasts a template-bound form of RNA polymerase (transcriptionally active chromosome, TAC) that appeared to preferentially transcribe rRNA sequences. Subsequent studies using TAC preparations from higher plants, however, did not reveal such preference (Briat et al. 1979; Reiss and Link 1985). Nevertheless, it was reported by Little and Hallick (1988) that antibodies raised against a soluble RNA polymerase preparation were capable of inactivating that enzyme, but did not affect the activity of the template-bound (TAC) form.

Two soluble forms of plastid RNA polymerase from peas were reported, one preferentially transcribing from the *rrn* promoter (Sun et al. 1986), the other from the *psbA* promoter (Lakhani et al. 1992). In this case, monoclonal antibodies raised against the first form also recognized polypeptides of the latter (Lakhani et al. 1992), suggesting that two related forms rather than entirely different polymerases were isolated. Two different enzyme forms were also obtained from maize chloroplasts (Zaitlin et al. 1989), one of which was capable of using relaxed templates, while the other was dependent on a supercoiled template. With few exceptions, a functional characterization of polypeptides found in most of these enzyme preparations remains to be done. Two (44- and 48-kDa) polypeptides of pea RNA polymerase were assigned as transcript-binding components (Khanna et al. 1991) and the largest (150-kDa) polypeptide was shown to be involved in DNA binding, both to *rrn* and *psbA* promoter templates (Khanna et al. 1992). In summarizing these data, it appears likely that various related forms of one single enzyme exist, rather than multiple unrelated classes of plastid RNA polymerases.

This conclusion is not necessarily at variance with recent findings that imply the existence of a second (nuclear-coded) plastid RNA polymerase. For instance, despite the absence of *rpo* genes from cloned ptDNA of the parasitic plant *Epifagus virginiana* (beechdrops), plastid transcripts were detected among total RNAs from that organism (Morden et al. 1991). It was concluded that plastid gene expression in *Epifagus* relies on the import of a nuclear-coded RNA polymerase. However, the presence of RNA sequences does not necessarily mean they were synthesized within the plastids containing the defective ptDNA. Based on ultrastructural studies, several different plastid types appear to exist in this parasitic plant (Walsh et al. 1980); it would be interesting to see, e.g., by in situ techniques if they differ with regard to their DNA organization and/or gene expression pattern.

Nevertheless, it is conceivable that subunits or accessory factors of the plastid polymerase are encoded by nuclear genes. This would be in accordance with the situation found for the photosynthetic apparatus, the ribosomes, and other multimeric protein complexes of the organelle, each of which is composed of both plastid and nuclear gene products (Bogorad 1991). In this regard, it is interesting to note the conspicuous absence of a plastid gene showing sufficient sequence similarity to qualify as the *rpoD* equivalent, i.e., coding for a polypeptide that acts similarly to the $sigma^{70}$ initiation factor of bacterial RNA polymerase. There is a limited degree of sequence similarity amongst various sigma factors from bacteria (Lonetto et al. 1992), including cyanobacteria (Brahamsha and Haselkorn 1991). It hence remains possible that a putative plastid *rpoD* gene (or other sigma factor-encoding gene) has thus far escaped detection, especially if it were organized as a split gene requiring extensive (*trans*-) splicing (Sugiura 1992). On the other hand, the apparent absence of *rpoD*-related sequences from ptDNA might be taken as an argument for the possible existence of nuclear genes for plastid sigma factors.

Surzycki and Shellenbarger (1976) partially purified a protein fraction containing a 51-kDa protein from *Chlamydomonas* cells. They termed this protein "sigma 2" as it seemed to carry out a reaction reminiscent of bacterial sigma, i.e., to stimulate the formation of open binary complexes by *E. coli* core enzyme in the presence of rifampicin. A 27.5-kDa polypeptide purified from maize chloroplasts (Jolly and Bogorad 1980) was shown to be required for preferential transcription of cloned chloroplast sequences (as compared to vector sequences) from supercoiled plasmids, if the appropriate form (Zaitlin et al. 1989) of plastid RNA polymerase was used. This polypeptide, termed S factor, was noted to act generally rather than on a specific set of chloroplast genes (Bogorad 1991). A 90-kDa polypeptide from spinach chloroplasts was detected by cross-reaction with antisera raised against *E. coli* $sigma^{70}$ and hence termed sigma-like factor (SLF) (Lerbs et al. 1988). Sigma-like activity was detected also in a protein fraction from mustard chloroplasts, using assays that detect the ability to confer sequence-specific DNA binding and transcription initiation in a system containing a plastid promoter and *E. coli* core RNA polymerase (Bülow and Link 1988). Further purification led to the separation of three distinct polypeptides of 67, 52, and 29 kDa, each of which revealed sigma-like activity (SLF^{67}, SLF^{52}, and SLF^{29}) (Tiller et al. 1991). The three factors differ in their affinity for several tested plastid promoters and in their ionic strength requirements. Furthermore, the same factors seem to be present also in etioplasts in comparable relative amounts (Tiller and Link 1993). The etioplast factors, however, show functional properties distinct from those of their chloroplast counterparts, including different promoter affinity and salt requirements (Tiller and Link 1993).

It has recently become clear that this is related to differential phosphorylation, with the factors from etioplasts being in a more highly phosphorylated form than those from chloroplasts. The etioplast factors bind more tightly to the plastid core enzyme but initiate transcription less efficiently than their chloroplast counterparts (K. Tiller and G. Link, unpubl. data).

Hence, the situation in the etioplast system seems to be somewhat similar to that observed for a mutant sigma factor from *B. subtilis*, which traps the RNA polymerase in a stable complex with promoter DNA in which it is unable to initiate transcription (Jones and Moran 1992). Phosphorylation of plastid sigma factors is likely to be the key mechanism that accounts for differential transcription from the *psbA* promoter and the associated promoter element switch (Link 1984; Eisermann et al. 1990). A TATA-binding protein (TBP)-associated phosphoprotein, Dr1, has been recently described, which seems to play an analogous role in nuclear class II gene transcription; in its phosphorylated form Dr1 interacts efficiently with TBP, thus inhibiting the formation of a transcription-competent complex (Inostroza et al. 1992).

In addition to the sigma-like factors, which do not bind to DNA per se but rather act together with other components of the transcription complex, several direct DNA-binding plastid proteins were reported. These include a 115-kDa protein from peas shown to interact with a 5′ region of the *rbcL* gene (Lam et al. 1988), polypeptides of 70- (Lerbs et al. 1983), 35- and 33-kDa from spinach (Baeza et al. 1991), as well as several DNA-binding polypeptides associated with TAC preparations from mustard (Bülow et al. 1987) and tobacco (Nemoto et al. 1988, 1990). The exact role of all these proteins remains to be established.

4 Conclusion

What might be the immediate and perhaps more long-term directions in plastid transcription research? As progress in a particular field depends on the development of new techniques, it is likely that the biolistic transfection approach (Blowers et al. 1990) will play an increasing role. Answers to the following questions can be expected by combining molecular-genetic and biochemical approaches: What is the functional architecture of plastid promoters? Are there additional regulatory regions that may be involved in the activation and/or repression of specific genes? What is the role of the multiple sigma-like factors and of the various DNA-binding proteins? Where are these factors encoded? Which protein kinase(s) and phosphatase(s) are responsible for the differential phosphorylation of the sigma-like factors? What are the signals for the plastid transduction pathway(s) via phosphorylation control of transcription factors? Are the plastid and nuclear transcription systems interconnected by common signal transduction mechanisms? Considering the rapid progress in the field, it is possible that answers to some of these questions are available already or will be very soon.

Acknowledgments. Cited work from the author's laboratory was supported by the Deutsche Forschungsgemeinschaft and the Fonds der Chemischen Industrie, FRG.

References

Baeza L, Bertrand A, Mache R, Lerbs-Mache S (1991) Characterization of a protein binding sequence in the promoter region of the 16S rRNA gene of the spinach chloroplast genome. Nucleic Acids Res 19:3577–3581

Baker NR, Barber J (eds) (1984) Chloroplast biogenesis. Topics in photosynthesis, vol 5. Elsevier, Amsterdam

Berends Sexton T, Jones JT, Mullet JE (1990) Sequence and transcriptional analysis of the barley ctDNA region upstream of *psbD-psbC* encoding *trnK*(UUU), *rps16*, *trnQ*(UUG), *psbK*, *psbI*, and *trnS*(GCU). Curr Genet 17:445–454

Blowers AD, Ellmore GS, Klein U, Bogorad L (1990) Transcriptional analysis of endogenous and foreign genes in chloroplast transformants in *Chlamydomonas*. Plant Cell 2:1059–1070

Bogorad L (1991) Replication and transcription of plastid DNA. In: Bogorad L, Vasil IK (eds) The molecular biology of plastids. Cell culture and somatic cell genetics, vol 7A. Academic Press, San Diego, pp 93–117

Bogorad L, Vasil IK (eds) (1991a) The molecular biology of plastids. Cell culture and somatic cell genetics, vol 7A. Academic Press, San Diego

Bogorad L, Vasil IK (eds) (1991b) The photosynthetic apparatus: molecular biology and operation. Cell culture and somatic cell genetics, vol 7B. Academic Press, San Diego

Bradley D, Gatenby AA (1985) Mutational analysis of the maize chloroplast ATPase-beta subunit gene promoter: the isolation of promoter mutants in *E. coli* and their characterization in a chloroplast in vitro transcription system. EMBO J 4:3641–3648

Brahamsha B, Haselkorn R (1991) Isolation and characterization of the gene encoding the principal sigma factor of the vegetative cell RNA polymerase from the cyanobacterium *Anabaena* sp. strain PCC 7129. J Bacteriol 173:2442–2450

Breathnach R, Chambon P (1981) Organization and expression of eukaryotic split genes coding for proteins. Annu Rev Biochem 50:349–383

Briat JF, Laulhere JP, Mache R (1979) Transcription activity of a DNA-protein complex isolated from spinach plastids. Eur J Biochem 98:285–292

Broyles SS, Moss B (1986) Homology between RNA polymerases of poxviruses, prokaryotes, and eukaryotes: nucleotide sequence and transcriptional analysis of vaccinia virus genes encoding 147-kDa and 22-kDa subunits. Proc Natl Acad Sci USA 83:3141–3145

Bülow S, Link G (1988) Sigma-like activity from mustard (*Sinapis alba* L.) chloroplasts conferring DNA-binding and transcription specificity to *E. coli* core RNA polymerase. Plant Mol Biol 10:349–357

Bülow S, Reiss T, Link G (1987) DNA-binding proteins of the transcriptionally active chromosome from mustard (*Sinapis alba* L.) chloroplasts. Curr Genet 12:157–159

Callahan FE, Ghirardi ML, Sopory SK, Mehta AM, Edelman M, Mattoo AK (1990) A novel metabolic from of the 32kDa-D1 protein in the grana-localized reaction center of photosystem II. J Biol Chem 256:15357–15360

Chen LJ, Rogers SA, Bennett DC, Hu MC, Orozco EM (1990) An in vitro transcription termination system to analyze chloroplast promoters: identification of multiple promoters for the spinach *atpB* gene. Curr Genet 17:55–64

Christopher DA, Kim M, Mullet JE (1992) A novel light-regulated promoter is conserved in cereal and dicot chloroplasts. Plant Cell 4:785–798

Cornelissen M, Vandewiele M (1989) Nuclear transcriptional activity of the tobacco plastid *psbA* promoter. Nucleic Acids Res 17:19–29

Costanzo MC, Fox TD (1990) Control of mitochondrial gene expression in *Saccharomyces cerevisiae*. Annu Rev Genet 24:91–113

Deng XW, Gruissem W (1987) Control of plastid gene expression during development: the limited role of transcriptional regulation. Cell 49:379–387

Dietrich G, Link G (1985) Transcriptional organization and possible function of mustard plastid DNA regions expressed in vivo. Curr Genet 9:683–692
Eisermann A, Tiller K, Link G (1990) In vitro transcription and DNA binding characteristics of chloroplast and etioplast extracts from mustard (*Sinapis alba*) indicate differential usage of the *psbA* promoter. EMBO J 9:3981–3987
Galli G, Hofstetter H, Birnstiel ML (1981) Two conserved sequence blocks within eukaryotic tRNA genes are major promoter elements. Nature 294:626–631
Geiduschek EP (1991) Regulation of expression of the late genes of bacteriophage T4. Annu Rev Genet 25:437–460
Gilmartin PM, Saroki L, Memelink J, Chua N-H (1990) Molecular light switches for plant genes. Plant Cell 2:369–378
Goodrich JA, McClure WR (1991) Competing promoters in prokaryotic transcription. TIBS 16:394–397
Gralla JD (1991) Transcriptional control. Lessons from an *E. coli* promoter data base. Cell 66:415–418
Gruissem W, Zurawski G (1985a) Identification and mutational analysis of the promoter for a spinach chloroplast transfer RNA gene. EMBO J 4:1637–1644
Gruissem W, Zurawski G (1985b) Analysis of promoter regions for the spinach chloroplast *rbcL*, *atpB* and *psbA* genes. EMBO J 4:3375–3383
Gruissem W, Elsner-Menzel C, Latshaw S, Narita JO, Schaffer MA, Zurawski G (1986) A subpopulation of spinach chloroplast tRNA genes does not require upstream promoter elements for transcription. Nucleic Acids Res 14:7541–7557
Haley J, Bogorad L (1990) Alternative promoters are used for genes within maize chloroplast polycistronic transcription units. Plant Cell 2:323–333
Hallick RB (1989) Proposals for the naming of chloroplast genes. II. Update to the nomenclature of genes for thylakoid membrane polypeptides. Plant Mol Biol 7:266–275
Hallick RB, Lipper C, Richards OC, Rutter WJ (1976) Isolation of a transcriptionally active chromosome from chloroplasts of *Euglena gracilis*. Biochemistry 15:3039–3045
Hanley-Bowdoin L, Chua N-H (1987) Chloroplast promoters. TIBS 12:67–70
Hawley DK, McClure WR (1983) Compilation and analysis of *Escherichia coli* promoter DNA sequences. Nucleic Acids Res 11:2237–2255
Hemmingsen SM, Woolford C, Van der Vies SM, Tilly K, Dennis DT, Georgopoulos CP, Hendrix RW, Ellis RJ (1988) Homologous plant and bacterial proteins chaperone oligomeric protein assembly. Nature 333:330–334
Herrmann RG, Westhoff P, Link G (1992) Chloroplast biogenesis in higher plants. In: Herrmann RG (ed) Cell organelles. Springer, Berlin Heidelberg New York, pp 275–349
Hoch B, Maier RM, Appel K, Igloi GL, Kössel H (1991) Editing of a chloroplast mRNA by creation of an initiation codon. Nature 353:178–180
Hu J, Bogorad L (1990) Maize chloroplast RNA polymerase: the 180- and 38-kilodalton polypeptides are encoded in chloroplast genes. Proc Natl Acad Sci USA 87:1531–1535
Hu J, Troxler RF, Bogorad L (1991) Maize chloroplast RNA polymerase: the 78-kilodalton polypeptide is encoded by the plastid *rpoC1* gene. Nucleic Acids Res 19:3431–3434
Hughes JE, Neuhaus H, Link G (1987) Transcript levels of two adjacent chloroplast genes during mustard (*Sinapis alba* L.) seedling development are under differential temporal and light control. Plant Mol Biol 9:355–363
Igloi GL, Kössel H (1992) The transcriptional apparatus of chloroplasts. Crit Rev Plant Sci 10:525–558
Inostroza JA, Mermelstein FH, Ha I, Lane WS, Reinberg D (1992) Dr1, a TATA-binding protein-associated phosphoprotein and inhibitor of class II gene transcription. Cell 70:477–489
Jolly SO, Bogorad L (1980) Preferential transcription of cloned maize chloroplast DNA sequences by maize chloroplast RNA polymerase. Proc Natl Acad Sci USA 77:822–826

Jones CH, Moran CP (1992) Mutant sigma factor blocks transition between promoter binding and initiation of transcription. Proc Natl Acad Sci USA 89: 1958–1962

Khanna NC, Lakhani S, Tewari KK (1991) Photoaffinity labelling of the pea chloroplast transcriptional complex by nascent RNA in vitro. Nucleic Acids Res 19:4849–4855

Khanna NC, Lakhani S, Tewari KK (1992) Identification of the template binding polypeptide in the pea chloroplast transcriptional complex. Nucleic Acids Res 20:69–74

Kirk JTO, Tilney-Bassett RAE (1967) The plastids. Freeman, London

Klein RR, Mullet JE (1987) Control of gene expression during higher plant chloroplast biogenesis. J Biol Chem 9:4341

Klein U, De Camp JD, Bogorad L (1992) Two types of chloroplast gene promoters in *Chlamydomonas reinhardtii*. Proc Natl Acad Sci USA 89:3453–3457

Kung SD, Lin CM (1985) Chloroplast promoters from higher plants. Nucleic Acids Res 13:7543–7549

Kustu S, North AK, Weiss DS (1991) Prokaryotic transcriptional enhancers and enhancer-binding proteins. TIBS 16:397–402

Lakhani S, Khanna NC, Tewari KK (1992) Two distinct transcriptional activities of pea (*Pisum sativum*) chloroplasts share immunochemically related functional polypeptides. Biochem J 286:833–841

Lam E, Hanley-Bowdoin L, Chua N-H (1988) Characterization of a chloroplast sequence-specific DNA binding factor. J Biol Chem 263:8288–8293

Lerbs S, Briat JF, Mache R (1983) Chloroplast RNA polymerase from spinach: purification and DNA-binding proteins. Plant Mol Biol 2:67–74

Lerbs S, Bräutigam E, Mache R (1988) DNA-dependent RNA polymerase of spinach chloroplasts: characterization of sigma-like and alpha-like polypeptides. Mol Gen Genet 211:459–464

Lidholm J, Gustafsson P (1992) A functional promoter shift of a chloroplast gene: a transcriptional fusion between a novel *psbA* gene copy and the *trnK*(UUU) gene in *Pinus contorta*. Plant J 2:875–886

Link G (1984) DNA sequence requirements for the accurate transcription of a protein-coding plastid gene in a plastid in vitro system from mustard (*Sinapis alba* L.). EMBO J 3:1697–1704

Link G (1988) Photocontrol of plastid gene expression. Plant Cell Environ 11: 329–338

Link G (1991) Photoregulated development of chloroplasts. In: Bogorad L, Vasil IK (eds) The photosynthetic apparatus: molecular biology and operation. Cell Culture and somatic cell genetics, vol 7B. Academic Press, San Diego, pp 365–386

Link G, Langridge U (1984) Structure of the chloroplast gene for the precursor of the M_r 32000 photosystem II protein from mustard (*Sinapis alba* L.). Nucleic Acids Res 12:945–958

Little MC, Hallick RB (1988) Chloroplast *rpoA*, *rpoB*, and *rpoC* genes specify at least three components of a chloroplast DNA-dependent RNA polymerase active in tRNA and mRNA transcription. J Biol Chem 263:14302–14307

Lonetto M, Gribskov M, Gross CA (1992) The sigma^{70} family: sequence conservation and evolutionary relationships. J Bacteriol 174:3843–3849

Marczynski GT, Schultz PW, Jaehning JA (1989) Use of yeast nuclear DNA sequences to define the mitochondrial RNA polymerase promoter in vitro. Mol Cell Biol 9:3193–3202

Masters BS, Stohl LL, Clayton DA (1987) Yeast mitochondrial RNA polymerase is homologous to those encoded by bacteriophages T3 and T7. Cell 51:89–99

Morden CW, Wolfe KH, DePamphilis CW, Palmer JD (1991) Plastid translation and transcription genes in a non-photosynthetic plant: intact, missing and pseudo genes. EMBO J 10:3281–3288

Mullet JE (1988) Chloroplast development and gene expression. Annu Rev Plant Physiol Plant Mol Biol 39:475–502

Nemoto Y, Kawano S, Nakamura S, Mita T, Nagata T, Kuroiwa T (1988) Studies on plastid-nuclei (nucleoids) in *Nicotiana tabacum* L. I. Isolation of proplastid-nuclei from cultured cells and identification of proplastid-nuclear proteins. Plant Cell Physiol 29:167–177

Nemoto Y, Kawano S, Kondoh K, Nagata T, Kuroiwa T (1990) Studies on plastid-nuclei (nucleoids) in *Nicotiana tabacum* L. III. Isolation of chloroplast-nuclei from mesophyll protoplasts and identification of chloroplast DNA-binding proteins. Plant Cell Physiol 31:767–776

Neuhaus H, Link G (1987) The chloroplast $tRNA^{Lys}$(UUU) gene from mustard (*Sinapis alba*) contains a class II intron potentially coding for a maturase-related polypeptide. Curr Genet 11:251–257

Neuhaus H, Link G (1990) The chloroplast *psbK* operon from mustard (*Sinapis alba* L.): multiple transcripts during seedling development and evidence for divergent overlapping transcription. Curr Genet 18:377–383

Neuhaus H, Scholz A, Link G (1989) Structure and expression of a split chloroplast gene from mustard (*Sinapis alba*): ribosomal protein gene *rps16* reveals unusual transcriptional features and complex RNA maturation. Curr Genet 15:63–70

Neuhaus H, Pfannschmidt T, Link G (1990) Nucleotide sequence of the chloroplast *psbI* and *trnS*-GCU genes from mustard (*Sinapis alba*). Nucleic Acids Res 18:368–368

Nickelsen J, Link G (1990) Nucleotide sequence of the mustard chloroplast genes *trnH* and *rps19′*. Nucleic Acids Res 18:1051–1051

Nickelsen J, Link G (1991) RNA-protein interactions at transcript 3′ ends and evidence for *trnK-psbA* cotranscription in mustard chloroplasts. Mol Gen Genet 228:89–96

Palmer JD (1991) Plastid chromosomes: structure and evolution. In: Bogorad L, Vasil IK (eds) The molecular biology of plastids. Cell culture and somatic cell genetics, vol 7A. Academic Press, San Diego, pp 5–42

Przybyl D, Fritzsche E, Edwards K, Kössel H, Falk H, Thompson JA, Link G (1984) The ribosomal RNA genes from chloroplasts of mustard (*Sinapis alba* L.): mapping and sequencing of the leader region. Plant Mol Biol 3:147–158

Raibaud O, Schwartz M (1984) Positive control of transcription initiation in bacteria. Annu Rev Genet 18:173–206

Rajasekhar VK, Sun E, Meeker R, Wu BW, Tewari KK (1991) Highly purified pea chloroplast RNA polymerase transcribes both rRNA and mRNA genes. Eur J Biochem 195:215–228

Rapp JC, Mullet JE (1991) Chloroplast transcription is required to express the nuclear genes *rbcS* and *cab*. Plastid DNA copy number is regulated independently. Plant Mol Biol 17:813–823

Rapp JC, Baumgartner BJ, Mullet J (1992) Quantitative analysis of transcription and RNA levels of 15 barley chloroplast genes. Transcription rates and mRNA levels vary over 300-fold; predicted mRNA stabilities vary 30-fold. J Biol Chem 267:21404–21411

Reiss T, Link G (1985) Characterization of transcriptionally active DNA-protein complexes from chloroplasts and etioplasts of mustard (*Sinapis alba* L.). Eur J Biochem 148:207–212

Reznikoff W, Siegele DA, Cowing DW, Gross CA (1985) The regulation of transcription initiation in bacteria. Annu Rev Genet 19:355–387

Rochaix J-D (1992) Control of plastid gene expression in *Chlamydomonas reinhardtii*. In: Herrmann RG (ed) Cell organelles. Springer, Berlin Heidelberg New York, pp 249–275

Staub JM, Maliga P (1992) Long regions of homologous DNA are incorporated into the tobacco plastid genome by transformation. Plant Cell 4:39–45

Sugiura M (1992) The chloroplast genome. Plant Mol Biol 19:149–168

Sun E, Shapiro DR, Wu BW, Tewari KK (1986) Specific in vitro transcription of 16S rRNA gene by pea chloroplast RNA polymerase. Plant Mol Biol 6:429–440

Sun E, Wu BW, Tewari KK (1989) In vitro analysis of the pea chloroplast 16S rRNA gene promoter. Mol Cell Biol 9:5650–5659

Surzycki SJ, Shellenbarger DL (1976) Purification and characterization of a putative sigma factor from *Chlamydomonas reinhardtii*. Proc Natl Acad Sci USA 73: 3961–3965
Taylor WC (1989) Regulatory interactions between nuclear and plastid genomes. Annu Rev Plant Physiol Plant Mol Biol 40:211–233
Thompson WF, White MJ (1991) Physiological and molecular studies of light-regulated nuclear genes in higher plants. Annu Rev Plant Physiol Plant Mol Biol 42:423–466
Thomson WW, Whatley JM (1980) Development of nongreen plastids. Annu Rev Plant Physiol 31:375–394
Tiller K, Link G (1993) Sigma-like transcription factors from mustard (*Sinapis alba* L.) etioplasts are similar in size to, but functionally distinct from, their chloroplast counter-parts. Plant Mol Biol 21:503–514
Tiller K, Eisermann A, Link G (1991) The chloroplast transcription apparatus from mustard (*Sinapis alba* L.): evidence for three different transcription factors which resemble bacterial sigma factors. Eur J Biochem 198:93–99
Tobin EM, Silverthorne J (1985) Light regulation of gene expression in higher plants. Annu Rev Plant Physiol 36:569–593
Vera A, Sugiura M (1992) Combination of in vitro capping and ribonuclease protection improves the detection of transcription start sites in chloroplasts. Plant Mol Biol 19:309–311
Vera A, Matsubayashi T, Sugiura M (1992) Active transcription from a promoter positioned within the coding region of a divergently oriented gene: the tobacco chloroplast *rpl32* gene. Mol Gen Genet 233:151–156
Walsh MA, Rechel EA, Popovich TM (1980) Observations on plastid fine-structure in the holoparasitic angiosperm *Epifagus virginiana*. Am J Bot 67 833–837
Weil JH (1987) Organization and expression of the chloroplast genome. Plant Sci 49:149–157
Wellington CL, Bauer CE, Beatty JT (1992) Photosynthesis gene superoperons in purple nonsulfur bacteria: the tip of the iceberg? Can J Microbiol 38:20–27
Westhoff P (1985) Transcription of the gene encoding the 51 kd chlorophyll a-apoprotein of the photosystem II reaction centre from spinach. Mol Gen Genet 201:115–123
Woodbury NW, Dobres M, Thompson WF (1989) The identification and localization of 33 pea chloroplast transcription initiation sites. Curr Genet 16:433–445
Zaitlin D, Hu J, Bogorad L (1989) Binding and transcription of relaxed DNA templates by fractions of maize chloroplast extracts. Proc Natl Acad Sci USA 86:876–880

4 AT-Rich Elements (ATREs) in the Promoter Regions of Nodulin and Other Higher Plant Genes: a Novel Class of *Cis*-Acting Regulatory Element?

Brian G. Forde

1 Introduction

Regions of DNA that are unusually rich in dA.dT base pairs play a key role in the structure and function of the eukaryotic genome. AT-rich sequences are found within origins of replication in both plants and animals (Eckdahl et al. 1989; Eckdahl and Anderson 1990), within the scaffold associated regions (SARs) that anchor chromatin to the nuclear matrix (Gasser and Laemmli 1986) and upstream of yeast (Struhl 1985) and mammalian (Fashena et al. 1992) promoters, where they serve as transcriptional activators.

The occurrence of extensive AT-rich sequences within these elements may partly be attributable to their unusual structural properties, such as their ability to produce bent DNA (Eckdahl and Anderson 1990) or to be readily unwound and unpaired in supercoiled DNA (Umek et al. 1988). Poly(dA.dT) stretches of 20 bp or more strongly inhibit nucleosome formation in vitro (Kunkel and Martinson 1981; Prunell 1982), and may therefore promote transcriptional activation by allowing access to *trans*-acting factors (Struhl 1985). However, it is also becoming clear that a number of nuclear proteins specifically recognize AT-rich DNA sequences. Some, such as the mammalian HMGI protein (Solomon et al. 1986), bind to any sequence of dAs and dTs, while others, such as the TATA-box binding protein TBP (Greenblatt 1991) and the homeodomain transcription factors *Antennapedia* and *Ultrabithorax* (Beachy et al. 1988), recognize short well-defined AT-rich sequences.

Little is known in higher plants about the nature and function of nuclear proteins that bind to AT-rich DNA sequences. However, a succession of studies in recent years has uncovered the existence in a variety of plant species of at least two groups of abundant nuclear proteins which recognize AT-rich sequences in the promoter regions of a diverse set of plant genes. This chapter will review the evidence relating to AT-rich binding sites in one coordinately regulated group of genes, the late nodulin genes of the Leguminosae, and then examine how information from other systems throws light on the possible role of these elements and their binding factors in

Biochemistry and Physiology Department, Rothamsted Experimental Station, Institute of Arable Crops Research, Harpenden, Herts AL5 2JQ, UK

Results and Problems in Cell Differentiation 20
L. Nover (Ed.)
Plant Promoters and Transcription Factors

regulating transcription. More general aspects of nodulin gene regulation have recently been reviewed (de Bruijn et al. 1990; de Bruijn and Schell 1993).

2 Nodulin Genes

During the development of the nitrogen-fixing symbiotic relationship between legumes and *Rhizobium* bacteria, each symbiont synthesizes a set of proteins known as nodulins, which are characteristic of the symbiotic state (see Verma et al. 1992 for a recent review). The 20 or more plant nodulins have been classified as either "early" or "late" according to the developmental stage at which they first appear. Amongst the late nodulins are the leghaemoglobins, whose function is to facilitate O_2 diffusion to the *Rhizobium* bacteroids. The regulation of the leghaemoglobin gene families in soybean and in the stem-nodulating legume *Sesbania rostrata* has been studied in detail at the molecular level. The other late nodulin genes that have received the most attention are the soybean N23 gene, which specifies a putative zinc finger protein of unknown function, and the *Phaseolus vulgaris gln-*γ gene, which codes for the γ subunit of a nodule-specific isoenzyme of glutamine synthetase (GS). Like the N23 and leghaemoglobin genes, *gln-*γ is expressed at very low levels in roots but is strongly induced in the developing nodule. In each of these coordinately regulated genes, as reviewed below, there is evidence for the occurrence of related AT-rich protein-binding sequences located at multiple sites upstream of the promoters.

2.1 Occurrence of AT-Rich Binding Sites in Nodulin Genes

2.1.1 Leghaemoglobin Genes

Jensen et al. (1988) identified two distinct regions upstream of the soybean *lbc*$_3$ promoter which strongly interacted in vitro with a nodule-specific nuclear factor, later designated NAT2 (Jacobsen et al. 1990). The two binding sites were mapped to the segments between −246 and −233 (Binding Site 1 = BS1) and between −176 and −161 (Binding Site 2 = BS2) (unless otherwise stated, nucleotide positions in this review are given relative to the transcription start site). Although both NAT2-binding sites are composed almost entirely of A and T base pairs, they are dissimilar in sequence. Binding site 2, but not the higher affinity binding site, BS1, is highly conserved at analogous positions in other soybean leghaemoglobin genes (Jensen et al. 1988) and in the *Srglb*$_2$ and *Srglb*$_3$ genes from *Sesbania* (Metz et al. 1988). The BS2 sequence in the *Srglb*$_3$ gene (BS2*) has been shown to compete efficiently with the *lbc*$_3$ BS2 sequence for binding to NAT2

(Metz et al. 1988). Although the BS1-analogous site is absent in *Srglb*$_3$, several additional binding sites for NAT2-like factors have been detected both upstream and downstream of the BS2* site (Metz et al. 1988; de Bruijn et al. 1989).

2.1.2 N23 Gene

At least five NAT2-binding sites were detected in the 1-kb promoter proximal region of the soybean N23 gene (Jacobsen et al. 1990). The two binding sites that were precisely mapped (at −344 and −407) are almost exclusively composed of A.T base pairs (see Fig. 1), but are otherwise dissimilar to each other and to the BS1- and BS2-binding sites in the *lbc*$_3$ promoter. These NAT2-binding sites were also found to interact in vitro with a second group of soybean nuclear factors (NAT1 and LAT1) that have many of the properties of HMGI-like proteins (see Sect. 2.2.1 below).

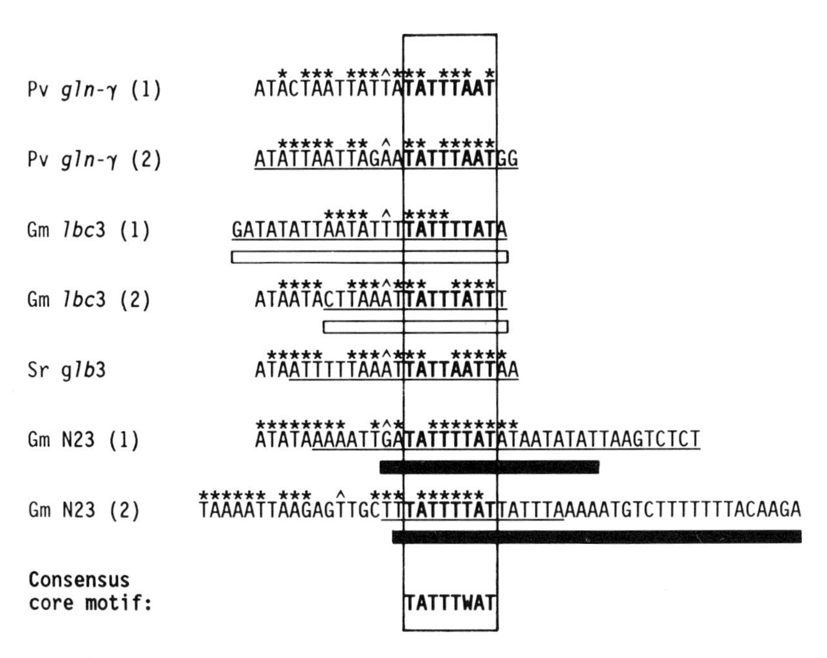

Fig. 1. Alignment of AT-rich binding sites from four nodulin genes illustrating a common 8-bp core motif (*boxed*). Where binding to an oligonucleotide sequence has been demonstrated, the relevant sequence has been *underlined*, while binding sites mapped by Exo III digestion or footprinting are indicated by *open* and *closed bars*, respectively. A region of imperfect dyad symmetry within each sequence is indicated by *asterisks*. The *gln-γ* sequences were reported by Forde et al. (1990), the *lbc*$_3$ and N23-binding sites were mapped by Jensen et al. (1988) and Jacobsen et al. (1990), respectively, and the *Srglb*$_3$ site (BS2*) was reported by Metz et al. (1988). *Pv* French bean; *Gm* soybean; *Sr Sesbania*. (Forde et al. 1990)

2.1.3 Glutamine Synthetase (*gln*-γ) Gene

Abundant nuclear factors in extracts from *P. vulgaris* roots and nodules (PRF-1 and PNF-1, respectively) bound in vitro to at least three sites in the region of the *gln*-γ promoter between −860 and −154 (relative to the ATG codon) (Forde et al. 1990). PNF-1 and PRF-1 bound strongly to a synthetic oligonucleotide containing the sequence of one copy of an AT-rich 21-bp imperfect repeat found at positions −516 and −466 (relative to the ATG codon). This AT-rich repeat has significant homology to the BS2-binding site in lbc_3 (Fig. 1) and it was subsequently shown that PNF-1 and PRF-1 also have a high affinity for the lbc_3 BS1 sequence. It was therefore concluded that PNF-1 is probably the homologue in *P. vulgaris* of NAT2.

2.2 Properties and Distribution of Proteins Binding to AT-Rich Sequences

2.2.1 HMGI-Like Proteins

HMG (high motility group) proteins are a heterogeneous group of eukaryotic chromosomal proteins which are rich in acidic and basic amino acids and appear to be preferentially associated with transcriptionally active chromatin. They are defined on an operational basis as being extractable from chromatin by 0.35 M NaCl and soluble in 2% trichloracetic acid (TCA) or 5% perchloric acid. The 11-kDa HMGI (as distinct from HMG1), and its isoform HMGY, differ from other HMG families in their preference for binding double-stranded AT-rich DNA (Bustin et al. 1990).

Jacobsen et al. (1990) noted that the same AT-rich sequences in the N23 gene that bound NAT2 were also recognized by another group of factors that formed faster moving DNA-protein complexes than NAT2 in the gel retardation assays. One such factor (NAT1) was present in both roots and nodules, while the other (LAT1) was detected in young leaves. In accordance with the operational definition of HMG proteins, NAT1 and LAT1 were released from chromatin by extraction with a low salt buffer and were soluble in 2% TCA. Furthermore, the binding of NAT1 and LAT1 to double-stranded AT-rich DNA could not be competed out by single-stranded poly(dA.dT) and was more sensitive to poly(dI.dC) than to poly (dG.dC), indicating the importance of minor groove contacts for NAT1/LAT1 binding, a further point of similarity to HMGI (Solomon et al. 1986). Finally, analysis of the binding affinity of LAT1 to mutated versions of the lbc_3 BS1 site showed that, like HMGI, LAT1 was relatively insensitive to changes that did not affect the AT-richness of the binding site (Jacobsen et al. 1990).

LAT1 was partially purified and found to consist of two heat-stable polypeptides of 21 and 23 kDa with identical N-terminal sequences that had no homology to mammalian HMGI (Jacobsen et al. 1990). Recently, a soybean cDNA clone (SB16) has been isolated for a 19-kDa polypeptide which binds to an AT-rich sequence in the soybean seed lectin gene (Laux and Goldberg 1991; T. Laux and R.B. Goldberg unpubl. results). This polypeptide has an unusual mosaic structure with an amino-terminal region that resembles histone H1 and a carboxy-terminal, DNA-binding domain that contains sequences homologous to HMGI(Y). It will be interesting to establish the relationship, if any, between the SB16 protein and LAT1/NAT1.

2.2.2 NAT2-Like Factors

Apart from the lower electrophoretic mobility of their DNA-protein complexes, NAT2-like factors differ in several other respects from the HMGI-like proteins. Firstly, although to some extent extractable with low salt, NAT2-like binding activity was completely lost after 2% TCA treatment and was also heat-sensitive (Forde et al. 1990; Jacobsen et al. 1990). Secondly, although the two types of factor share common recognition sequences in the nodulin genes, and both appear to interact with the minor groove in the DNA helix, the NAT2-like factors are more discriminating in their affinity for different AT-rich sequences (Forde et al. 1990; Jacobsen et al. 1990).

In soybean, NAT2 is absent in leaves and only barely detectable in roots (Jensen et al. 1988; Metz et al. 1988), but appears to be a highly abundant nuclear protein in soybean nodules (estimated as 0.3% of the protein content, assuming a molecular weight of 50 kDa; Jensen et al. 1988). In a developmental study, the period of most rapid accumulation of binding activity in nodules was between 5 and 11 days after rhizobial infection.

In *P. vulgaris* unnodulated roots contain significant amounts of the NAT2-like binding factor, PRF-1, which differs only slightly from the nodule form, PNF-1, having a marginally higher electrophoretic mobility and differing relative affinities for some AT-rich binding sites (Forde et al. 1990). Similarly, NAT2-like factors giving complexes with different mobility to the nodule forms were identified in *Sesbania* roots and in leaves of both *Sesbania* and alfalfa (Metz et al. 1988). Thus, in many legume species, factors related to NAT2 appear to be widely distributed throughout the plant and may represent either alternatively modified forms of the nodule factor or distinct gene products.

Forde et al. (1990), comparing AT-rich binding sites in nodulin genes, identified a shared 8-bp core motif with the consensus sequence TATTTa/tAT (Fig. 1). Mutational analysis of an oligonucleotide containing the sequence of one of the two 21-bp AT-rich repeats in *gln-*γ confirmed that PNF-1 has quite stringent sequence requirements for binding, even when the AT-rich base composition of the binding site was maintained, and provided evidence for the importance of the core motif. A search of other late nodulin genes, for which sufficient upstream sequences are available, reveals that potential binding sites for NAT2 also occur at −190 in the soybean N35 (uricase II) gene (D.P.S. Verma, unpubl.; Genbank accession no. M10594), and in the soybean N20 gene at −440 and −330 (Sandal et al. 1987; Richter et al. 1991), but it is not known whether these or any other sequences in the N35 or N20 genes do bind NAT2.

2.3 Functional Analysis of AT-Rich Binding Sites

The occurrence of conserved binding sites for abundant nuclear proteins in the 5′-flanking region of the late nodulin genes raises the question of what role, if any, these sequences play in regulating transcription. Experiments which address this question have so far been carried out with *lbc₃*, *Srglb₃*, N23 and *gln*-γ. In each case these have involved a deletion analysis of the 5′-flanking region of the gene, combined with the construction of a number of hybrid promoters in which the ability of nodulin promoter sequences to activate expression from a heterologous promoter has been tested in transgenic plants.

2.3.1 Leghaemoglobin Genes

The 5′ deletion analysis of the *lbc*$_3$ gene, using transgenic *Lotus corniculatus* plants, revealed two positive regulatory elements (PEs), one between −1100 and −950, and another weaker one between −230 and −170, as well as organ-specific and negative elements between −139 and −49 (Stougaard et al. 1987). As the promoter proximal PE lies between and overlaps the BS1 and BS2 NAT2-binding sites, this suggested their possible involvement in enhancing transcription. However, later studies found that the weak PE and either or both binding sites could be removed from the 2-kb *lbc*$_3$ promoter without any major effect on the level of reporter gene activity in transgenic nodules (Stougaard et al. 1990). Although this does not rule out a role for the BS1 and BS2 binding sites in the activity of the weak PE, it did lead the authors to conclude that the two NAT2-binding sites and the weak PE were dispensable components of the *lbc*$_3$ promoter. When Jacobsen et al. (1990) detected additional NAT2 binding sites further upstream in the *lbc*$_3$ promoter, they suggested that these might be able to substitute for BS1 and BS2 in the deletion mutants, leaving open the possibility that interaction with NAT2 is necessary for *lbc*$_3$ expression.

In *Srglb*$_3$ two positive elements have also been identified by 5′ deletion analysis (Szabados et al. 1990). The distal element (−1601/−670 relative to the ATG codon) behaved like a general transcriptional activator when placed upstream of the minimal 35S and nopaline synthetase (*nos*) promoters, since it was able to stimulate promoter activity in both orientations in roots and nodules of *L. corniculatus* and in leaves, stems and roots of tobacco. The proximal element (−429/−162) performed similarly except that in *L. corniculatus* it functioned only in the correct orientation. The proximal PE contains the BS2* NAT2-binding site (between −220 and −200), so that these results contrast with those of Stougaard et al. (1990), who were unable to obtain activation of the −90 CaMV 35S promoter using a selection of restriction fragments from the *lbc₃* promoter, including one carrying both BS1 and BS2. It has recently been found that oligonucleotide monomers and multimers corresponding to the *Srglb₃* BS2* sequence are able to activate minimal plant promoters when fused to them in *cis*, an effect which was most pronounced in roots (de Bruijn and Schell 1993). Confirmation of a regulatory role for this AT-rich element has been obtained from mutagenesis experiments which showed that point mutations within the BS2* sequence of the complete 1.9-kb *Srglb₃* promoter reduced promoter activity by up to 70% (K. Szczyglowski, L. Szabados, S. Fujimoto, P. Welters, F. de Bruijn unpubl. results). In the same series of experiments, a deletion that precisely removed most of the BS2* sequence surprisingly reduced promoter activity by only 25%, a result which could indicate that complex multiple interactions are involved in the functioning of the BS2* element.

2.3.2 N23 Gene

Jørgensen et al. (1988) identified two positive elements in the N23 promoter by 5′ deletion analysis, one at −344/−293 and the other at −247/−165. The first of these coincides with the more proximal of the two mapped NAT2-binding sites (at −344 to −329; see Sect. 2.1.2). However, subsequent studies showed that deletion of the NAT2-binding site from the Δ(−344) promoter had no significant effect on expression in transgenic nodules (Jørgensen et al. 1991). The positive elements were redefined as PE-A (−320 to −298) and PE-B (−257 to −165) and shown to constitute a two-component enhancer that was able to confer nodule-specific expression on the minimal 35S promoter. It was concluded that the adjacent NAT2-binding site in N23 is dispensable for enhancer activity.

2.3.3 Glutamine Synthetase (*gln-γ*) Gene

At least two positive elements were detected in the 2-kb *gln-γ* promoter, one in the region between −2000 and −978 (relative to the ATG codon), another between −597 and −354 (Shen et al. 1992). The latter positive element contains at least two PNF-1/PRF-1 binding sites (the 21-bp AT-rich repeats at −516 and −446). When the sequences from −516 to −343 or from −474 to −293 (containing two and one of the 21-bp repeats, respectively) were inserted at the −90 position of the 35S promoter, they were each able to stimulate expression in *L. corniculatus* nodules, but only if they were in the correct orientation. The same results were obtained when the constructs were tested in a transient expression system consisting of electroporated tobacco mesophyll protoplasts, indicating the presence of a general transcriptional activator. Thus the region of the *gln-γ* promoter that includes the 21-bp AT-rich repeats has regulatory properties very similar to the BS2*-containing proximal PE of the *Srglb₃* gene (see Sect. 2.3.1), but whether the AT-rich repeats themselves are required for this activity has not been determined.

3 Other Plant Genes

The studies described above have established that two distinct types of nuclear protein interact with AT-rich upstream elements in nodulin genes, the HMGI-like factors and the apparently novel NAT2-like factors. The properties of HMGI-like factors are fairly well defined, based on studies in mammalian systems (see above), but those of the NAT2-like factors are much less so. In Table 1 is compiled a list of the features that characterize the latter group and distinguish it from the HMGI-like proteins. In the following discussion the factors that fit these criteria will be referred to as ATBFs (AT-rich Binding Factors), as proposed by Czarnecka et al. (1992), to distinguish them from the HMGI-like factors.

3.1 Occurrence of ATBF-Binding Sites in Non-Nodulin Genes

Using the criteria in Table 1 as a guide, a list of ATBF-type interactions within plant promoters has been compiled (Table 2). In addition to the nodulin genes, ATBF-type interactions have been observed with a diverse group of plant genes, both developmentally regulated (e.g., seed-specific) and inducible (e.g., photoregulated and heat-shock induced). Although there is considerable diversity in the AT-rich sequences contained within the binding sites (Table 2), in most cases the motifs do conform to the consensus $WA(T)_{3-5}(A)_{1-2}T$.

Table 1. Properties of ATBF-type nuclear factors

1. Bind in vitro to AT-rich sequences upstream of the TATA box (*typically multiple sites 150 bp or more from the cap site*)
2. Abundant (*readily detectable in gel retardation assays*)
3. Heat-sensitive (*unlike HMGI*)
4. Insoluble or unstable in 2% TCA (*unlike HMGI*)
5. Multiple complexes or broad band seen in gel retardations (*unlike HMGI*)
6. Show sequence specificity in binding, as demonstrated by competition experiments or by ability to obtain DNase I footprints (*unlike HMGI*)
7. Target sequence conforms to the motif $WA(T)_{3-5}(A)_{1-2}T$ (where W = A or T) or similar (Forde et al. 1990)
8. Associated with a much faster moving complex in gel retardations (*due to HMGI-type binding at the same site*)
9. Binding effectively competed out by a well-characterized ATBF-binding site (e.g., *lbc*$_3$ BS1)

Table 2. ATBFs, their binding sites and their possible regulatory role

Gene	Location of binding sites[a]	AT-rich motifs[b]	ATBF-type binding factor[c]	Occurrence of binding factor	Evidence for a regulatory role[d]	References[e]
Nodulin genes						
Soybean *lbc₃*	−246 to −223 −176 to −161	AATATTTTATTTTATA ATTATTTATTTA	NAT2[1,2,3,4,5,6,7,8]	Nodule-specific in soybean	Binding sites closely associated with a PE, but non-essential	a,b,c,d
Soybean N23	−407 to −374 −344 to −329	ATATTTTATATAA TTATTTTATTATTTAA (rev)	NAT2	As above	Binding sites closely associated with a PE, but non-essential	c,d,e,f
Sesbaunia rostrata glb₃	−570 to −483† −220 to −220†	? TTATTAATTAA	Unnamed[1,2,7,8,9]	In leaves as well as nodules of *S. rostrata* and alfalfa	Binding sites closely associated with PEs; 20-bp BS2* is able to activate a heterologous promoter	g,h,i
Phaseolus vulgaris gln-γ	−516 to −496† −466 to −446†	ATATTTAAT ATATTTAAT	PNF-1, PRF-1[1,2,3,4,5,6,7,8,9]	PNF-1 in *P. vulgaris* nodules, PRF-1 in roots	Binding sites closely associated with a PE	j,k
Seed-specific genes						
Soybean seed lectin *Le1*	−184 to −173 −165 to −126	AATTTAATTAA AATTTAAAATTTAAT (rev)	Unnamed[1,2,6,7,9]	60-kDa protein developmentally regulated in soybean embryogenesis	Correlation between activity of binding factor and transcriptional activity of *Le1*	l
Soybean Kunitz trypsin inhibitor	−298 to −261	TATTTAATTATAAATT (rev)	As for *Le1*	As above	nd	l,m
P. vulgaris phytohaemagglutinin *lec2*	−550 to −310	TATTTTAATATAAATT (rev)	Unnamed[1,2,5,7,9]	In developing *P. vulgaris* cotyledons (mid-mature)	Binding site closely associated with a PE	m
Soybean β-conglycinin α′ subunit	−788 to −769 −674 to −643	ATATTTATATTTTAAT TATATTGTT	SEF1[1,2,3,5,6,7]	In developing soybean embryos	Binding sites can be deleted without effect (but see ref. x)	n,o
Soybean β-conglycinin β subunit	−553 to −442	?	SEF1	–	Binding site closely associated with a PE	o,p
P. vulgaris β-phaseolin	−682 to −628	AATATTTTAATT (rev) TTAATTTTAA	Unnamed[1,2,5,6,7,9]	In developing *P. vulgaris* cotyledons and hypocotyls; carrot somatic embryos	55-bp binding site able to activate heterologous promoter	q,r,s
Sunflower helianthinin HaG3D	−705 to −654	AATAAATTTATTTT	Unnamed[1,2,6,7,9]	In developing sunflower embryos and hypocotyls	Binding site closely associated with a PE	r
Sunflower helianthinin HaG3A	−1514 to −1463 −702 to −653	? GATAAATTTATTTT	As for HaG3D	–	nd	r

Table 2. *Continued*

Gene	Location of binding sites[a]	AT-rich motifs[b]	ATBF-type binding factor[c]	Occurrence of binding factor	Evidence for a regulatory role[d]	References[e]
Photoregulated genes						
Pea *rbcS-3.6*	−566 to −533	ATATTTTTAATTATTTTTATT	AT-1[1,2,3,5,6,7,8]	40–45 kDa proteins in etiolated pea plumules	nd	t
Tomato *rbcS-3A*	−1021 to −994	ATATTTTTATT	AT-1	–	AT-1 box is within a strong PE	t
Nicotiana plumbaginifolia cab-E	−1177 to −1150 −1139 to −1112 −1064 to −1035	TTATTTTTAAA ATATTTTTATT ATATTTTTATC	AT-1	–	Three AT-1 boxes located within a negative regulatory element	t,u
Other genes						
Soybean heat shock *Gmhsp 17.5E*	−907 to −830 −153 to −120	ATAATATTAATATTATATT ATTATATTTATTTTT (rev)	Unnamed[1,2,4,5,6,7,8,9]	46–69 kDa proteins in soybean plumules and roots	33-bp AT-rich motif activates a heterologous promoter; binding activity increases after heat shock	v,w,x
Mesembryanthemum crystallinum PEP carboxylase *Ppc1*	−205 to −187 −158 to −128	TTATTTTTAG (rev) TTATTAATTAA	PCAT1[1,2,3,4,5,6,7]	In leaves but not roots	Binding activity increased by salt stress	y
Carrot extensin	−311 to −271	ATTTTTATAATT ATTTTTTTTTATTT	EGBF-1, EGBF-2[1,2,5,6,7]	EGBF-1 and EGBF-2 are specific to root phloem and root xylem, respectively	EGBF-1 binding activity negatively correlated with extensin expression	z
Potato class I patatin B33 gene	−775 to −739 −505 to −479 −175 to −139	TATTATATAATA TATTATATAATA TATTATTTATTA (rev)	Unnamed[1,2,7]	In potato tubers and leaves	Binding sites interspersed with PEs	a′
Pea ferredoxin *Fed-1*	−393 to −256	AATATTTTAAT GTTAATTTAAT	BC2[1,2,3,5,7,8]	In pea leaf buds	nd	b′

[a] Locations of binding sites are given relative to the transcription initiation site, with the exception of those marked †, which are relative to the ATG codon.
[b] rev = sequence occurs in the reverse orientation.
[c] Numbers indicate which of the ATBF-like properties listed in Table 1 have been ascribed to each factor.
[d] nd = not determined; PE = positive regulatory element.
[e] References: a, Stougaard et al. (1987); b, Jensen et al. (1988); c, Jacobsen et al. (1990); d, Stougaard et al. (1990); e, Jørgensen et al. (1988); f, Jørgensen et al. (1991); g, Metz et al. (1988); h, Szabados et al. (1990); i, de Bruijn and Schell (1993); j, Forde et al. (1990); k, Shen et al. (1992); l, Jofuku et al. (1987); m, Riggs et al. (1989); n, Allen et al. (1989); o, Lessard et al. (1991); p, P.A. Lessard, T. Fujiwara, R.N. Beachy (unpubl. results); q, Bustos et al. (1989); r, Jordano et al. (1989); s, Guiltinan et al. (1989); t, Datta and Cashmore (1989); u, Schindler and Cashmore (1990); v, Czarnecka et al. (1990); w, Czarnecka et al. (1992); x, Fox et al. (1992); y, Cushman and Bohnert (1992); z, Holdsworth and Laties (1989); a′, Liu et al. (1990); b′, Pedersen et al. (1991).

3.2 Properties and Distribution of ATBFs

Binding factors of the ATBF type are widespread amongst dicotyledonous species of plants, having been detected not only in legumes, but also in solanaceous species, in sunflower (Asteraceae) and in *M. crystallinum* (Aizoaceae) (Table 2). Within these species the ATBFs have been found in a variety of plant organs, including leaves, developing embryos, roots, hypocotyls and potato tubers, as well as nodules. Evidence for the occurrence of ATBF-like interactions with monocot genes is surprisingly limited up to now. An abundant factor from immature maize kernels (designated MNP1) was found to interact at multiple AT-rich sites upstream of the maize sucrose synthase gene (Werr et al. 1988). While the consensus for these AT-rich binding sequences is quite different from those identified in dicots, it is nevertheless possible that MNP1 is a member of the ATBF family.

In the absence of amino acid sequences, the evidence that ATBF-like fators from diverse dicotyledonous sources are related to one another is based on the general similarities in their properties (as summarized in Tables 1 and 2). Of particular interest are the competition experiments that have demonstrated, on the one hand, that the ATBF(s) in a given nuclear extract are able to bind with high affinity to well-characterized AT-rich binding sites in other genes, often from other species and, on the other hand, that a given ATBF-binding site is recognized by factors present in a variety of different species and tissues (Metz et al. 1988; Datta and Cashmore 1989; Jordano et al. 1989; Riggs et al. 1989; Forde et al. 1990; Schindler and Cashmore 1990; Czarnecka et al. 1992).

In the few studies in which estimates have been made of the molecular weights of the ATBF-type factors, these have been in the range 40–69 kDa (Jofuku et al. 1987; Schindler and Cashmore 1990; Czarnecka et al. 1992), significantly greater than the 21–32 kDa estimated for the HMGI-like plant proteins (Jacobsen et al. 1990; Czarnecka et al. 1992).

3.3 Evidence from Non-Nodulin Genes that AT-Rich Elements Have a Regulatory Function

AT-Rich Binding Sites Are Frequently Associated with Positive Regulatory Elements. As has been noted for each of the nodulin genes (Sect. 2.3), binding sites for ATBFs are often found within or close to regions of the promoter that have a positive effect on gene expression. This general correlation has been reported for the *P. vulgaris lec2* gene, the soybean β-conglycinin β subunit gene, the sunflower helianthinin HaG3D gene, and the tomato *RbcS-3A* gene (see Table 2). In the case of the *N. plumbaginifolia Cab-E* gene, three binding sites for the AT-1 binding factor are located within a negative regulatory element (Datta and Cashmore 1989). However, the far upstream AT-rich binding sites in the β-conglycinin α′ subunit gene could be removed without apparent effect on gene expression in transgenic tobacco (Chen et al. 1986; Allen et al. 1989), while in the B33 patatin gene several AT-rich binding sites were interspersed with, but did not coincide with, four positive elements located between −1512 and −228 (Liu et al. 1990).

Oligonucleotides Containing AT-Rich Motifs Can Act as General Transcriptional Activators. When a 55-bp synthetic oligonucleotide containing the AT-rich binding motif from the seed-specific β-phaseolin gene was fused to the minimal 35S promoter, expression in transgenic tobacco was enhanced not only in seeds but also in vegetative tissues, particularly roots (Bustos et al. 1989), an effect that was stronger when the binding site was in the correct rather than the reverse orientation. Similarly, Czarnecka et al. (1992), using a transgenic sunflower tumour system, found that a 33-bp sequence containing AT-rich motifs from two binding sites upstream of the soybean heat-shock gene *Gmhsp 17.5E* stimulated expression from a maize *Adh1* promoter up to 24-fold in anaerobically stressed cells, the degree of activation being linearly related to the number of copies of the oligomer inserted in tandem and inversely proportional to its distance from the start of transcription. In a transient expression assay using electroporated maize protoplasts, a trimer of the same 33-bp sequence and a pentamer of an AT-rich binding site from the soybean β-conglycinin α′ subunit gene were able to stimulate expression from the *mas 2′* mannopine synthase promoter 7.5-fold and 3-fold, respectively (Fox et al. 1992).

Although these experiments clearly demonstrate the ability of AT-rich sequences to act as weak transcriptional activators, they have so far fallen short of demonstrating conclusively that the interaction with ATBFs and/or HMGI-like proteins is involved in the regulatory effect. This will necessitate experiments in which mutated AT-rich binding sites are assayed for their ability to bind ATBFs and HMGI-like proteins in vitro and the results correlated with their ability to activate gene expression in vivo. Such experiments would rule out the alternative possibilities that the structural properties of the AT-rich DNA sequences, or their interactions with other binding proteins, are important for their positive regulatory effect.

Correlation Between ATBF-Binding Activity and Transcriptional Activity of the Target Gene. A correlation between the abundance of a binding protein and the transcriptional activity of its target gene may provide circumstantial evidence for its involvement in gene regulation. Such a correlation has been noted for the 60-kDa protein that binds to AT-rich motifs upstream of the soybean seed lectin gene, its binding activity being temporally regulated in developing embryos in a similar way to the *Le1* gene itself (Jofuku et al. 1987. Similarly, the binding activities of the ATBF-like proteins that bind upstream of the soybean heat-shock *Gmhsp17.5E* gene and the *M. crystallinum Ppc1* gene are increased by the same stress factors (heat and salt, respectively) that induce expression of the target genes (Czarnecka et al.

1990; Cushman and Bohnert 1992). The binding activity of EGBF-1 in carrot roots is *reduced* by two treatments (ethylene and wounding) that stimulate extensin gene expression, suggesting that the EGBF-1 binding site in the extensin gene might act as a silencer (Holdsworth and Laties 1989).

4 Conclusions

The information reviewed above indicates that NAT2 and related nuclear factors in legume root nodules belong to a diverse family of abundant DNA-binding proteins (ATBFs) that have an affinity for certain AT-rich sequences. These ATBFs seem to occur in most, if not all (dicotyledonous) plant tissues and have binding sites in a wide variety of plant genes. The broad band that characterizes the ATBF complexes in gel retardation analysis suggests that multiple ATBFs occur within a given organ, and differences in the mobility of these complexes and their binding site preferences from one organ to another (e.g., Metz et al. 1988; Forde et al. 1990; Czarnecka et al. 1992) indicate that the complement of ATBF proteins differs between cell types. This multiplicity of factors within one extract may explain why it has often been difficult to define precisely the binding site requirements of ATBFs. Whether the isoforms of ATBF are separate gene products or modified versions of a single protein is not yet known. A cDNA clone has recently been isolated for a tobacco protein that binds to an AT-rich general transcriptional activator (Box VI) in the pea *rbcS-3A* gene (Lam et al. 1990). The encoded protein contains putative zinc-finger DNA-binding motifs and hybridizes to multiple transcripts of 1.5–4.5 kb expressed in roots, stems and leaves. Although the nucleotide sequence recognized by this protein and its location (at −45) are not typical of ATBFs, it is nevertheless possible that the binding protein is a member of the ATBF family.

The function of the AT-rich elements (ATREs) that serve as binding sites for both ATBFs and HMGI-like proteins is still unclear. While there is a close association between the ATREs and positive elements in a wide variety of nodulin and non-nodulin plant genes, when this has been analyzed in more detail, it has been found that the AT-rich binding sites can be deleted without significantly affecting gene expression. These observations, together with the abundance of ATBFs in nuclear extracts and the widespread occurrence of their binding sites in the upstream regions of plant genes, suggest that despite their ability to act as weak transcriptional activators when fused to heterologous promoters, ATREs may not be conventional *cis*-acting elements.

Possible clues to the role of ATREs and their binding proteins may come from what is known about HMGI-like proteins and their functions in other organisms. In animals, HMGI has been implicated in the positioning of nucleosomes on DNA (Strauss and Varshavsky 1984) and in the stimulation of transcription of ribosomal RNA (Yang-Yen and Rothblum 1988), lymphotoxin (Fashena et al. 1992) and interferon-β genes (Thanos and Maniatis 1992). It has been proposed that the role of HMGI in these diverse functions is to serve as an accessory factor which acts by altering the structure of naked DNA or chromatin, allowing and/or facilitating the binding of other proteins (Fashena et al. 1992; Thanos and Maniatis 1992). If HMGI-like proteins have a similar function in plants, then the association between ATREs and positive elements in many plant promoters may not be fortuitous. Thus ATREs may act in concert with more conventional *cis*-acting elements (such as the two-component enhancer in the soybean N23 gene; see Sect. 2.3.2) by influencing the accessibility of the DNA to *trans*-acting factors. Chromatin structure and its degree of unfolding or "openness" is known to be an important factor in gene regulation (Felsenfeld 1992). In yeast an abundant nuclear protein, GRF2, binds to the GAL upstream activating sequence and creates a nucleosome-free region of about 230 bp (Chasman et al. 1990). Like the ATREs, the GRF2 binding site on its own is a weak positive element, but when combined with a neighbouring weak activator it stimulates transcription up to 170-fold, an effect that is strongly dependent on the distance between the two elements. If the ATBFs and/or the HMGI-like proteins that bind to ATREs play a role analogous to GRF2, we would expect to observe synergistic interactions between ATREs and neighbouring positive (or negative) regulatory elements. The absence of any published evidence for such interactions may arise from the use of transgenic plants as an assay system for *cis*-acting elements. As foreign DNA is inserted into the host genome more or less at random and the gene constructs are flanked by prokaryotic sequences from the vector DNA (usually the *Agrobacterium* Ti or Ri plasmid), the context of the foreign gene is very different from that of the endogenous gene, a situation which is likely to affect profoundly the chromatin structure in the vicinity of the promoter. Several studies have found that T-DNA inserted into the plant genome is unusually sensitive to DNase I and micrococcal nuclease, indicating either a disorganized chromatin structure or an "open" conformation (Shafer et al. 1984; Coates et al. 1987; Weising et al. 1990). Thus *coarse* control of gene expression, which involves changes in chromatin structure and in accessibility of the DNA to transcription factors and RNA polymerases (Spiker 1988), may not be amenable to investigation using standard techniques. Testing the hypothesis that ATREs function as auxiliary gene activators in higher plants may call for alternative approaches, such as the mutation of ATREs in situ by homologous recombination, or antisense manipulation of the synthesis of ATBFs and HMGI-like proteins in transgenic plants.

Acknowledgements. I am very grateful to Professors R.B. Goldberg, R.N. Beachy and F.J. de Bruijn for providing access to unpublished data. I also wish to acknowledge the contribution of my late wife and colleague Dr. Janice Forde, without whose encouragement and support this review would not have been written and to whom it is dedicated.

References

Allen RD, Bernier F, Lessard PA, Beachy RN (1989) Nuclear factors interact with a soybean β-conglycinin enhancer. Plant Cell 1:623–631

Beachy PA, Krasnow MA, Gavis ER, Hogness DS (1988) An *Ultrabithorax* protein binds sequences near its own and the *Antennapedia* P1 promoters. Cell 55:1069–1081

Bustin M, Lehn DA, Landsman D (1990) Structural features of the HMG chromosomal proteins and their genes. Biochim Biophys Acta 1049:231–243

Bustos MM, Guiltinan MJ, Jordano J, Begum D, Kalkan FA, Hall TC (1989) Regulation of β-glucuronidase expression in transgenic tobacco plants by an A/T-rich, *cis*-acting sequence found upstream of a French bean β-phaseolin gene. Plant Cell 1:839–853

Chasman DI, Lue NF, Buchman AR, LaPointe JW, Lorch Y, Kornberg RD (1990) A yeast protein that influences the chromatin structure of UAS_G and functions as a powerful auxiliary gene activator. Genes Dev 4:503–514

Chen Z-L, Schuler MA, Beachy RN (1986) Functional analysis of regulatory elements in a plant embryo specific gene. Proc Natl Acad Sci USA 83:8560–8564

Coates D, Taliercio EW, Gerlrin SB (1987) Chromatin structure of integrated T-DNA in crown gall tumors. Plant Mol Biol 8:159–168

Cushman JC, Bohnert HJ (1992) Salt stress alters A/T-rich DNA-binding factor interactions within the phosphoenolpyruvate carboxylase promoter from *Mesembryanthemum crystallinum*. Plant Mol Biol 20:411–424

Czarnecka E, Fox PC, Gurley WB (1990) In vitro interaction of nuclear proteins with the promoter of soybean heat shock gene Gmhsp 17.5E. Plant Physiol 94:935–943.

Czarnecka E, Ingersoll JC, Gurley WB (1992) AT-rich promoter elements of soybean heat shock gene Gmhsp 17.5E bind two distinct sets of nuclear proteins in vitro. Plant Mol Biol 19:985–1000

Datta N, Cashmore AR (1989) Binding of a pea nuclear protein to promoters of certain photoregulated genes is modulated by phosphorylation. Plant Cell 1:1069–1077

de Bruijn FJ, Schell J (1993) Regulation of plant genes specifically induced in developing and mature nitrogen-fixing nodules: *cis*-acting elements and *trans*-acting factors. In: Verma DPS (ed) Control of plant gene expression. CRC Press, Boca Raton, p 241

de Bruijn FJ, Felix G, Grunenberg B, Hoffmann HJ, Metz B, Ratet P, Simons-Schreier A, Szabados L, Welters P, Schell J (1989) Regulation of plant genes specifically induced in nitrogen-fixing nodules: role of *cis*-acting elements and *trans*-acting factors in leghemoglobin gene expression. Plant Mol Biol 13:319–325

de Bruijn FJ, Szabados L, Schell J (1990) Chimeric genes and transgenic plants are used to study the regulation of genes involved in symbiotic plant-microbe interactions (nodulin genes). Dev Genet 11:182–196

Eckdahl TT, Anderson JN (1990) Conserved DNA structures in origins of replication. Nucleic Acids Res 18:1609–1612

Eckdahl TT, Bennetzen JL, Anderson JN (1989) DNA structures associated with autonomously replicating sequences from plants. Plant Mol Biol 12:507–516

Fashena SJ, Reeves R, Ruddle NH (1992) A poly (dA-dT) upstream activating sequence binds high-mobility group I protein and contributes to lymphotoxin (tumor necrosis factor-β) gene regulation. Mol Cell Biol 12:894–903

Felsenfeld G (1992) Chromatin as an essential part of the transcriptional mechanism. Nature 355:219–224

Forde BG, Freeman J, Oliver JE, Pineda M (1990) Nuclear factors interact with conserved A/T-rich elements upstream of a nodule-enhanced glutamine synthetase gene from French bean. Plant Cell 2:925–939

Fox PC, Vasil V, Vasil IK, Gurley WB (1992) Multiple ocs-like elements required for efficient transcription of the mannopine synthase gene of T-DNA in maize protoplasts. Plant Mol Biol 20:219–233

Gasser SM, Laemmli UK (1986) Cohabitation of scaffold binding regions with upstream/enhancer elements of three developmentally regulated genes of *D. melanogaster*. Cell 46:521–530
Greenblatt J (1991) Roles of TFIID in transcriptional initiation by RNA polymerase II. Cell 66:1067–1070
Guiltinan MJ, Thomas JC, Nessler CL, Thomas TL (1989) Expression of DNA binding proteins in carrot somatic embryos that specifically interact with a *cis* regulatory element of the French bean phaseolin gene. Plant Mol Biol 13:605–610
Holdsworth MJ, Laties GC (1989) Site-specific binding of a nuclear factor to the carrot extensin gene is influenced by both ethylene and wounding. Planta 179:17–23
Jacobsen K, Laursen NB, Jensen EO, Marcker A, Poulsen C, Marcker KA (1990) HMG I-like proteins from leaf and nodule nuclei interact with different AT motifs in soybean nodulin promoters. Plant Cell 2:85–95
Jensen EO, Marcker KA, Schell J, de Bruijn FJ (1988) Interaction of a nodule specific, *trans*-acting factor with distinct DNA elements in the soybean leghaemoglobin lbc_3 5′ upstream region. EMBO J 7:1265–1271
Jofuku KD, Okamuro JK, Goldberg RB (1987) Interaction of an embryo DNA binding protein with a soybean lectin gene upstream region. Nature 328:734–737
Jordano J, Almoguera C, Thomas TL (1989) A sunflower helianthinin gene upstream sequence ensemble contains an enhancer and sites of nuclear protein interaction. Plant Cell 1:855–866
Jørgensen J-E, Stougaard J, Marcker A, Marcker KA (1988) Root nodule specific gene regulation: analysis of the soybean nodulin N23 gene promoter in heterologous symbiotic systems. Nucleic Acids Res 16:39–50
Jørgensen JE, Stougaard J, Marcker KA (1991) A two-component nodule-specific enhancer in the soybean N23 gene promoter. Plant Cell 3:819–827
Kunkel GR, Martinson HG (1981) Nucleosomes will not form on double-stranded RNA or over poly (dA).poly (dT) traits in recombinant DNA. Nucleic Acids Res 9:6869–6888
Lam E, Kano-Murakami Y, Gilmartin P, Niner B, Chua N-H (1990) A metal-dependent DNA-binding protein interacts with a constitutive element of a light-responsive promoter. Plant Cell 2:857–866
Laux T, Goldberg RB (1991) A plant DNA binding protein shares highly conserved sequence motifs with HMG-box proteins. Nucleic Acids Res 19:4769
Lessard PA, Allen RD, Bernier F, Crispino JD, Fujiwara T, Beachy RN (1991) Multiple nuclear factors interact with upstream sequences of differentially regulated β-conglycinin genes. Plant Mol Biol 16:397–413
Liu XJ, Prat S, Willmitzer L, Frommer WB (1990) *Cis* regulatory elements directing tuber-specific and sucrose-inducible expression of a chimeric class I patatin promoter/GUS-gene fusion. Mol Gen Genet 223:401–406
Metz BA, Welters P, Hoffman HJ, Jensen EØ, Schell J, de Bruijn J (1988) Primary structure and promoter analysis of leghemoglobin genes of the stem-nodulated tropical legume *Sesbania rostrata*: conserved coding sequences, *cis*-elements and *trans*-acting factors. Mol Gen Genet 214:181–191
Pedersen TJ, Arwood LJ, Spiker S, Guiltinan MJ, Thompson F (1991) High mobility group chromosomal proteins bind to AT-rich tracts flanking plant genes. Plant Mol Biol 16:95–104
Prunell A (1982) Nucleosome reconstitution on plasmid-inserted poly(dA).poly(dT). EMBO J 1:173–179
Richter HE, Sandal NN, Marcker KA, Sengupta-Gopalan C (1991) Characterization and genomic organization of a highly expressed late nodulin gene subfamily in soybeans. Mol Gen Genet 229:445–452
Riggs CD, Voelker TA, Chrispeels MJ (1989) Cotyledon nuclear proteins bind to DNA fragments harboring regulatory elements of phytohemagglutinin genes. Plant Cell 1:609–621

Sandal NN, Bojsen K, Marcker KA (1987) A family of nodule specific genes from soybean. Nucleic Acids Res 15:1507–1519

Schindler U, Cashmore AR (1990) Photoregulated gene expression may involve ubiquitous DNA binding proteins. EMBO J 9:3415–3427

Shafer W, Weising K, Kahl G (1984) T-DNA of a crown gall tumor is organized into nucleosomes. EMBO J 3:373–376

Shen W-J, Williamson MS, Forde BG (1992) Functional analysis of the promoter region of a nodule-enhanced glutamine synthetase gene from *Phaseolus vulgaris* L. Plant Mol Biol 19:837–846

Solomon MJ, Strauss F, Varshavsky A (1986) A mammalian high mobility group protein recognizes any stretch of six A.T base pairs in duplex DNA. Proc Natl Acad Sci USA 83:1276–1280

Spiker S (1988) Histone variants and high mobility group non-histone chromosomal proteins of higher plants: their potential for forming a chromatin structure that is either poised for transcription or transcriptionally inert. Physiol Plant 75:200–213

Stougaard J, Sandal NN, Grøn A, Kühle A, Marcker KA (1987) 5′ analysis of the soybean leghaemoglobin *lbc₃* gene: regulatory elements required for promoter activity and organ specificity. EMBO J 6:3565–3569

Stougaard J, Jørgensen JE, Christensen T, Kühle A, Marcker KA (1990) Interdependence and nodule specificity of *cis*-acting regulatory elements in the soybean leghaemoglobin *lbc₃* and N23 gene promoters. Mol Gen Genet 220:353–360

Strauss F, Varshavsky A (1984) A protein binds to a satellite DNA repeat at three specific sites that would be brought into mutual proximity by DNA folding in the nucleosome. Cell 37:889–890

Struhl K (1985) Naturally occurring poly(dA-dT) sequences are upstream promoter elements for constitutive transcription in yeast. Proc Natl Acad Sci USA 82:8419–8423

Szabados L, Ratet P, Grunenberg B, de Bruijn FJ (1990) Functional analysis of the *Sesbania rostrata* leghemoglobin glb_3 gene 5′-upstream region in transgenic *Lotus corniculatus* and *Nicotiana tabacum* plants. Plant Cell 2:973–986

Thanos D, Maniatis T (1992) The high mobility group protein HMG I(Y) is required for NF-κB-dependent virus induction of the human IFN-β gene. Cell 71:777–789

Umek RM, Eddy MJ, Kowalski D (1988) DNA sequences required for unwinding prokaryotic and eukaryotic replication origins. Cancer Cells 6:473–478

Verma DPS, Hu C-A, Zhang M (1992) Root nodule development: origin, function and regulation of nodulin genes. Physiol Plant 85:253–265

Weising K, Bohn H, Kahl G (1990) Chromatin structure of transferred genes in transgenic plants. Dev Genet 11:233–247

Werr W, Springer B, Schürmann, Bellmann R (1988) Multiple interactions between nuclear proteins of *Zea mays* and the promoter of the *Shrunken* gene. Mol Gen Genet 212:342–350

Yang-Yen HF, Rothblum LI (1988) Purification and characterization of a high-mobility-group-like DNA binding protein that stimulates mRNA synthesis in vitro. Mol Cell Biol 8:3406–3414

5 In Vitro Transcription of Plant Nuclear Genes

Patrick Schweizer

1 Introduction

Higher plants, like other eukaryotes, contain three RNA polymerases (pol I, II and III) that catalyze transcription from the respective nuclear class I, II and III promoters. In addition to RNA polymerase, general transcription factors are needed for correct initiation at class I (reviewed by Reeder 1990), class II (reviewed by Roeder 1991) or class III (reviewed by Gabrielsen and Sentenac 1991) promoters. The development of efficient and faithful animal in vitro transcription systems has enabled the functional study of, e.g., the general class II transcription factors (reviewed by Roeder 1991; Lu et al. 1992) or of several gene-specific *trans*-acting factors and coactivators (see e.g., Goldenberg et al. 1988; Bagchi et al. 1990a,b; Corthésy et al. 1990; Johnson and Krasnow 1990; Kelleher et al. 1990; Lin et al. 1990; Ohkuma et al. 1990; Pugh and Tjian 1990; Workman et al. 1990; Croston et al. 1991; Flanagan et al. 1991).

In the last few years, an increasing number of DNA-binding proteins that probably regulate transcription of nuclear genes have been identified from plants (see accompanying reviews in this volume; Katagiri et al. 1989; Scharf et al. 1990; Oeda et al. 1991; Tabata et al. 1991; Weisshaar et al. 1991; Dehesh et al. 1992; Gilmartin et al. 1992; Perisic and Lam 1992). The functional analysis of these proteins would be considerably facilitated by the use of plant in vitro transcription systems. However, the development of plant systems accurately transcribing nuclear genes in vitro has been much more difficult than the development of animal transcription systems, several of which have been established since 1980 (Manley et al. 1980; Dignam et al. 1983).

"Accurate" in vitro transcription requires that several criteria are fulfilled. First, the site of in vitro initiation must be identical or very close (within less than 10 bp) to the in vivo initiation site. Second, transcription should be dependent on at least one basic promoter element, e.g., the TATA box (see e.g., Wobbe and Struhl 1990) or the initiator, a less well-defined element at the in vivo cap site of pol II-transcribed genes (Smale and Baltimore 1989; Greenfield and Smale 1992). Third, the same class

Institute de Biologie Végétale, Université de Fribourg, rte Albert-Gockel 3, CH-1700 Fribourg, Switzerland

Results and Problems in Cell Differentiation 20
L. Nover (Ed.)
Plant Promoters and Transcription Factors

of RNA polymerase (I, II or III) should transcribe a given promoter in vitro and in vivo. Fulfillment of these minimal criteria will allow the study of basal transcription. However, in vitro transcription systems will only have their full impact, if they fulfil two additional criteria: (1) responsiveness to upstream located *cis*-acting elements and/or (2) responsiveness to *trans*-acting factors that are not involved in basal transcription.

This chapter gives an overview of the progress made with in vitro transcription of nuclear plant genes, although the body of publications in this area is still very limited and covers almost exclusively pol II-based transcription. I will furthermore emphasize problems often encountered with plant systems and will describe strategies for circumventing such problems.

2 Transcription in Isolated Nuclei and Chromatin

In vitro transcription in isolated plant nuclei has become a standard method for the analysis of transcriptional regulation. In this assay, isolated nuclei are incubated with salts and nucleotide triphosphates, one usually being radiolabeled, with subsequent counting of incorporation of the radiolabel or hybridization of radiolabeled RNA. However, the signals obtained by this so-called run-on transcription are thought to reflect only in vitro elongation of in vivo initiated transcripts (e.g., Mösinger et al. 1985; Chappell and Hahlbrock 1986; Somssich et al. 1986; Schweizer et al. 1989).

In addition to measuring in vitro elongation of RNA, attempts have been made to detect in vitro initiation of transcription in isolated nuclei and chromatin. The rationale behind these approaches is to isolate all components necessary for accurate initiation as complexes bound to nuclei or chromatin. The advantage of such systems would be a better reflection of the in vivo situation, because the endogenous histone-DNA complex referred to as chromatin serves as template rather than naked exogenous DNA. A major disadvantage of transcribing naked DNA is an elevated basal level of transcription, probably due to the absence of histone cores and histone H1 (see e.g., Laybourn and Kadonaga 1991). Reconstituted chromatin systems also respond much more strongly to *trans*-acting factors due to the combined effects of true activation and anti-repression (Croston et al. 1991; Laybourn and Kadonaga 1991). In vitro initiated, gene-specific transcripts are usually detected by DNA blot hybridization (Hipskind and Reeder 1980; Washington and Stallcup 1982; Mauro and Verma 1987; Mennes et al. 1992), although other detection methods like RNA protection (Zhang-Keck and Stallcup 1988) or PCR (Schweizer and Hahlbrock 1993) are possible. However, these detection methods do not discriminate between in vivo and in vitro initiated transcripts, and discrimination between these two RNA pools, prior to detection, is a major difficulty associated with nuclear or chromatin systems.

2.1 Use of Known Initiation Inhibitors

Transcription signals from in vitro initiation are sensitive to initiation inhibitors, the best studied being the polyanionic polysaccharide sulfate heparin (Cox 1973; Dynan and Burgess 1979; Teissère et al. 1979) and the anionic detergent sarcosyl (Hawley and Roeder 1985). The easiest way to detect in vitro initiation is to measure heparin or sarcosyl-sensitive transcription. The difference between signals obtained with and without the inhibitor could be referred to as in vitro initiation (Hipskind and Reeder 1980; Schweizer 1988; Furter and Hall 1989). Unfortunately, this method has several weak points. First, heparin also affects the in vitro elongation rates. Depending on the experimental conditions, inhibition (Schweizer 1988; Mennes et al. 1992) or stimulation (Hentschel and Tata 1978; Bouman et al. 1979; Mösinger et al. 1987; Schweizer 1988) of elongation takes place, resulting in superimposed effects which make the interpretation almost impossible. The same ambiguity holds true for sarcosyl (Schweizer 1988). Second, under conditions where heparin and sarcosyl inhibited total transcription, no inhibition of gene-specific signals in isolated wheat nuclei was observed (Schweizer 1988). Possibly, more specific inhibition like functional oligonucleotide competition experiments, using e.g., TATA box sequences, would produce better results.

2.2 In Vitro Inducible Transcription

The study of in vitro inducible transcription in isolated nuclei would be a very elegant way to circumvent the problem of background transcription from in vivo initiated polymerases. In this approach, *trans*-activators are added to isolated nuclei. Using the endogenous transcription machinery of the nuclei, these *trans*-activators might then stimulate transcription from promoters containing the respective *cis*-elements, provided the activators enter the nuclei and the transcription machinery is intact enough to respond to the activators. For example, activation of overall and gene-specific transcription in isolated nuclei by adding auxin-binding proteins (Sakai et al. 1986; Kikuchi et al. 1989), *Rhizobium* extract (Mauro and Verma 1987), or native phytochrome (Mösinger et al. 1987) has been reported. However, stimulation of in vitro transcription in such systems does not discriminate between stimulated in vitro initiation, stimulated elongation rate, or release of pausing RNA polymerase that initiated in vivo. There is evidence from several promoters that the holdback of transcriptionally engaged pol II at the site of initiation is an existing mechanism of gene regulation (Rougvie and Lis 1988; Strobl and Eick 1992).

2.3 Affinity Labeling of in Vitro Reinitiated Transcripts

Probably the most promising approach to measure in vitro initiation in isolated nuclei is the separation of in vitro initiated transcripts from bulk RNA by affinity chromatography. In this experimental system, one or several nucleoside triphosphates are replaced by a β-thiol or γ-thiol analog. Ideally, the thiol group will remain at the 5′ end of newly initiated transcripts only if the thio-nucleotide is the initiating nucleotide, otherwise it will be cleaved off. The thio-labeled RNA is then purified by affinity chromatography on mercury sepharose (or agarose) columns. Usually, the thioaffinity-labeled RNA is internally ^{32}P-labeled, and can be used for hybridization experiments after purification. Such reinitiation systems have been described for yeast (Ide 1981), animals (Yi-Chi Sun et al. 1979; Hipskind and Reeder 1980; Zhang-Keck and Stallcup 1988), and plants (Kikuchi et al. 1989; Mennes et al. 1992). As with the other nuclear systems described earlier, several problems go along with this system, too. First, gene-specific signals from the hybridization of affinity-labeled in vitro transcripts are usually at the limit of detection, because only 3 to 5% of the total incorporated radioactivity is in the affinity-labeled RNA. This problem can be solved by using PCR amplification of affinity-labeled RNA (Schweizer and Hahlbrock 1993). Second, the γ-thio affinity label can also be transferred posttranscriptionally to in vivo initiated RNA by polynucleotide kinase activity (Hipskind and Reeder 1980; Washington and Stallcup 1982). This artifact is all the more treacherous because the initiation inhibitor heparin also inhibits polynucleotide kinases from several organisms (Wu 1971; Levin and Zimmermann 1976). The use of β-thio nucleotides would eliminate this problem (Washington and Stallcup 1982). We compared affinity labeling of parsley chalcone synthase transcripts in isolated parsley nuclei by γ-thio ATP and β-thio ATP (Schweizer and Hahlbrock 1993). We found that the γ-thio affinity label was transferred posttranscriptionally and that no affinity labeling by the β-thio affinity label occurred. Therefore, up to now, no successful affinity labeling of plant in vitro transcripts by β-thio nucleotides has been reported, and results from γ-thio affinity labeling are not conclusive.

3 Transcriptionally Competent Extracts

The approach to obtain a functional plant in vitro transcription system that has attracted most attention is the establishment of transcriptionally competent extracts. This seemed straightforward, especially after the establishment of soluble animal systems based on crude whole-cell extracts (Manley et al. 1980) or crude nuclear extracts (Dignam et al. 1983). The existing systems seem simple, resembling an enzyme assay (RNA polymerase plus substrates, including DNA template), plus cofactors, including general transcription factors). Accurately initiated in vitro transcripts are detected by direct radioactive labeling and "run-off" transcription from linearized templates (Manley et al. 1980) or G-free transcription from supercoiled templates containing a G-free cassette downstream from the promoter (Sawadogo and Roeder 1985). Alternative, although less sensitive and more complicated, detection methods are primer extension or RNA protection with nonradioactive in vitro transcripts (Cooke and Penon 1990; Roberts and Okita 1991). For mostly unknown reasons, it has turned out to be very difficult to obtain accurate, transcriptionally competent plant extracts. The limited success made so far is summarized below, while some possible reasons for these difficulties are discussed in Section 5.

There are two reports on partly successful in vitro transcription in a plant system derived from whole-cell extracts. While a crude extract from tobacco cells was incapable of accurately transcribing from the cauliflower mosaic virus (CaMV) 19S and 35S promoters, the combination of two fractions resulted in the accurate initiation from the 19S promoter, but not from the 35S promoter (Cooke and Penon 1990). On the other hand, accurate transcription initiation of rDNA in a crude extract from embryonic axes of *Vicia faba* was reported (Yamashita et al. 1993).

The preparation of nuclear or chromatin extracts has the advantage of removing possible vacuolar or cytoplasmic inhibitors, but the disadvantage of losing necessary nuclear factors during nuclei or chromatin isolation. There are conflicting results from in vitro transcription with nuclear or chromatin extracts. Crude nuclear extracts from wheat germ were unable to transcribe from the mammalian adenovirus-2 major late promoter (Ackerman et al. 1987; Flynn et al. 1987). The failure of the wheat germ extract to transcribe was thought to reflect the presence of transcription inhibitors (see Sect. 5.1). In contrast, crude nuclear extracts from rice, wheat, and tobacco cells were capable of accurately transcribing from the wheat gliadin and from a truncated CaMV 35S promoter (Roberts and Okita 1991). Furthermore, a wheat germ chromatin extract (WGCE) accurately transcribed from the TC7 promoter (on T-DNA) and from a CaMV 35S minimal promoter (Yamazaki and Imamoto 1987, Yamazaki et al. 1990a). Most promisingly, transcription in the WGCE has been reported to be stimulated by the added recombinant plant DNA-binding protein TGA1a (Yamazaki et al. 1990b). The stimulation was dependent on the presence of as-1, the binding site of TGA1a. We have analyzed the nucleotide sequence requirements for accurate initiation in the WGCE and have found that the WGCE transcribed in an initiator-, but not in a TATA box- or CCAAT box-dependent manner (Schweizer and Mösinger submitted). The same has been observed with the rice nuclear extract mentioned above (Okita 1991). Therefore, these systems are only partly functional, and further improvement will be necessary. The initiator dependence of transcription in the WGCE and the rice nuclear extract prompted me to compare the cap-site sequences of the promoters, from which accurate in vitro transcription by plant extracts was reported, to sequences, from which there was no or inefficient transcription (Fig. 1). Some homologies between the plant cap-site sequences seem to exist, but the degree of similarity is not correlated with the efficiency of transcription in vitro. However, there is a tendency in efficiently transcribed cap-site sequences for absence of G-residues. It is tempting to speculate that A-, T-rich and A-, C-rich sequences are a common characteristic of plant class II initiators, because two of such G-free cap-site sequences, from the parsley chalcone synthase and wheat gliadin promoter, function as initiators in vitro (Okita 1991; Schweizer and Mösinger submitted). The putative plant initiators are different from animal initiators

Promoter	Cap-site Sequence (-6 to +11)	In vitro Transcr.	Reference
CHS Inr	AAAATAACAACACAAAT	+	Schweizer and Mösinger (submitted)
CaMV 19S	AAAATCAGACCTCCAAG	+	Cooke and Penon (1990)
Glia (Inr)	ATCCTTCCTCACCCATC	+	Roberts and Okita (1991)
TC7	TTACACAAACACACCTC	+	Yamazaki and Imamoto (1987)
PR2m Inr	TTTTCTTaCACCAAATT	(+)	Schweizer and Mösinger (submitted)
CaMV 35S	GAGGACACGCTGAAATC	– (+)	Cooke and Penon (1990); Roberts and Okita (1991)
PI2	ACACTCTTCACCCCAAA	(+)	Schweizer and Mösinger (submitted)
AdML Inr	GTCCTCACTCTCTTCCG	–	Flynn et al. (1987); Smale and Baltimore (1989)

Fig. 1. Comparison of cap-site sequences used for in vitro transcription in plant extracts. Nucleotide sequences of the nontranscribed strand from positions −6 to +11, relative to the major in vivo transcription start site (marked by a *bent arrow*), are shown. The sequences are aligned for best fit, without allowing gaps. *Boxed* sequences are identical to the CHS cap-site sequence. *Inr* Initiator; *CHS* parsley chalcone synthase promoter; *CaMV 19S* cauliflower mosaic virus 19S promoter; *Glia* wheat gliadin promoter; *TC7* TC7 promoter on T-DNA; *PR2m* mutated parsley pathogenesis-related protein 2 promoter (the mutated residue is marked by a *lowercase letter*); *CaMV 35S* cauliflower mosaic virus 35S promoter; *PI2* potato proteinase-inhibitor 2 promoter; *AdML* adenovirus-2 major late promoter; +, efficiently transcribed in vitro; (+), weakly transcribed in vitro; −, not transcribed in vitro

(Greenfield and Smale 1992), and the initiator of the adenovirus-2 major late promoter did not direct efficient transcription in a plant system (Fig. 1).

In general, the plant transcriptionally competent extracts described up to now will have to be further characterized. Their ability to transcribe from a larger number of plant promoters and detailed nucleotide sequence requirements for initiation would have to be examined. Another point to consider is the possible need for topological strain on supercoiled DNA templates (Handa et al. 1989; Sekiguchi and Kmiec 1989; Mizutani et al. 1991).

4 Reconstituted Plant Systems

Since RNase, DNase, phosphatase, proteinase, or other enzyme activities, together with unidentified inhibitors, may have a detrimental effect on in vitro transcription in crude plant extracts, reconstituted systems from highly purified or recombinant components are an attractive alternative. In studies on mammalian in vitro transcription, the use of reconstituted systems based on pol II and partly to highly purified general transcription factors has become a standard method (see e.g., Carcamo et al. 1991; Dynlacht et al. 1991; Zhu and Prywes 1992). The minimum factor requirements for basal transcription from TATA box-containing class II promoters are pol II and the general transcription factors TFIIA, TFIIB, TBP (TATA box-binding protein, a component of TFIID), TFIIE, TFIIF, TFIIH, and TFIIJ (Flores et al. 1992; Lu et al. 1992). For initiator-mediated basal transcription, an additional factor, TFII-I, seems to be necessary (reviewed by Roeder 1991). For activated transcription, TAFs (forming TFIID together with TBP) and/or other proteins called "coactivators" are required (Kelleher et al. 1990; Pugh and Tjian 1990; Dynlacht et al. 1991; Flanagan et al. 1991; Meisterernst et al. 1991; Zhu and Prywes 1992). Since several of these general class II transcription factors copurify, five fractions of partly purified factors, formerly called TFIIA to TFIIF, plus pol II, are sufficient for basal and activated in vitro transcription (reviewed by Sawadogo and Sentenac 1990).

Standardized protocols for the purification of enzymatically active plant pol I, II, and III have existed for several years (reviewed by Cooke et al. 1984; see also by Bakó et al., this Vol.). A crude reconstituted system consisting of transcriptionally inactive chromatin and pea and cauliflower pol II has been described (Tomi et al. 1981). The authors reported initiation by the plant RNA polymerases at specific sites that were distinct from initiation sites used by *E. coli* RNA polymerase, although no gene-specific data seem to be available from this system. Other approaches for reconstituted systems based on pol II and partly purified general class II transcription factors were partly successful (Cooke and Penon 1990) or unsuccessful (Flynn et al. 1987; see also Sect. 3).

Until now, only two class II general transcription factors, but no class I or class III factors, have been identified and isolated from plants. The class II factor TFIIA has been purified from wheat germ (Burke et al. 1990). Wheat TFIIA could functionally replace human (HeLa) TFIIA in vitro. cDNAs encoding the TATA box-binding subunit of TFIID (TBP) have been isolated from *Arabidopsis* (Gasch et al. 1990), potato (Holdsworth et al. 1992), maize (Haass and Feix 1992), and wheat (Kawata et al. 1992). In contrast to *Drosophila*, human, and yeast, *Arabidopsis* and maize contain two TBP-encoding genes. It is not yet known whether the two plant TBP isoforms have different functions. Recombinant TBPs from *Arabidopsis* could functionally replace human (HeLa) TFIID (Gasch et al. 1990). The crystal structure of TBP-2 from *Arabidopsis* has also been determined (Nikolov et al. 1992).

In conclusion, to obtain a functional reconstituted plant system for in vitro transcription of class II promoters, three more factors, TFIIB, TFIIE, and TFIIF, will have to be (partially) purified or obtained as recombinant products using heterologous probes to clone the genes. Even if all these gene products can be cloned from plants and overexpressed, they will probably not be sufficient to reconstitute transcription. As mentioned above, other general factors and coactivators copurify with fractions TFIIA, TFIID, and TFIIF (see e.g., Dynlacht et al. 1991; Flores et al. 1992; Zhu and Prywes 1992). There are several lines of evidence indicating interchangeability of the general class II transcription factors between plants and animals (Burke et al. 1990; Gasch et al. 1990). Therefore, complementation of incomplete plant systems with the respective animal or yeast factors might result in a functional, partly plant-derived system. Such a "hybrid" system may be an acceptable compromise to start with when developing plant systems.

5 Problems Associated with Plant Systems

The limited progress made with development of plant in vitro transcription systems seems to reflect some severe problems associated with the plant material as a source of such systems. In vitro transcription systems are multifactor systems and therefore more sensitive to degradation, loss of components, or inhibition than, e.g., a "simple" one-enzyme assay. The loss or inhibition of one factor might be enough to completely inactivate the system. The release of lytic enzyme activities or inhibitory secondary metabolites from plant vacuoles during extract preparation is certainly a major concern (Boller 1982; Boller and Wiemken 1986). Other problems might arise from the high amounts of heterochromatin in nuclei of wheat and other plant species (Bennett and Smith 1976), which might be associated with inhibitors of transcription, the most evident being histones (Croston et al. 1991; Further and Hall 1991). A third problem might be the often low starting protein concentration in plant extracts, which might destabilize sensitive factors or drastically reduce recovery in concentration or fractionation procedures (Deutscher 1990). Finally, nonspecific transcription by plant RNA polymerases, the best studied being pol II, might obscure signals from accurate in vitro initiation.

5.1 Endogenous Inhibitors of Transcription

The presence in plant extracts of inhibitors of accurate transcription has been demonstrated in mixing experiments with functional yeast pol III (Furter and Hall 1991) or animal pol II (Flynn et al. 1987; Henfrey et al. 1989) transcription systems. Moreover, the antiviral activity of *Phyllanthus* sp. extracts, a small herb used often in folk medicine, has been attributed to its capacity to inhibit viral DNA polymerase and reverse transcriptase in vitro (Unander 1991). We found that the WGCE described in Section 3 also contains an activity inhibiting TATA box-dependent transcription by pol II in a *Drosophila* extract (Schweizer and Mösinger submitted). The inhibiting activity in the WGCE is heat labile and inhibits initiation complex formation, but not elongation. An inhibiting activity in wheat germ nuclear extracts did not affect elongation in an active yeast pol III extract, but seemed to inhibit initiation complex formation by interaction with transcription factor(s) (Further and Hall 1991).

There is evidence from fractionated wheat germ nuclear extracts that the transition of pol II from initiation to elongation is inhibited (Flynn et al. 1987), although one of the wheat germ fractions increased the processivity of purified pol II (de Mercoyrol et al. 1989). It is now known that the C-terminal heptapeptide repeat (CTD) of pol II is highly phosphorylated by the general transcription factor TFIIH during transition from the initiation to the elongation mode (Lu et al. 1992). A CTD-specific kinase, which might represent TFIIH, has also been partially purified from wheat germ (Guilfoyle 1989). It therefore could be speculated that the block in early elongation was due to contamination of the wheat germ nuclear extract with phosphatase, thus interfering with TFIIH function. However, in a mammalian transcription system, the adenovirus-2 major late promoter was also accurately transcribed by pol IIB lacking the CTD (Kim and Dahmus 1989). Moreover, the CTD was not required for transcription factor Spl to function in vitro (Zehring and Greenleaf 1990).

It is a well-known fact that plant extracts often contain elevated levels of lytic enzyme activities, some of which might inhibit transcription. However, DNase, RNase, or proteinase contamination was not the main reason for inhibition of transcription in the wheat germ nuclear extract mentioned earlier (Ackerman et al. 1987). Wheat germ extracts might be an exception in this respect, because wheat germ is a vacuole-free tissue.

Although the described transcription inhibitors from wheat germ are not vacuolar, the vacuole might disturb the function of transcription systems prepared from other sources like cultured cells. A solution to this problem might be the preparation of transcriptionally competent extracts from evacuolized cells. Transcription of parsley chalcone synthase promoter constructs in such a system, prepared from evacuolized parsley protoplasts, resulted in promising signals (H. Frohnmeyer pers. comm.).

Irrespective of the origin of an inhibitor, fractionation and reconstitution of plant extracts seems to be the most straightforward approach to obtain inhibitor-free systems. This approach has led to a partially functional (Cooke and Penon 1990) and to inhibitor-free, but very inefficient (Flynn et al. 1987), plant extracts. A laborious but more targeted approach would be the identification and specific removal or inactivation (e.g., by antibodies) of inhibitors. With one exception (Further and Hall 1991), no information on such an approach is currently available. Recently, a simple method for preparation of an inhibitor-free, transcriptionally active extract from *Schizosaccharomyces pombe* has been described, and the authors suggest using this method also for preparation of plant extracts (Flanagan et al. 1992).

5.2 Nonspecific Transcription

Nonspecific background signals are a major problem in crude transcriptionally competent extracts from plants (Henfrey et al. 1989; Roberts and Okita 1991; Schweizer and Hahlbrock unpubl.). Initiation of pol II at ends (Reines et al. 1989; Yamazaki et al. 1990a) and nicks (Dynan and Burgess 1979; Lewis and Burgess 1980) of added DNA templates is one reason for this background transcription. Nonspecific initiation in an extract from wheat shoot nuclei was repressed by the addition of active human (HeLa) extract, indicating that sufficient amounts of general transcription factors repress this artifact (Henfrey et al. 1989). Another possible reason for the often observed smear of background signals in plant extracts is the premature termination at sites where the nascent RNA forms secondary structures (see e.g., Bengal and Aloni 1989; Kessler et al. 1989; Resnekov and Aloni 1989). In animals, the transcription elongation factor TFIIS allows pol II to read through such attenuation sites (Reines et al. 1989; Yoo et al. 1991). This factor might have been missing or present in insufficient amounts in the plant extracts used up to now. We have found that another problem associated with soluble plant systems is template-independent background transcription, i.e., transcription on genomic DNA which contaminates the extract (Schweizer and Hahlbrock unpubl.). We found that this background was not eliminated under G-free conditions and was partly resistant to RNase T1 digestion, indicating covalent linkage of RNA to the contaminating genomic DNA at nicks (Lewis and Burgess 1980) and extension of short transcripts until the first G-residue.

An elegant method to circumvent the problem of nonspecific background signals has been described by Roberts and Okita (1991). They affinity-purified in vitro transcripts with a poly(A) tail on oligo(dT), thereby revealing specific signals that were hidden behind the nonspecific background.

We used another approach to cope with the problem of background signals by studying transcription on immobilized DNA templates (Arias and Dynan 1989; Lin and Green 1991). This approach would possibly also circumvent the problem of inhibitors of early elongation (Flynn et al. 1987). The experimental system was based on linear templates which were biotinylated at the upstream end and bound to streptavidin-agarose. The immobilized templates were incubated with parsley nuclear and/or chromatin extracts, and washed extensively before the purified initiation complexes were allowed to start elongation by addition of the nucleoside triphosphates. Using different constructs of the parsley PR2 promoter (van de Löcht et al. 1990), we indeed found in vitro run-off transcripts of the expected size that were previously hidden behind nonspecific background, but the signals were not dependent on any promoter sequences down to position −9 relative to the in vivo start site (Schweizer and Hahlbrock unpubl.). Furthermore, they were only partially (approximately 75%) inhibited by α-amanitin, indicating no strict specificity for pol II. The major transcript was obviously from end-to-end transcription, and we could not obtain clear evidence for correct initiation by primer extension. We abandoned this system, because we could not exclude the possibility that the correctly sized transcripts were aborted antisense transcripts from initiation at the downstream end of the linear DNA templates. However, promising results have been recently obtained by using immobilized DNA templates and whole-cell or nuclear extracts from suspension-cultured soybean cells (Arias et al. 1993). The applicability of this system for transcription of a wide range of plant nuclear promoters, as well as TATA-box dependence of transcription from class II promoters, remains to be shown.

6 Heterologous Transcription Systems

It is tempting to use heterologous transcription systems from human (see e.g., Dignam et al. 1983), animals (see e.g., Parker and Topol 1984), or yeast (see e.g., Ponticelli and Struhl 1990) instead of continuing the laborious search for functional plant systems. Functional (see references cited in Sect. 4) and sequence information (Haass and Feix 1992; Nikolov et al. 1992) indicates that the general transcription machinery needed for basal transcription is highly conserved between plants, animals, and fungi. This might not be the case at the level of finely tuned transcriptional regulation involving interactions between *trans*-acting factors and the general transcription machinery, although no evidence against compatibility of plant and animal systems or factors has been reported up to now.

Recognition of plant promoters by the animal transcription machinery does not seem to be a problem. Accurate in vitro transcription from the

CaMV 35S promoter (Henfrey et al. 1989), a 15-kDa zein-gene promoter (Boston and Larkins 1986), a legumin-gene promoter (Evans et al. 1985), and from several *Arabidopsis* $tRNA^{Tyr}$-gene promoters (Stange et al. 1991) in a human (HeLa) cell lysate was reported. A CaMV 35S minimal promoter was also accurately transcribed in a *Drosophila* extract (Schweizer and Mösinger submitted).

There is one report on successful *trans*-activation by a plant DNA-binding protein, TGA1a from tobacco, in a human in vitro transcription system (Katagiri et al. 1990). This example demonstrates that it is in principle possible to test gene-specific plant DNA-binding proteins in animal systems. Although this example encourages us to use heterologous transcription systems, it remains open whether it was only a "lucky" exception, reflecting similarity of TGA1a to an animal *trans*-factor.

7 Conclusion

Up to now, no exhaustively tested plant in vitro transcription system capable of accurately initiating transcription from many plant nuclear promoters has been established. I believe that reconstitution of an active plant system from partly or highly purified fractions is the most promising approach. A difficult question will be the choice of an appropriate starting material, which should contain as few inhibitors as possible, but all general transcription factors of a given class. The WGCE might be a promising system to start such a fractionation and reconstitution approach. The availability of reconstituted human (HeLa) systems based on pol II, and purified or recombinant general class II transcription factors, would certainly help to functionally define plant fractions needed for accurate pol II transcription. In this approach, plant fractions would have to be tested for complementation of defined animal systems lacking specific factors. On the other hand, complementation of plant systems with defined general transcription factors from animals, or yeast, would help to define limiting, inactivated, or missing general plant transcription factors, once inhibitors have been removed. Cooperation between animal and plant molecular biologists is therefore highly desirable. Furthermore, the identification of protein-based plant transcription inhibitors might give insight into the general mechanisms of gene repression.

Acknowledgements. I would like to thank Drs. Egon Mösinger, Elizabeth Furter-Graves, and Rolf Furter for critical reading of the manuscript.

References

Ackerman S, Flynn PA, Davis EA (1987) Partial purification of plant transcription factors. I Initiation. Plant Mol Biol 9:147–158

Arias JA, Dynan WS (1989) Promoter-dependent transcription by RNA polymerase II using immobilized enzyme complexes. J Biol Chem 264:3223–3229
Arias JA, Dixon RA, Lamb CJ (1993) Dissection of the functional architecture of a plant defense gene promoter using a homologous in vitro transcription initiation system. Plant Cell 5:485–496
Bagchi MK, Tsai SY, Weigel NL, Tsai MJ, O'Malley BW (1990a) Regulation of in vitro transcription by progesterone receptor. J Biol Chem 265:5129–5134
Bagchi MK, Tsai SY, Tsai MJ, O'Malley BW (1990b) Identification of a functional intermediate in receptor activation in progesterone-dependent cell-free transcription. Nature 345:547–550
Bengal E, Aloni Y (1989) A block of transcription elongation by RNA polymerase II at synthetic sites in vitro. J Biol Chem 264:9791–9798
Bennett MD, Smith JB (1976) Nuclear DNA amounts in angiosperms. Philos Trans R Soc (Lond) B 274:227–274
Boller T (1982) Enzymatic equipment of plant vacuoles. Physiol Veg 20:247–257
Boller T, Wiemken A (1986) Dynamics of vacuolar compartmentation. Annu Rev Plant Physiol 37:137–164
Boston RS, Larkins BA (1986) Specific transcription of a 15-kilodalton zein gene in HeLa cell extracts. Plant Mol Biol 7:71–79
Bouman H, Mennes AM, Libbenga KR (1979) Transcription in nuclei isolated form tobacco tissues. FEBS Lett 101:369–372
Burke C, Yu XB, Marchitelli L, Davis EA, Ackerman S (1990) Transcription factor IIA of wheat and human function similarly with plant and animal viral promoters. Nucleic Acids Res 18:3611–3620
Carcamo J, Buckbinder L, Reinberg D (1991) The initiator directs the assembly of a transcription factor IID-dependent transcription complex. Proc Natl Acad Sci USA 88:8052–8056
Chappell J, Hahlbrock K (1986) Salt effects on total and gene-specific in-vitro transcriptional activity of isolated plant nuclei. Plant Cell Rep 5:398–402
Cooke RM, Penon P (1990) In vitro transcription from cauliflower mosaic virus promoters by a cell-free extract from tobacco cells. Plant Mol Biol 14:391–405
Cooke RM, Durand R, Job C, Penon P, Teissere M, Job D (1984) Enzymatic properties of plant RNA polymerases. Plant Mol Biol 3:217–225
Corthésy B, Claret FX, Wahli W (1990) Estrogen receptor level determines sex-specific in vitro transcription from the *Xenopus* vitellogenin promoter. Proc Natl Acad Sci USA 87:7878–7882
Cox RF (1973) Transcription of high-molecular-weight RNA from hen-oviduct chromatin by bacterial and endogenous form-B RNA polymerases. Eur J Biochem 39:49–61
Croston GE, Kerrigan LA, Lira LM, Marshak DR, Kadonaga JT (1991) Sequence-specific antirepression of histone H1-mediated inhibition of basal RNA polymerase II transcription. Science 251:643–649
Dehesh K, Hung H, Tepperman JM, Quail PH (1992) GT-2: a transcription factor with twin autonomous DNA-binding domains of closely related but different target sequence specificity. EMBO J 11:4131–4144
De Mercoyrol L, Job C, Ackerman S, Job D (1989) A wheat-germ nuclear fraction required for selective initiation in vitro confers processivity to wheat-germ RNA polymerase II. Plant Sci 64:31–38
Deutscher MP (1990) Guide to protein purification. In: Deutscher MP (ed) Methods in enzymology, vol 182. Academic Press, San Diego, pp 83–89
Dignam JD, Lebovitz RM, Roeder RG (1983) Accurate transcription initiation by RNA polymerase II in a soluble extract from isolated mammalian nuclei. Nucleic Acids Res 11:1475–1489
Dynan WS, Burgess RR (1979) In vitro transcription by wheat germ ribonucleic acid polymerase II: effects of heparin and role of template integrity. Biochemistry 18:4581–4588
Dynlacht BD, Hoey T, Tjian R (1991) Isolation of coactivators associated with the TATA-binding protein that mediate transcriptional activation. Cell 66:563–576

Evans IM, Bown D, Lycett GW, Croy RRD, Boulter D, Gatehouse JA (1985) Transcription of a legumin gene from pea (*Pisum sativum* L.) in vitro. Planta 165:554–560
Flanagan PM, Kelleher RJ, Sayre MH, Tschochner H, Kornberg RD (1991) A mediator required for activation of RNA polymerase II transcription in vitro. Nature 350:436–438
Flanagan PM, Kelleher RJ III, Tschochner H, Sayre MH, Kornberg RD (1992) Simple derivation of TFIID-dependent RNA polymerase II transcription systems from *Schizosaccharomyces pombe* and other organisms, and factors required for transcriptional activation. Proc Natl Acad Sci USA 89:7659–7663
Flores O, Lu H, Reinberg D (1992) Factors involved in specific transcription by mammalian RNA polymerase II. J Biol Chem 267:2786–2793
Flynn PA, Davis EA, Ackerman S (1987) Partial purification of plant transcription factors. II. An in vitro transcription system is inefficient. Plant Mol Biol 9:159–169
Furter R, Hall BD (1989) Specific transcription and reinitiation of class III genes in wheat embryo nuclei and chromatin. Plant Mol Biol 12:567–577
Furter R, Hall BD (1991) Substances in nuclear wheat germ extracts which interfere with polymerase III transcriptional activity in vitro. Plant Mol Biol 17:773–785
Gabrielsen OS, Sentenac A (1991) RNA polymerase III (C) and its transcription factors. Trends Biochem 16:412–416
Gasch A, Hoffmann S, Horikoshi M, Roeder RG, Chua NH (1990) *Arabidopsis thaliana* contains two genes for TFIID. Nature 346:390–394
Gilmartin PM, Memelink J, Hiratsuka K, Kay SA, Chua NH (1992) Characterization of a gene encoding a DNA binding protein with specificity for a light-responsive element. Plant Cell 4:839–849
Goldenberg CJ, Luo Y, Fenna M, Baler R, Weinmann R, Voellmy R (1988) Purified human factor activates heat shock promoter in a HeLa cell-free transcription system. J Biol Chem 263:19734–19739
Greenfield AO'S, Smale ST (1992) Roles of TATA and initiator elements in determining the start site location and direction of RNA polymerase II transcription. J Biol Chem 267:1391–1402
Guilfoyle TJ (1989) A protein kinase from wheat germ that phosphorylates the largest subunit of RNA polymerase II. Plant Cell 1:827–836
Haass MM, Feix G (1992) Two different cDNAs encoding TFIID proteins of maize. FEBS Lett 301:294–298
Handa H, Watanabe H, Suzuki Y, Hirose S (1989) Effect of DNA supercoiling on in vitro transcription from the adenovirus early region 4. FEBS Lett 249:17–20
Hawley DK, Roeder RG (1985) Separation and partial characterization of three functional steps in transcription initiation by human RNA polymerase II. J Biol Chem 260:8163–8172
Henfrey RD, Proudfoot LMF, Covey SN, Slater RJ (1989) Identification of an inhibitor of transcription in extracts prepared from wheat shoot nuclei. Plant Sci 64:91–98
Hentschel CC, Tata JR (1978) Template-engaged and free RNA polymerases during *Xenopus* erythroid cell maturation. Dev Biol 65:496–507
Hipskind RA, Reeder RH (1980) Initiation of ribosomal RNA chains in homogenates of oocyte nuclei. J Biol Chem 255:7896–7906
Holdsworth MJ, Grierson C, Schuch W, Bevan M (1992) DNA-binding properties of cloned TATA-binding protein from potato tubers. Plant Mol Biol 19:455–464
Ide GJ (1981) Nucleoside 5′-(γ-S)triphosphates will initiate transcription in isolated yeast nuclei. Biochemistry 20:2633–2638
Johnson FB, Krasnow MA (1990) Stimulation of transcription by an *Ultrabithorax* protein in vitro. Genes Dev 4:1044–1052
Katagiri F, Lam E, Chua NH (1989) Two tobacco DNA-binding proteins with homology to the nuclear factor CREB. Nature 340:727–729
Katagiri F, Yamazaki KI, Horikoshi M, Roeder RG, Chua NH (1990) A plant DNA-binding protein increases the number of active preinitiation complexes in a human in vitro transcription system. Genes Dev 4:1899–1909

Kawata T, Minami M, Tamura T, Sumita K, Iwabuchi M (1992) Isolation and characterization of a cDNA clone encoding the TATA box-binding protein (TFIID) from wheat. Plant Mol Biol 19:867–873
Kelleher RJ, Flanagan PM, Kornberg RD (1990) A novel mediator between activator proteins and the RNA polymerase II transcription apparatus. Cell 61:1209–1215
Kessler M, Ben-Asher E, Aloni Y (1989) Elements modulating the block of transcription elongation at the adenovirus 2 attenuation site. J Biol Chem 264:9785–9790
Kikuchi M, Imaseki H, Sakai S (1989) Modulation of gene expression in isolated nuclei by auxin-binding proteins. Plant Cell Physiol 30:765–773
Kim WY, Dahmus ME (1989) The major late promoter of adenovirus-2 is accurately transcribed by RNA polymerases IIO, IIA and IIB. J Biol Chem 264:3169–3176
Laybourn PJ, Kadonaga JT (1991) Role of nucleosomal cores and histone H1 in regulation of transcription by RNA polymerase II. Science 254:238–245
Levin CJ, Zimmermann SB (1976) A deoxyribonucleic acid kinase from nuclei of rat liver. J Biol Chem 251:1767–1774
Lewis MK, Burgess RR (1980) Transcription of simian virus 40 DNA by wheat germ RNA polymerase II. J Biol Chem 255:4928–4936
Lin YS, Green MR (1991) Mechanism of action of an acidic transcriptional activator in vitro. Cell 64:971–981
Lin YS, Carey M, Ptashne M, Green MR (1990) How different eucaryotic transcriptional activators can cooperate promiscuously. Nature 345:359–361
Lu H, Zawel L, Fisher L, Egly JM, Reinberg D (1992) Human general transcription factor IIH phosphorylates the C-terminal domain of RNA polymerase II. Nature 358:641–645
Manley JL, Fire A, Cano A, Sharp PA, Gefter ML (1980) DNA-dependent transcription of adenovirus genes in a soluble whole-cell extract. Proc Natl Acad Sci USA 77:3855–3859
Mauro VP, Verma DPS (1987) Transcriptional activation in nuclei from uninfected soybean of a set of genes in symbiosis with *Rhizobium*. Mol Plant-Microbe Interact 1:46–51
Meisterernst M, Roy AL, Lieu HM, Roeder RG (1991) Activation of class II gene transcription by regulatory factors is potentiated by a novel activity. Cell 66:981–993
Mennes AM, Quint A, Gribnau JH, Boot CJM, van der Zaal EJ, Maan AC, Libbenga KR (1992) Specific transcription and reinitiation of 2,4-D-induced genes in tobacco nuclei. Plant Mol Biol 18:109–117
Mizutani M, Ohta T, Watanabe H, Handa H, Hirose S (1991) Negative supercoiling of DNA facilitates an interaction between transcription factor IID and the fibroin gene promoter. Proc Natl Acad Sci USA 88:718–722
Mösinger E, Batschauer A, Schäfer E, Apel K (1985) Phytochrome control of in vitro transcription of specific genes in isolated nuclei from barley (*Hordeum vulgare*). Eur J Biochem 147:137–142
Mösinger E, Batschauer A,Vierstra R, Apel K, Schäfer E (1987) Comparison of the effects of exogenous native phytochrome and in-vivo irradiation on in-vitro transcription in isolated nuclei from barley (*Hordeum vulgare*). Planta 170:505–514
Nikolov DB, Hu SH, Lin J, Gasch A, Hoffmann A, Horikoshi M, Chua NH, Roeder RG, Burley SK (1992) Crystal structure of TFIID TATA-box binding protein. Nature 360:40–46
Oeda K, Salinas J, Chua NH (1991) A tobacco bZip transcription activator (TAF-1) binds to a G-box-like motif conserved in plant genes. EMBO J 10:1793–1802
Ohkuma Y, Horikoshi M, Roeder RG, Desplan C (1990) Engrailed, a homeodomain protein, can repress in vitro transcription by competition with the TATA box-binding protein transcription factor IID. Proc Natl Acad Sci USA 87:2289–2293
Okita TW (1991) Development of in vitro transcription systems for both monocot and dicot genes. Progress report to Proj No WNP02690 at Washington State

University, Pullmann, WA 99164. Current Research Information System, US Department of Agriculture (Databank File: CRIS/USDA) (Abstr)

Parker CS, Topol J (1984) A *Drosophila* RNA polymerase II transcription factor contains a promoter-region-specific DNA-binding activity. Cell 26:357–369

Perisic O, Lam E (1992) A tobacco DNA binding protein that interacts with a light-responsive box II element. Plant Cell 4:831–838

Ponticelli AS, Struhl K (1990) Analysis of *Saccharomyces cerevisiae* his3 transcription in vitro: biochemical support for multiple mechanisms of transcription. Mol Cell Biol 10:2832–2839

Pugh BF, Tjian R (1990) Mechanism of transcriptional activation by Spl: evidence for coactivators. Cell 61:1187–1197

Reeder RH (1990) rRNA synthesis in the nucleolus. Trends Genet 6:390–395

Reines D, Chamberlin MJ, Kane CM (1989) Transcription elongation factor SII (TFIIS) enables RNA polymerase II to elongate through a block to transcription in a human gene in vitro. J Biol Chem 264:10799–10809

Resnekov O, Aloni Y (1989) RNA polymerase II is capable of pausing and prematurely terminating transcription at a precise location in vivo and in vitro. Proc Natl Acad Sci USA 86:12–16

Roberts MW, Okita TW (1991) Accurate in vitro transcription of plant promoters with nuclear extracts prepared from cultured plant cells. Plant Mol Biol 16:771–786

Roeder RG (1991) The complexities of eukaryotic transcription initiation: regulation of preinitiation complex assembly. Trends Biochem 16:402–408

Rougvie AE, Lis JT (1988) The RNA polymerase II molecule at the 5′ end of the uninduced hsp 70 gene of *D. melanogaster* is transcriptionally engaged. Cell 54:795–804

Sakai S, Seki J, Imaseki H (1986) Stimulation of RNA synthesis in isolated nuclei by auxin-binding proteins-I and -II. Plant Cell Physiol 27:635–643

Sawadogo M, Roeder RG (1985) Factors involved in specific transcription by human RNA polymerase II: analysis by a rapid and quantitative in vitro assay. Proc Natl Acad Sci USA 82:4394–4398

Sawadogo M, Sentenac A (1990) RNA polymerase B (II) and general transcription factors. Annu Rev Biochem 59:711–754

Scharf KD, Rose S, Zott W, Schöff F, Nover L (1990) Three tomato genes code for heat stress transcription factors with a region of remarkable homology to the DNA-binding domain of yeast HSF. EMBO J 9:4495–4501

Schweizer P (1988) Induced local resistance of winter wheat (*Triticum aestivum* L.) to *Erysiphe graminis* f. sp. *tritici*: cDNA cloning and in vitro transcription of induced wheat mRNA. Thesis, University of Bern, Switzerland

Schweizer P, Hahlbrock K (1993) Posttranscriptional transfer of γ-thio affinity label to RNA in isolated parsley nuclei. Plant Mol Biol 21:943–947

Schweizer P, Hunziker W, Mösinger E (1989) cDNA cloning, in vitro transcription and partial sequence analysis of mRNAs from winter wheat (*Triticum aestivum* L.) with induced resistance to *Erysiphe graminis* f. sp. *tritici*. Plant Mol Biol 12:643–654

Sekiguchi JM, Kmiec EB (1989) DNA superhelicity enhances the assembly of transcriptionally active chromatin in vitro. Mol Gen Genet 220:73–80

Smale ST, Baltimore D (1989) The "initiator" as transcription control element. Cell 57:103–113

Somssich IE, Schmelzer E, Bollmann J, Hahlbrock K (1986) Rapid activation by fungal elicitor of genes encoding "pathogenesis-related" proteins in cultured parsley cells. Proc Natl Acad Sci USA 83:2427–2430

Stange N, Beier D, Beier H (1991) Expression of nuclear $tRNA^{Tyr}$ genes from *Arabidopsis thaliana* in HeLa cell and wheat germ extracts. Plant Mol Biol 16:865–875

Strobl LJ, Eick D (1992) Hold back of RNA polymerase II at the transcription start site mediates down-regulation of c-myc in vivo. EMBO J 11:3307–3314

Tabata T, Nakayama T, Mikami K, Iwabuchi M (1991) HBP-1a and HBP-1b: leucine zipper-type transcription factors of wheat. EMBO J 10:1459–1467

Teissère M, Durand R, Ricard J (1979) Transcription in vitro of cauliflower mosaic virus DNA by RNA polymerase I, II and III purified from wheat embryos. Biochem Biophys Res Commun 89:526–533

Tomi H, Sasaki Y, Kamikubo T (1981) A comparison between homologous and heterologous RNA polymerase recognition sites during in vitro chromatin transcription. J Biochem 90:1705–1714

Unander DW (1991) Callus induction in *Phyllanthus* species and inhibition of viral DNA polymerase and reverse transcriptase by callus extracts. Plant Cell Rep 10:461–466

van de Löcht U, Meier I, Hahlbrock K, Somssich IE (1990) A 125 bp promoter fragment is sufficient for strong elicitor-mediated gene activation in parsley. EMBO J 9:2945–2950

Washington LD, Stallcup MR (1982) A comparison of nucleoside (β-S)triphosphates and nucleoside (γ-S)triphosphates as suitable substrates for measuring transcription initiation in preparations of cell nuclei. Nucleic Acids Res 10:8311–8322

Weisshaar B, Armstrong GA, Block A, da Costa e Silva O, Hahlbrock K (1991) Light-inducible and constitutively expressed DNA-binding proteins recognizing a plant promoter element with functional relevance in light responsiveness. EMBO J 10:1777–1786

Wobbe CR, Struhl K (1990) Yeast and human TATA-binding proteins have nearly identical DNA sequence requirements for transcription in vitro. Mol Cell Biol 10:3859–3867

Workman JL, Roeder RG, Kingston RE (1990) An upstream transcription factor, USF (MLTF) facilitates the formation of preinitiation complexes during in vitro chromatin assembly. EMBO J 9:1299–1308

Wu R (1971) Inhibition of polynucleotide kinase by agar, dextran sulfate and other polysaccharide sulfates. Biochem Biophys Res Commun 43:927–934

Yamashita J, Nakajima T, Tanifuji S, Kato A (1993) Accurate transcription initiation of *Vicia faba* rDNA in a whole cell extract from embryonic axes. Plant J 3:187–190

Yamazaki KI, Imamoto F (1987) Selective and accurate initiation of transcription at the T-DNA promoter in a soluble chromatin extract from wheat germ. Mol Gen Genet 209:445–452

Yamazaki KI, Chua NH, Imaseki H (1990a) Accurate transcription of plant genes in vitro using a wheat germ-chromatin extract. Plant Mol Biol Rep 8:114–123

Yamazaki KI, Katagiri F, Imaseki H, Chua NH (1990b) TGA1a, a tobacco DNA-binding protein, increases the rate of initiation in a plant in vitro transcription system. Proc Natl Acad Sci USA 87:7035–7039

Yi-Chi Sun I, Johnson EM, Allfrey VG (1979) Initiation of transcription of ribosomal deoxyribonucleic acid sequences in isolated nuclei of *Physarum polycephalum*: studies using nucleoside 5′-(γ-S)triphosphates and labelled precursors. Biochemistry 18:4572–4580

Yoo D, Yoon H, Baek K, Jeon C, Miyamoto K, Ueno A, Agarwal K (1991) Cloning, expression and characterization of the human transcription elongation factor, TFIIS. Nucleic Acids Res 19:1073–1079

Zehring WA, Greenleaf AL (1990) The carboxyl-terminal repeat domain of RNA polymerase II is not required for transcription factor Sp1 to function in vitro. J Biol Chem 265:8351–8353

Zhang-Keck ZJ, Stallcup MR (1988) Optimized reaction conditions and specific inhibitors for initiation of transcription by RNA polymerase II in nuclei from cultured mammalian cells. J Biol Chem 263:3513–3520

Zhu H, Prywes R (1992) Identification of a coactivator that increases activation of transcription by serum response factor and GAL4-VP16 in vitro. Proc Natl Acad Sci USA 89:5291–5295

B: Promoters Regulated by Stress and Developmental Signals

6 Heat Stress Promoters and Transcription Factors

Klaus-Dieter Scharf[1,2], Tilo Materna[1], Eckardt Treuter[2], and Lutz Nover[1,2]

1 Introduction: Heat Stress Response and Stress Protein Families

The initial description of heat stress (hs)-induced gene activity using polytene chromosomes of *Drosophila* salivary glands (Ritossa 1962) was followed 12 years later by the detection of the corresponding heat stress proteins (HSP) (Tissieres et al. 1974) and progress toward cloning the *Drosophila* hs genes (for summaries, see Ashburner and Bonner 1979; Schlesinger et al. 1982). The explosive development of molecular stress biology in the following decade extended the investigations to all types of living organisms. In all cases the heat stress (hs) response was found to comprise a highly complex but transient reprogramming of cellular activities necessary to protect cells from extensive damage and to provide optimum conditions for recovery after the stress period. Results have been summarized in many reviews and books (Lindquist and Craig 1988; Morimoto et al. 1990; Nover et al. 1990; Nover 1991; Vierling 1991; Gething and Sambrook 1992).

Two main aspects of the heat stress response are currently the center of interest:

1. The structure and function of the newly formed heat stress proteins (HSPs), which are essential for many aspects of induced stress tolerance. Seven stress protein families were defined, each of them with highly conserved representatives in bacteria, yeast, plants and animals (Table 1). Remarkably, constitutive and stress-induced members of these families have been characterized as indispensible components of the protein processing machineries in different parts of the cell. As "molecular chaperones" they participate in protein folding, assembly, topogenesis and degradation (see reviews by Gething and Sambrook 1992; Hartl et al. 1992).

---

[1] Lehrstuhl Zellbiologie, Biozentrum, J.-W.-Goethe-Universität, Marie-Curie-Straße 9, D-60439 Frankfurt, FRG
[2] Institut für Pflanzenbiochemie, Weinberg 3, D-06120 Halle, FRG

Throughout the chapter genes are symbolized by lower case italicited letters (*hsp*, *hsf*, *gus*), whereas the corresponding proteins are symbolized by upper case letters (HSP, HSF, GUS).

Results and Problems in Cell Differentiation 20
L. Nover (Ed.)
Plant Promoters and Transcription Factors

2. The signal transduction mechanisms involved in the regulation of the hs genes. In addition to hs, a considerable number of chemical stressors act as inducers of HSP synthesis, including heavy metals, amino acid analogs and certain antibiotics. These agents lead to mistranslation or abortion of protein fragments as well as to protein denaturation, indicating that the stress-induced accumulation of denatured proteins is central to the signal transduction mechanisms (see model in Fig. 10; Sect. 7.3). At the end of the signal transduction chain is the specific receptor element (HSE) in the promoter of the hs genes and the corresponding transcription factor (HSF) binding to it. Both will be the focus of this review. We will present data not only on plants, but we will frequently include results obtained with other eukaryotes. The experimental basis for this

Table 1. Survey of eukaryotic HSP families

Organism (No. of genes)	Cellular localization			Functional characteristics[a]
	Cytoplasm/nucleus	ER	Chloroplasts (c)/ mitochondria (m)	
HSP110 family	(Ec-ClpA/B)			
Mammals (2)	**HSP110**	–	–	In *E. coli* ATP-binding subunits of high mol.wt. protease complex[b]
Plants (>2)	ClpB	–	ClpC	
Yeast (>2)	**HSP104**	–	**ClpBm**	
HSP90 family (Ec-HtpG)				
Mammals (3)	**HSP89α,β**	GRP94	–	In mammals characterized as inhibitory subunit of steroid hormone receptors, complexed with the HSP56 immunophilin[c]
Plants (1)	**HSP80**	?	–	
Yeast (2)	**HSP90**, HSC90	?	–	
HSP70 family (Ec-DnaK)				
Mammals (7–8)	**HSP70 A, B**, HSC72	GRP78	HSC70m	ATP-binding chaperones, essential for folding, intracellular distribution and assembly/disassembly of proteins
Plants (>5)	**HSP70**, HSC70	GRP78	HSC70c, **HSP68m**	
Yeast (8)	**SSA 1–4**, SSB 1,2	KAR2	SSC1m	
HSP60 family (Ec-GroEL)[d]				
Mammals (>2)	TCP1	–	**HSP60** (Cpn60)m	ATP-binding, heptameric chaperone essential for assembly of protein complexes
Plants (>3)	TCP1	–	HSP60c, **HSP60m**	
Yeast (>2)	TCP1	–	HSP60m	
HSP40 family (Ec-DnaJ)[e]				
Mammals (1)	**HDJ-1**	?	?	In *E. coli* characterized as substrate-binding subunit of the DnaK complex
Plants (?)	**ANJ1**	?	?	
Yeast (4)	YDJ1, SIS1	SEC63	**SCJ1**	
HSP20 family (Ec-IbpA,IbpB)[f]				
Mammals (2)	**HSP25, αB-crystallin**[f]	?	?	Form oligomeric complexes of 200–500 kDa, which aggregate under hs conditions (hs granules)
Plants (15–20)	**HSPs 15–18**	**HSP18/19**	**HSP21/22c**	
Yeast (1)	**HSP26**	–	–	
HSP10 family (Ec-GroES)[g]				
Mammals (1)	–	–	Cpn10m	Heptameric complex interacts with corresponding proteins of HSP60 family
Plants (1–2)	–	–	Cpn10m, Cpn12c	
Yeast	–	–	?	

Table 1. *Continued*

Organism (No. of genes)	Cellular localization			Functional characteristics[a]
	Cytoplasm/nucleus	ER	Chloroplasts (c)/ mitochondria (m)	
HSP8.5 family (lacking in *E. coli*)[h]				
Mammals (several)	**Ubiquitin**	–	–	Tag for protein degradation by ATP-dependent protease complex
Plants (several)	**Ubiquitin**	–	–	
Yeast (several)	**Ubiquitin**	–	–	

Members of each stress protein family are characterized by their size (apparent molecular weight in kDa), sequence homology and in many cases functional similarity (see last column). The *E. coli* proteins corresponding to the particular stress protein family are indicated in parentheses. For additional information see Georgopoulos (1992); Nover (1991); Nover et al. (1990); Vierling (1991) and references given in footnotes.

[a] Family-specific characteristics are indicated in short. However, in many cases, the allocation of a given stress gene product to the family is based on sequence homology only and not on its functional analysis. Proteins with a pronounced heat stress induction of synthesis are printed in boldface type. Others are constitutively expressed (e.g., HSC72) or their synthesis increases under conditions of glucose deficiency (e.g., GRP78).

[b] Sequencing and characterization of the tomato ClpA/B and the yeast HSP104 led to the definition of a new HSP110 family with a highly variable size (756 to 926 amino acid residues). The mammalian HSP110 is a nucleolar stress protein (for reviews and references, see Nover 1991; Squires and Squires 1992).

[c] See Tai et al. (1992); Georgopoulos (1992).

[d] The GroEL type of stress proteins is localized in the organelles. Cytoplasmic homologs of mammalian plant and yeast cells (TCP-1 proteins) were described recently (Frydman et al. 1992; Lewis et al. 1992; Mori et al. 1992). They form large heterooligomeric ring complexes of 970 kDa which are active in protein folding.

[e] The existence and possible role of eukaryotic representatives of the DnaJ (HSP40) family are just emerging (Blumberg and Silver 1991; Luke et al. 1991; Raabe and Manley 1991; Caplan et al. 1992; Zhu et al. 1993).

[f] Chaperone-like activities were recently described for αB-crystallin (Horwitz 1992) and mouse HSP25 (Buchner et al. 1993). The *E. coli* HSP20 homologs were dected as inclusion body-associated proteins IbpA/B (Allen et al. 1992; Chuang and Blattner (1993) J Bacteriol 175:5242–5252)

[g] See Hartman et al. (1992a,b).

[h] Among the numerous ubiquitin-coding genes of eukaryotes the polyubiquitin genes are stress-induced. Interaction with the HSP110 as part of the high molecular weight protease complex is very likely (Jentsch et al. 1991; Jentsch 1992).

is the remarkable conservation of most parts of the hs response extending from the structure and function of stress proteins to the promoter structure and essential parts of the transcription factor.

2 Heat Stress Promoters

Sequence comparison led Pelham (1982) to the definition of a palindromic 14 nucleotide motif (CTnGAAnnTTCnAG) conserved in the upstream regions of *hsp* genes of *Drosophila*. Introduction of a synthetic oligonucleotide of this type into the herpes simplex virus thymidine kinase promoter gave the first hs-inducible reporter construct tested in a mammalian cell (Bienz and Pelham 1982). The universal significance of this heat stress promoter element (HSE) in practically all eukaryotes was soon recognized by sequencing more and more promoters of hs genes (Bienz 1984, 1986; Schöffl et al. 1984; Kay et al. 1986) and by expression of corresponding reporter constructs in many different systems (for summaries, see Nover 1987, 1991). The only exceptions so far are two hs-inducible genes in yeast (Kobayashi and McEntee 1990, 1993; Wieser et al. 1991) and the hs genes of parasitic protozoa (Glass et al. 1986; Swindle et al. 1988).

The concise analysis of hs promoters resulted in a simplified, more general structure of the HSE, which in fact represents multimers of five nucleotide modules with alternating, palindromic orientation (Nover 1987; Amin et al. 1988; Cuniff et al. 1991; Xiao et al. 1991; Lis and Wu 1992):

–(aGAAn)(nTTCt)(aGAAn)(nTTCt)–

Three consecutive modules are evidently the minimum combination required for an active hs promoter. When taking the ($a_1G_2A_3A_4n_5$) as the basis module, only the G is invariant, whereas the extent of conservation of the A residues decreases from $A_3 > A_4 > a_1$ (Nover 1987, 1991; Lis and Wu 1992).

HSE motifs are frequently found in TATA-proximal positions, but they act equally well as enhancer elements from positions far upstream of the TATA box (Bienz and Pelham 1986; Thomas and Elgin 1988). In appropriate promoter configurations (see the *Drosophila hsp70* promoter in Fig. 1, No. 1) the HSE confers both high selectivity of hs induction and high overall efficiency of transcription initiation. Frequently, additional regulatory motifs have been identified by sequence comparison, deletion and/or footprint analysis (Fig. 1). The complexity of stress promoters reflects the multiplicity of metabolic and developmental situations under which the corresponding genes are expressed. From the examples with typical patterns of regulatory elements in hs-inducible promoters (Fig. 1), three will be discussed in more detail:

1. The *Drosophila hsp70* promoter (No. 1) is the simplest prototype of a hs-inducible promoter. It was widely used for gene constructs expressed in many different eukaryotic cells (see summaries in Nover 1987, 1991). Its high effectivity and selectivity (hs-inducibility) depends solely on the TATA-proximal region with a combination of two HSE motifs (bp −47 to −80). The distal HSE motif centered at bp −180 has no influence. The length of the spacer of 10 bp between the two TATA-proximal HSE motifs is critical (Amin et al. 1987) for the cooperativity of interactions between transcription factors bound to both HSE motifs (Shuey and Parker 1986; Cunniff et al. 1991; Xiao et al. 1991; Lis and Wu 1992).
2. In contrast to the *Drosophila hsp70* gene the promoter of the rat *hsc73* gene is much more complex (No. 2 in Fig. 1). HSC73 is the constitutively expressed, dominant chaperone of the HSP70 family in rat cells. A multiplicity of regulatory elements provides the basis for permanently high gene expression. Full activity with moderate hs-inducibility (about twofold) requires at least 300 bp of the promoter region. Though detailed footprint analyses are lacking, it is tempting to speculate that under hs conditions the HSE motifs function as maintenance elements replacing

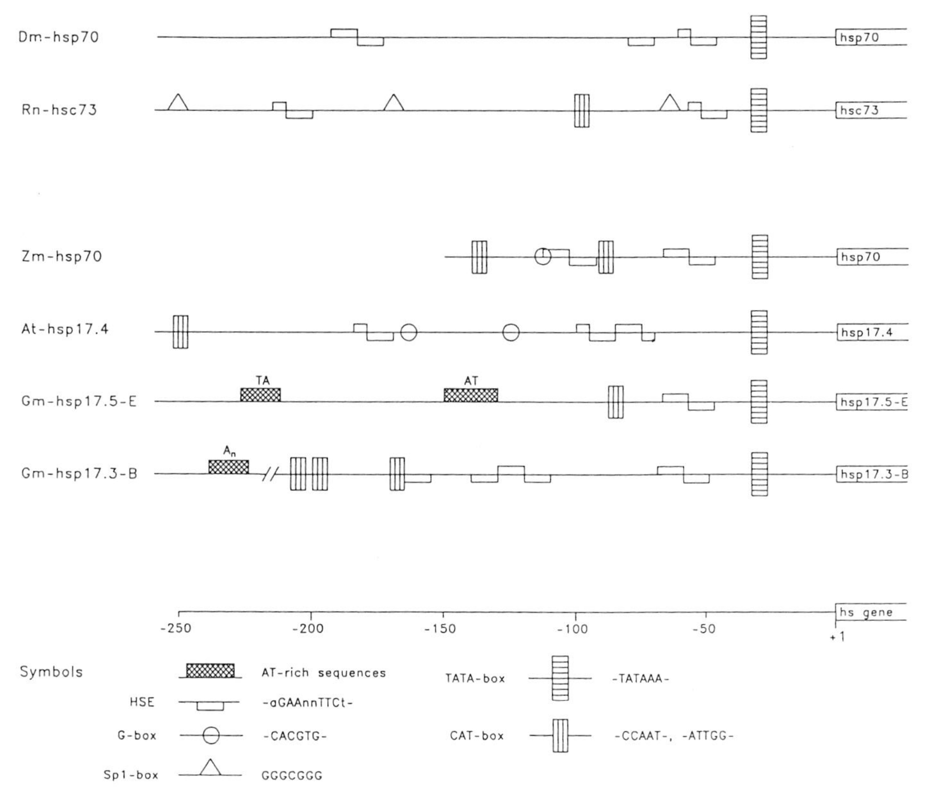

Fig. 1. Basic structure of hs-inducible promoters. Explanations of *symbols* used for regulatory elements with the corresponding consensus sequences are given at the *bottom*. For further information, see text and reviews of Gurley and Key (1991); Nover (1991); and Katagiri and Chua (1992). Sequence information obtained from the following sources: *Drosophila* (*Dm*) *hsp70* (Amin et al. 1987); rat (*Rn*) *hsc73* (Sorger and Pelham 1987); maize (*Zm*) *hsp70* (Rochester et al. 1986); *Arabidopsis* (*At*) *hsp17.4* (Takahashi and Komeda 1989); soybean (*Gm*) *hsp17.5-E* (Czarnecka et al. 1989); *hsp17.3-B* (Baumann et al. 1987)

other regulatory elements, e.g., the Sp1 or CAT box, in the assembly of the transcription complex.

3. A tripartite structure is characteristic of the two soybean *hsp17* promoters (Fig. 1, Nos. 5, 6). TATA-proximal heat stress elements are combined with one to three CAT boxes further upstream and distal A/T-rich, so-called simple sequences. Deletion of the latter results in a considerable decrease in promoter activity without loss of hs-inducibility (Baumann et al. 1987; Czarnecka et al. 1989). Reconstruction of the *hsp17.3-B* promoter by combining various parts with an inactive reporter construct demonstrated that the distal poly(dA) motif and the CAT boxes represent quantitative elements which contribute to the overall efficiency of the promoter. Only one TATA-proximal HSE motif is required for hs-inducibility (Rieping and Schöffl 1992).

3 Cloning of Heat Stress Transcription Factor (*hsf*) Genes

The identification and functional analysis of the eukaryotic heat stress promoter element HSE (see Sect. 2) were rapidly followed by the first evidence for the existence of a HSE-binding protein in *Drosophila* (Parker and Topol 1984a,b; Wu 1984a,b, 1985; Topol et al. 1985; Shuey and Parker 1986). Gel retardation assays and footprint analyses with crude and partially purified fractions were used to characterize the putative heat stress transcription factor (HSF). A critical observation was the reversible activation/deactivation of the *Drosophila* HSF by repeated cycles of stress and recovery in the presence of cycloheximide (Zimarino and Wu 1987). Heat stress treatments or chemical stressors (2,4-dinitrophenol, salicylate) were equally effective for activation. Evidently, under control conditions, the HSF exists in a preformed but inactive state not capable of HSE binding. Similar results were reported for chicken and mammalian cells (Kingston et al. 1987; Sorger et al. 1987; Zimarino et al. 1990) as well as for tomato cell cultures (Scharf et al. 1990).

An important milestone was the cloning of the yeast *hsf1* gene by Sorger and Pelham (1988) and Wiederrecht et al. (1988). Both research groups purified the yeast HSF to homogeneity, followed by the generation of antibodies and immunoscreening of a cDNA expression library (see Table 2). Unfortunately, these yeast *hsf* clones could not be used as probes to screen for similar genes in unrelated eukaryotes. Instead, to obtain the *Drosophila hsf* gene Clos et al. (1990) purified *Drosophila* HSF and used synthetic oligonucleotides derived from microsequencing data as probes for screening a *Drosophila* cDNA library. An alternative route was followed by our group in the search for the plant *hsf* clones. We avoided the possible complications of HSF purification by direct screening for HSE-binding activity (Southwestern screening) in a lambda gt11 cDNA expression library prepared with randomly primed cDNA (Scharf et al. 1990). The remarkable result was the isolation of three different *hsf* genes.

Meanwhile other *hsf* clones were isolated from two other yeasts, and from chicken, man and mouse. Results and references are summarized in Table 2. The functional significance of the *hsf* clones is revealed by conserved sequence elements (see Sect. 4) and a number of activity tests performed with derived recombinant proteins (Table 2). The capability for specific DNA binding can be tested by footprinting (FP), retardation assays (RA) and Southwestern blot (SW) analysis. HSE-dependent transcription activity was studied using in vitro transcription systems (IT) and *trans*-activation (TA) assays (see Sect. 6). Particularly interesting is a gene replacement test (GR) with *hsf*-defective yeast strains. In addition to *Saccharomyces* and *Kluyveromyces hsf* clones, two of the three tomato clones (*hsf8* and *hsf30*) and the *Drosophila hsf* can substitute for the defective chromosomal *hsf1* gene of *Saccharomyces cerevisiae* (Scharf et al. unpubl.).

Table 2. Survey of HSF

Organism (HSF)	Gene cloning procedure	ORF (codons)	Mr (kDa)	MW (kDa)	mRNA (kb)	Activity assays[a] FP	RA	SW	IT	TA	GR	Reference[b]
Yeasts												
Saccharomyces cerevisiae (Sc-HSF1)	HSF purification, antibody screening	833	130	93.2		–	–	–	–	+	+	a
						+	–	+	–	+	+	
Kluyveromyces lactis (K1-HSF)	Cross-hybridization with Sc-HSF1-derived oligonucleotide	677		75.4		–	–	–	–	–	+	b
Schizosaccharomyces pombe (Sp-HSF)	Cross-hybridization with degenerate primers	609	108			–	–	–	–	–	–	c
Plant												
Lycopersicon peruvianum (tomato)	Southwestern screening with HSE concatemer											d
(Lp-HSF8)		527	70	56	2.0	–	+	+	–	+	+	
(Lp-HSF24)		301	n.d.	33.3	1.5	–	+	+	–	+	–	
(Lp-HSF30)		351	n.d.	39	1.4	–	–	+	–	–	+	
Insect												
Drosophila melanogaster (Dm-HSF)	HSF purification, microsequencing, oligonucleotide screening	691	110	77.3		–	+	–	+	–	+	e
Vertebrates												
Homo sapiens (human) (Hs-HSF1)	PCR amplification with degenerate primers derived from Dm-HSF	529	83	57.3	2.0	+	+	–	+	–	–	f
(Hs-HSF2)	HSF purification, microsequencing, oligonucleotide screening	536	87	60.3	2.7	+	+	–	–	–	–	g

Table 2. *Continued*

Organism (HSF)	Gene cloning procedure	ORF (codons)	Mr (kDa)	MW (kDa)	mRNA (kb)	Activity assays[a]						Reference[b]
						FP	RA	SW	IT	TA	GR	
Mus musculus (mouse)												
(Mm-HSF1)	Cross-hybridization with Hs-HSF1 fragment	503	75	54.8	2.0	+	+	–	–	–	–	h
(Mm-HSF2)		517	72	58.2	2.1	+	+	–	–	–	–	
Gallus domesticus (chicken)												
(Gd-HSF1)	Cross-hybridization with mouse HSF clones	491	65		1.9	–	+	–	–	–	–	i
(Gd-HSF2)		564	80		2.8	–	+	–	–	–	–	
(Gd-HSF3)		467	64		4.5	–	+	–	–	–	–	

[a] To identify isolated clones, tests for activity were performed in vitro with recombinant or native HSF or in vivo (transactivation assays): FP, footprint analysis (DNase, methylation interference); RA, gel retardation assay; SW, Southwestern assay; IT, in vitro transcription; TA, transactivation assay in transgenic yeast or transformed tobacco protoplasts; GR, replacement of the defective yeast *hsf1* gene by the *hsf* genes or cDNA from homologous or heterologous sources.

[b] References are: a, Sorger and Pelham (1988), Wiederrecht et al. (1988); b, Jakobsen and Pelham (1991); c, Gallo et al. (1993), d, Scharf et al. (1990, 1993); e, Clos et al. (1990); f, Rabindran et al. (1991); g, Schuetz et al. (1991); h, Sarge et al. (1991); i, Nakai and Morimoto (1993).

4 Characteristics of Transcription Factor Clones and Proteins

Results of sequence analysis of 14 *hsf* genes or cDNA clones of yeasts, *Drosophila*, chicken, mammals and plants are summarized in Fig. 2 and Table 2. Conserved parts in all HSF are the DNA-binding domain in the N-terminal half and, downstream from it, characteristic patterns of two to three hydrophobic heptad repeats of the Leu-zipper type. The C-terminal activation domains are highly variable, even when HSFs of the same organism, e.g. of tomato, chicken, human or mouse, are compared.

4.1 The DNA-Binding Domain

As to be expected from the conservation of the HSE motif and the effective function of various hs promoter constructs in heterologous expression systems (Nover 1987, 1991), all HSF contain similar DNA-binding domains of about 93–100 amino acid residues (Fig. 3). This new type of DNA-binding domain was first defined by Wiederrecht et al. (1988) analyzing deletion clones of the yeast *hsf* gene. Interestingly, our direct approach to cloning the tomato *hsf* genes by Southwestern screening (see Sect. 3) resulted in the identification of a 380-bp fragment of the *hsf24* gene. It represents a minimum clone coding for the 93 amino acid residues of the DNA-binding domain plus a few additional residues of the C-terminal part (Scharf et al. 1990).

Though the functional significance remains to be established, a number of structural features of the DNA-binding domain of eukaryotic HSF are noteworthy (Fig. 3):

1. Three conserved boxes (I to III) are separated by two less conserved regions of 15 and 2–14 amino acid residues, respectively. Brackets in Fig. 2 indicate the position of a duodecapeptide motif in the most conserved part of box II which has an intriguing homology to motifs from the DNA-binding regions of two *E. coli* sigma factors, sigma 70 and sigma 32 (Stragier et al. 1985; Clos et al. 1990) as shown below:

HSF	$\mathrm{V\ R\ A\ L\ N}^{\mathrm{T}}_{\mathrm{M}}\mathrm{Y\ G}^{\mathrm{F\ R}}_{\mathrm{WH}}\mathrm{K}^{\mathrm{I}}_{\mathrm{V}}$	
sigma 70	$\underline{\mathrm{I\ R}}\ \mathrm{Q}\ \underline{\mathrm{I}}\ \mathrm{E\ A\ K\ A\ L}\ \underline{\mathrm{R\ R\ L}}$	(599)
sigma 32	$\underline{\mathrm{V\ R}}\ \mathrm{Q}\ \underline{\mathrm{L}}\ \mathrm{E\ K\ N\ A\ M}\ \underline{\mathrm{K\ K\ L}}$	(279)

2. The net charge is positive with a concentration of basic amino acid residues in the central box II and at the C-terminus of box III (see Fig. 3B).
3. Structural predictions for all HSF indicate three short regions with potential for α-helix formation, whereas the N-terminus (box I) and the

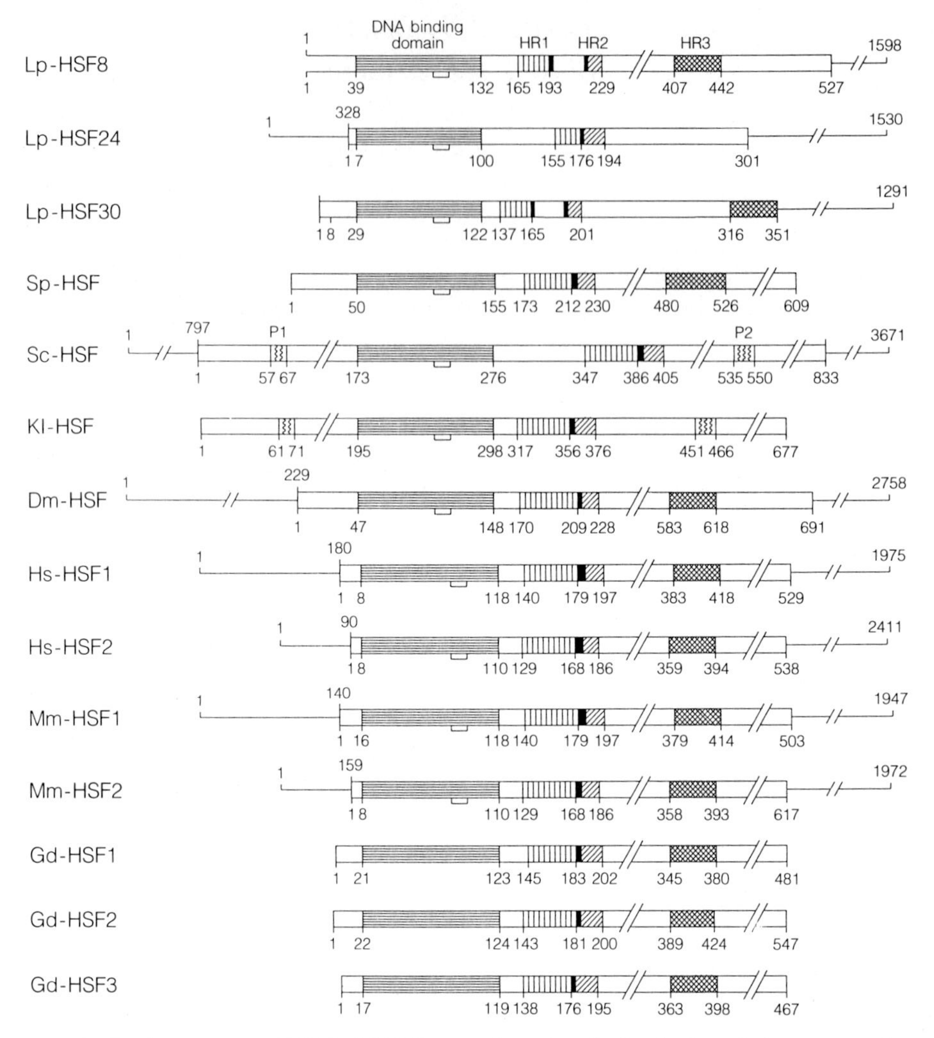

Fig. 2. Basic structure of HSF. All HSF are aligned at the beginning of the conserved DNA-binding domain (for sequences, see Fig. 3). Other conserved parts are the heptad hydrophobic repeats (*HR1*, *HR2* and *HR3*; for sequences, see Fig. 4), the Q-rich linker between HR1 and HR2 (*black bars*) and short peptide motifs (*P1*, *P2*) found only in the two yeast HSF. The *bracket* under the DNA-binding domain denotes the region with homology to the *E. coli* sigma factor (see Sect. 4.1). For references to sequence data and for explanation of the two-letter code for organisms, see Table 2

most highly conserved part in box II are dominated by β-sheets (Fig. 3B).

4. There are 21 invariant positions (capital letters of consensus sequence) and 12 additional positions with maximum two deviations among the 14 HSF sequenced so far (lowercase letters). Ten out of these 33 most

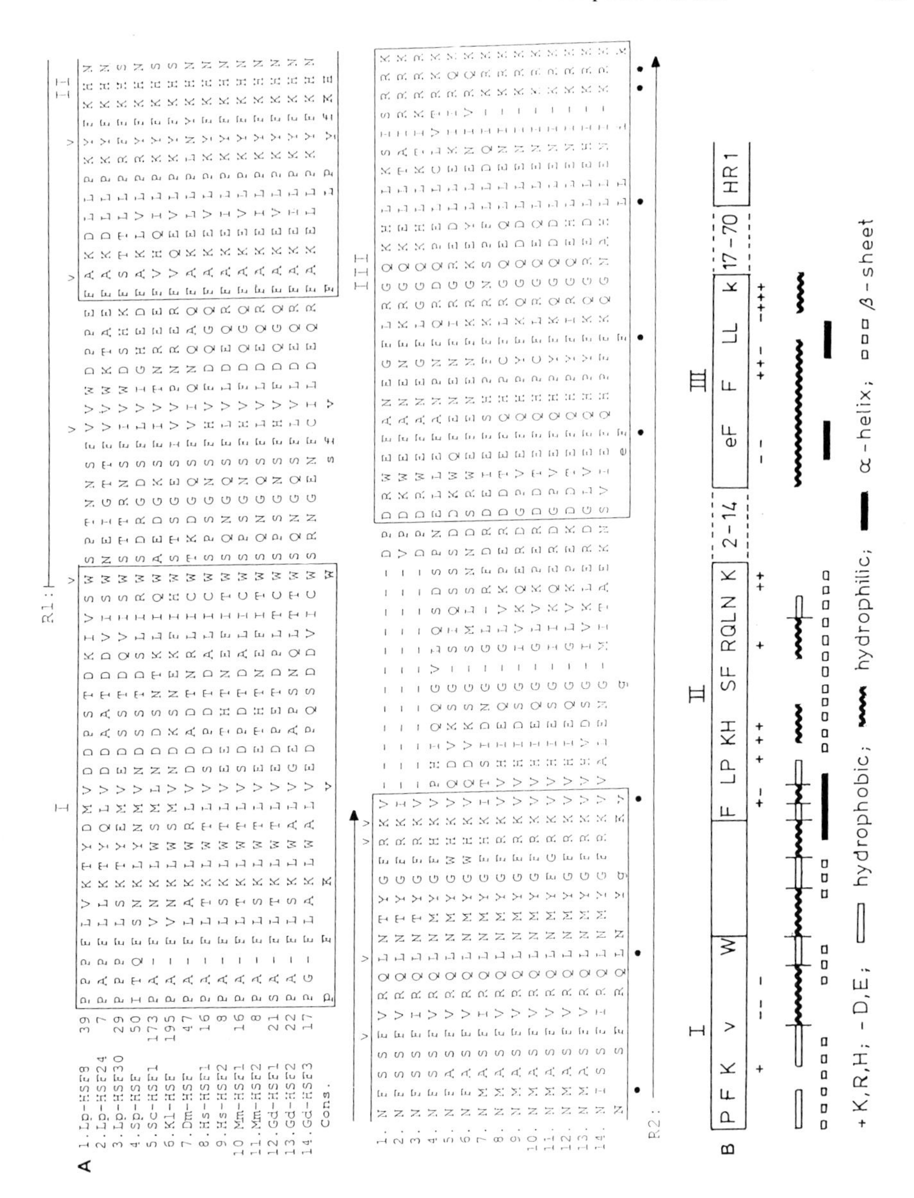

Fig. 3. Sequence comparison (*above*) and structural features (*below*) of the DNA-binding domain. The three most conserved boxes *I–III* are separated by two less conserved regions. Invariant (*capital letters*) and highly conserved residues (*lowercase letters*) are marked in the consensus sequence. *Long arrows* (*R1* plus *arrowheads*; *R2* plus *dots*) emphasize repeat patterns of hydrophobic, aromatic and basic amino acid residues (see text). The block diagram (*below*) represents structural features with charge distributions, hydrophilicity and predictions of secondary structure (explanations are given at the bottom). For better orientation the conserved boxes *I–III* and the beginning of the first hydrophobic repeat (*HR1*) are indicated (see details given in Fig. 1)

conserved positions are occupied by aromatic amino acid residues (F, Y, W). Some of the highly conserved or invariant hydrophobic amino acid residues of the DNA-binding domain form a pattern of two overlapping repeats, each ending with two basic residues (see arrows in Fig. 3A):

Repeat 1:	W(7X)	V (7X)	A (6X)	Y (5X)	F (6X)	L (5X)	R	K
		I	V	F	M			H
Repeat 2:	F (6X)	L (7X)	V (6X)	F (4X)	F (6X)	L (5X)	R	K
	M		I				Q	R

A similar density and pattern of aromatic amino acid residues are also found in the DNA-binding domain of other classes of transcription factors, e.g., the yeast flocculent suppressor (Fujita et al. 1989) or representatives of the class of fork-head proteins (Weigel and Jäckle 1990; Grossniklaus et al. 1992). Aromatic interactions may be crucial to form the stable hydrophobic cores of DNA-binding domains, an effect well documented for steroid hormone receptors (Burley and Petsko 1985; Härd et al. 1990; Schwabe et al. 1990; Luisi et al. 1991).

4.2 Leucine Zipper-Type Hydrophobic Repeats

Irrespective of their particular receptor element, activity and/or binding specificity of many transcription factors depend on the formation of homo- or heterooligomers. A universal oligomerization element was defined by Landschulz et al. (1988). It is created by a repetitive heptad pattern of leucine residues, which in the α-helical conformation are positioned on the same site of the protein domain. Interaction of two or three structures of this kind forms a stable hydrophobic pocket (leucine zipper). Detailed investigations of this interesting type of oligomerization domain revealed that the helices are wrapped around each other in a coiled-coil structure (Hu et al. 1990; O'Shea et al. 1991; Rasmussen et al. 1991; Lovejoy et al. 1993).

For all HSF, stress-dependent regulation of activity and high affinity binding to DNA are intimately connected with the formation of oligomers (see Sect. 7.2). A conserved heptad repeat pattern of hydrophobic amino acids (HR1, HR2) is found downstream of the DNA-binding domain (Fig. 4). The extended HR1 region is hydrophilic. It is separated by a Q-rich linker from the short hydrophobic HR2, which is actually a short-nested double zipper with a central invariant phenylalanine. With the exception of two yeast HSFs and the tomato HSF24, a third hydrophobic repeat (HR3) is found in the C-terminal parts of all other HSF sequenced so far (Figs. 2 and 4). Both, HR1 and HR2 belong to the class of Leu-Zippers with a regular pattern of hydrophilic residues (K, R, Q, D, E, S) in the fourth position (see dots in Fig. 4). There is evidence that hydrogen bonds between the polar side chains and the peptide carbonyls of the other subunit as well

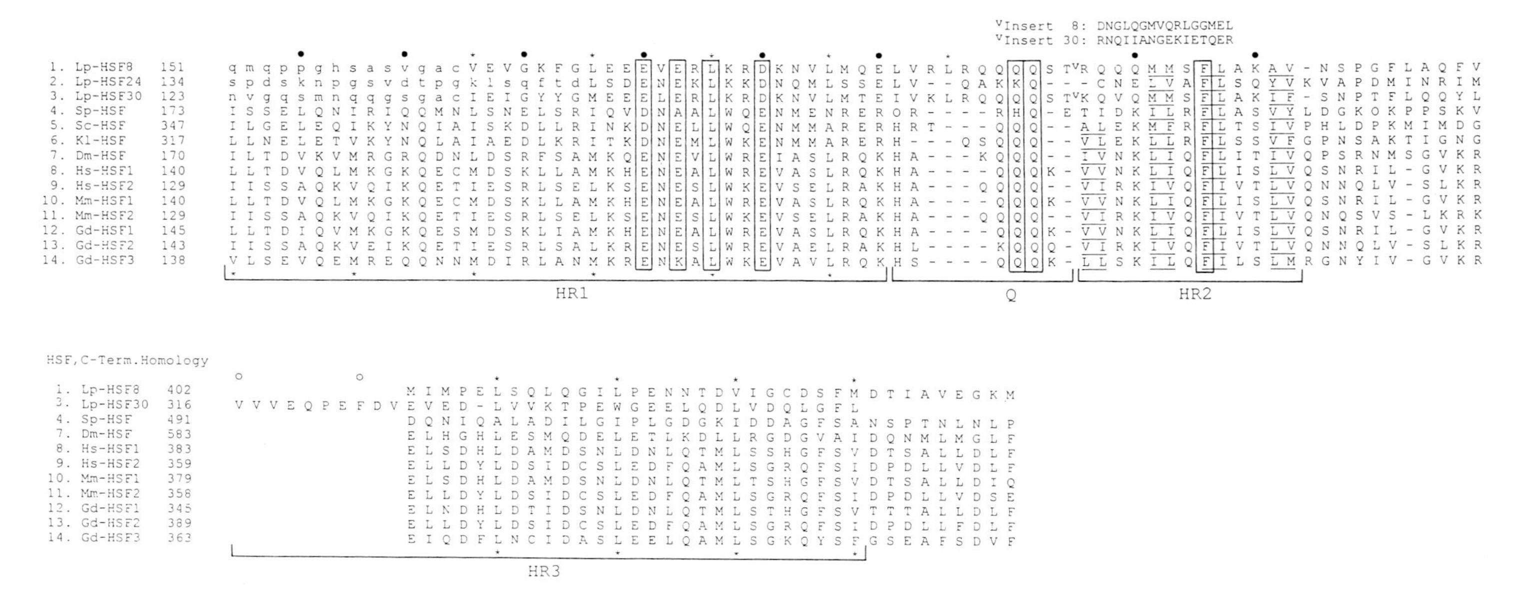

Fig. 4. Sequence comparison of the hydrophobic repeats HR1–3. Two heptad repeats of hydrophobic amino acids (*stars* in *HR1*, *underlined residues* in *HR2*) are separated by a Q-rich motif. In both cases the fourth position is occupied by hydrophilic residues (*dots*). In addition, most HSF contain a third hydrophobic repeat (*HR3*) in their C-terminal domains (see Fig. 1). Highly conserved or invariant amino acid residues in the HR1/HR2 part are *boxed*

as charge interactions may be decisive for the stability and specificity of oligomerization mediated by two Leu-zipper-type hydrophobic repeats (Paluh and Yanofsky 1991; O'Shea et al. 1991; Schuermann et al. 1991; Tropsha et al. 1991).

In contrast to many other transcription factors with hydrophobic repeats of the Leu-zipper type, the HSFs evidently form trimeric complexes upon hs activation. This is well documented for HSF from yeast (Sorger and Nelson 1989; Peteranderl and Nelson 1992), *Drosophila* (Clos et al. 1990; Rabindran et al. 1993) and man (Baler et al. 1993). Recently, crystal structure analysis of a synthetic 33-residue peptide, derived from the GCN4 Leu-zipper domain, revealed such a triple-stranded coiled-coil structure with helices oriented up-up-down (Lovejoy et al. 1993).

4.3 The C-Terminal Activation Domain

Sequence conservation in the C-terminal parts of HSF is low. There is a general enrichment of sequence elements with T, S, P, Q, N, G residues, indicating a high flexibility in this region of the molecule. Many HSF contain a third hydrophobic heptad repeat (Fig. 2, HR3). Extended parts with dominant negative charges (E, D residues) are generally lacking except in the tomato HSF8 and HSF30. Attempts to define functional parts important for activity control of the tomato HSF will be discussed in Section 6.2. However, it is worthwhile summarizing C-terminal sequence elements shared between HSF8 and HSF30 but lacking in HSF24 (Fig. 5). The C-terminal part of HSF24 contains no HR3 but two short clusters of acidic (−) and basic (+) residues as well as a subterminal tryptophan (W) motif (see Sect. 6.2). In contrast, the C-terminal parts of the other two tomato HSF are intriguingly complex but similar. It appears that HSF30 is a structurally related, smaller and hs-expressed version of HSF8. Common to both are the additional hydrophobic repeat (HR3), a short basic cluster (+) upstream of HR2 and an extended acidic region (−) towards the C-terminus. The whole C-terminus is covered by a patchwork of conserved positions starting with HR1. Three homology patterns (A, B, C) can be defined. The larger size of HSF8 is mainly due to an insertion of 100 amino acid residues between region A and HR3 and to a C-terminal extension. Regions B and C are part of the activity control elements, each characterized by a conserved W-residue (see Sect. 6.2). Remarkably, the C-terminal extension of HSF8 is related to the C-terminus of the *Drosophila* HSF:

Lp-HSF8 (478)- P S (2X) D (6X) I (3X) SE (6X) G (4X) Q (7X) Q (3X) T (X) I (3X) KH (X) I*
Dm-HSF (643)- P S (2X) E (6X) M (3X) SD (6X) G (3X) Q (7X) Q (3X) S (X) L (3X) RH (X) L*

The significance of this homology is unclear. At least, deletion of this C-terminal part does not impair the function of HSF8 as hs-inducible activator. (Treuter et al. 1993).

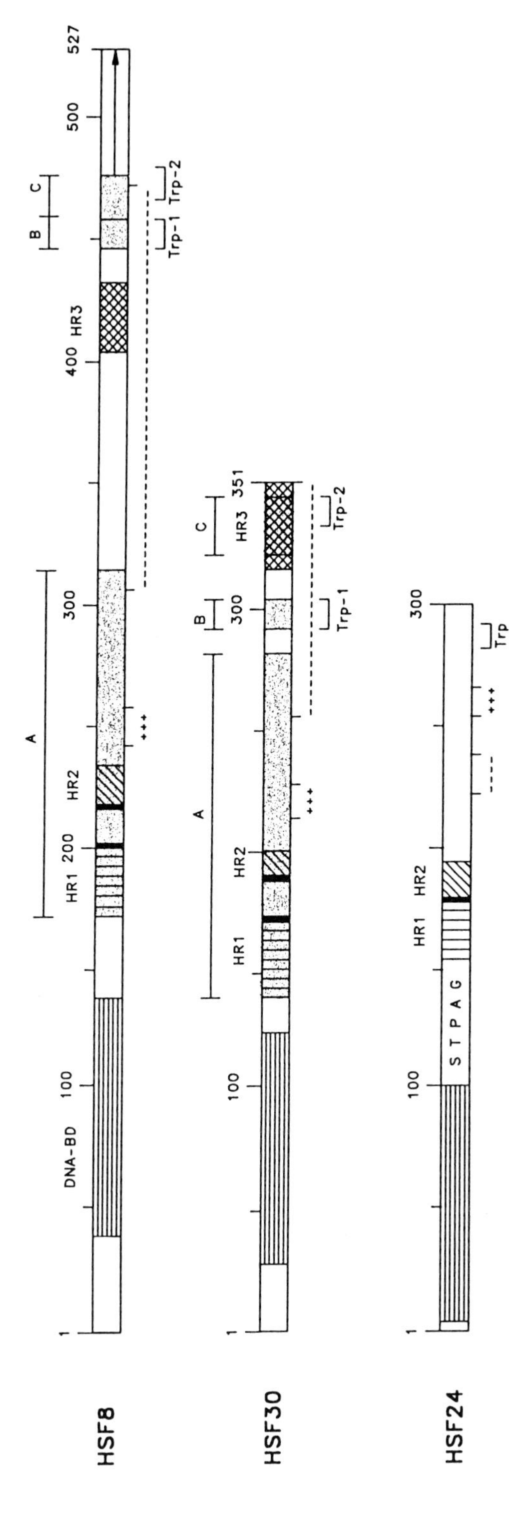

Fig. 5. Details of structural elements in the C-terminal activation domains of tomato HSF. For explanations, see Figs. 2–4. In addition to the DNA-binding domain and the heptad hydrophobic repeats, *HSF8* and *HSF30* are characterized by a pattern of conserved amino acid residues (*A*, *B*, *C*). *Trp-1*, *Trp-2* are the activity control elements defined by deletion analysis (Fig. 9), which are part of boxes *B* and *C*. Charged regions are indicated by + for basic amino acid residues (K, R, H) and − for acidic residues (D, E), respectively

5 Stress-Induced Expression of Tomato HSFs

The surprising multiplicity of *hsf* genes in tomato is connected with hs-inducibility of two of them, HSF24 and HSF30 (Scharf et al. 1990). Thus, in contrast to all other systems studied so far, the changing composition of the hs transcription complex may contribute to the regulatory fine tuning of stress gene expression in tomato. Using different hs regimes and comparing levels of HSF mRNAs with those of two standard heat stress proteins, i.e., of HSP70 and HSP17, the phenomenon is documented in Fig. 6 (Materna unpubl. results). Figure 6A contains results on the influence of temperature and time, whereas Fig. 6B presents changing mRNA levels during a long-term hs and recovery regime. As described earlier (Nover et al. 1986, 1989), this preinduction protocol is connected with the generation of a thermotolerant state. The following results are worth noting:

1. No HSF30 mRNA is detectable under control conditions. Its level increases rapidly but transiently at 33–35 °C. This increase is much more pronounced and permanent at 37–39 °C. From these expression data HSF30 appears to be a high temperature HSF.
2. The low level of HSF24 mRNA at 25 °C rises during heat stress with a temperature optimum of 35–37 °C. The response to a direct, severe hs (39 °C) is reduced (Fig. 6A) unless preceded by a preinduction period (Fig. 6B).
3. In contrast to the former results, the amount of mRNA of the constitutively expressed HSF8 declines at the beginning of a hs treatment but recovers relatively soon at 33–35 °C (Fig. 6A). Recovery at high temperatures is delayed but can be improved in thermotolerant (preinduced) cells (Fig. 6B).
4. If the time scales of mRNA accumulation and decay are compared (Fig. 6A,B), a sequence from HSF30 (early) to HSF24 (middle) to HSP17/HSP70 (delayed) can be derived. Though it is tempting to speculate on a sequence of inductive events during the stress response, much further work is needed before any conclusions can be drawn. One important problem is the lack of information concerning HSF protein levels. Despite the relatively high levels of mRNAs, there are evidently only trace amounts of the corresponding proteins present in tomato cell cultures (unpubl. observ.). Thus, the possibility cannot be ruled out that the strong hs-induced increase in HSF30 and HSF24 mRNA levels is not connected with corresponding changes in the protein levels.

6 Functional Analysis of *hsf* Clones in Tobacco Protoplasts

6.1 *Trans*-Activation vs. *Trans*-Repression Assays

The functional characterization of tomato *hsf* clones is based on cotransformation of *Nicotiana plumbaginifolia* (tobacco) protoplasts with *hsf* ex-

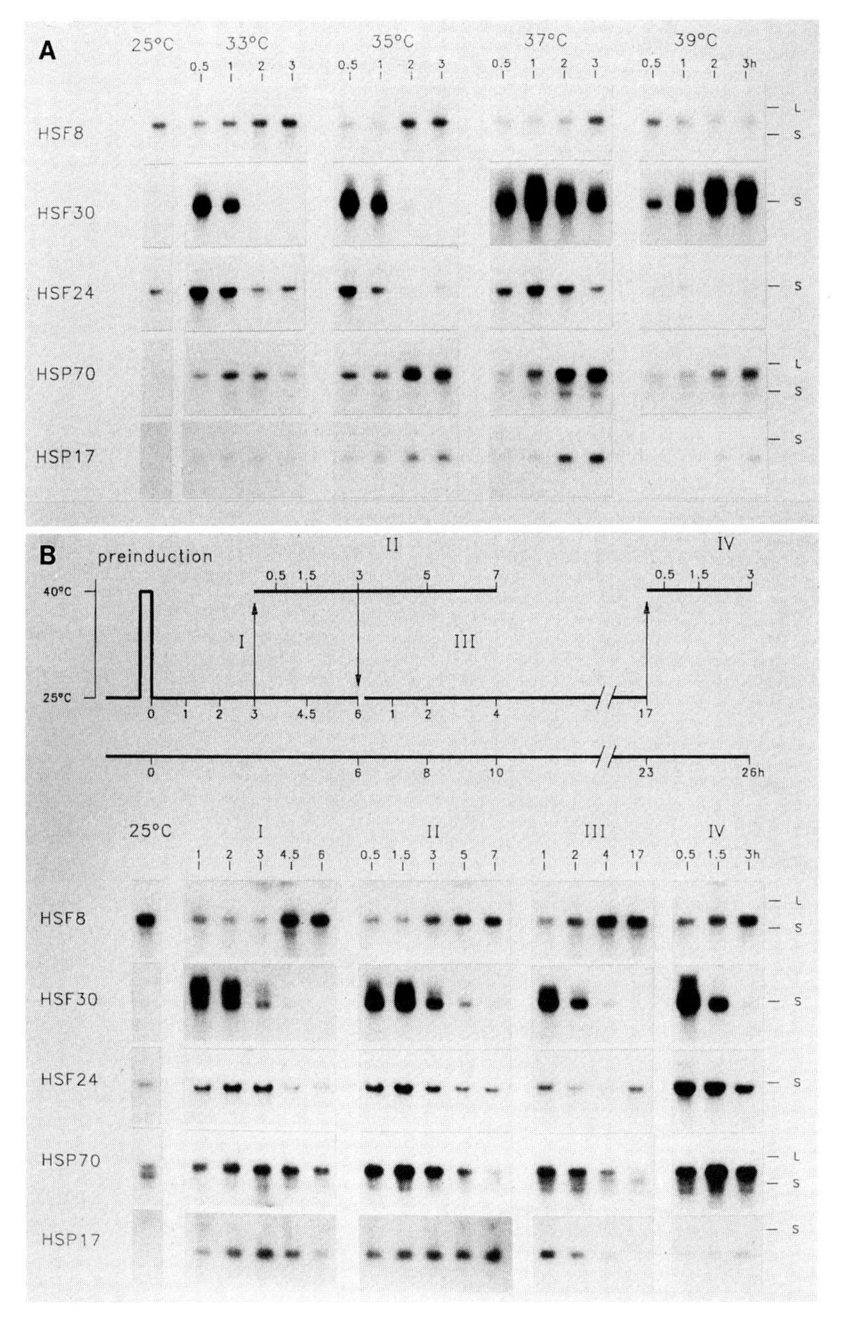

Fig. 6. Northern analysis of HSF and HSP mRNA levels (T. Materna, unpubl.). Total RNA was prepared from tomato cell cultures after the indicated hs treatments. For detection of the transcript levels Northern blots were hybridized with the indicated cDNA fragments. For comparison, HSP70 and HSP17 mRNAs were included in the analysis. For size markers, see *bars* in the right-hand margin for the large (*L*) and small (*S*) rRNA. Transcript sizes are (kb): 2.0 for HSF8, 1.4 for HSF30, 1.5 for HSF24, 2.2 for HSP70 and 0.9 for HSP17. Using 10^6 cpm/ml of phosphate-labeled probe for the hybridization, the exposition time was about 2 days for all probes except HSF8 which required about a tenfold longer exposure. **A** Transcript levels at different temperatures; **B** changing transcript levels in the course of a 26-h heat stress and recovery regime (see pictograph for details of temperature shifts)

A: HSF expression plasmid

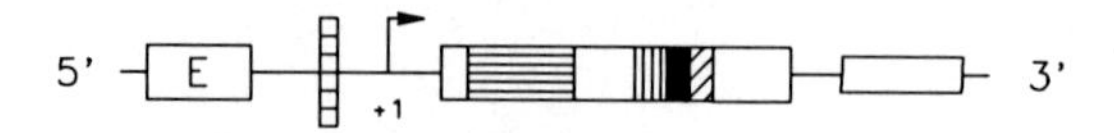

B: Constitutive gus reporter

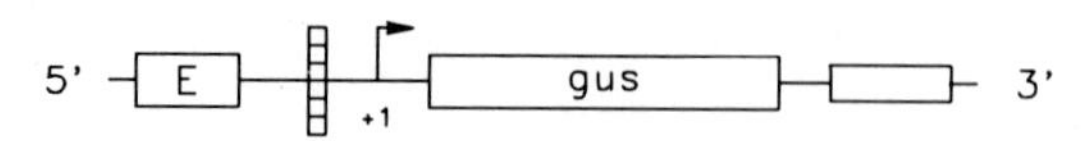

C: HS-inducible gus reporter
(Activator test construct)

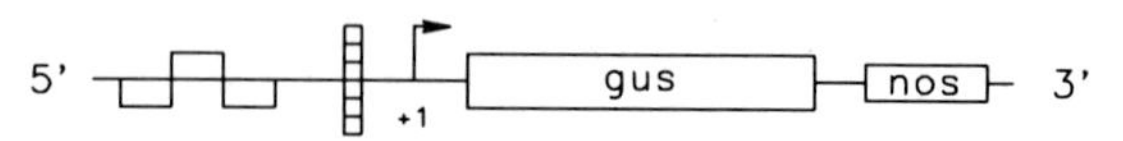

D: HS-repressible gus reporter
(Repressor test construct)

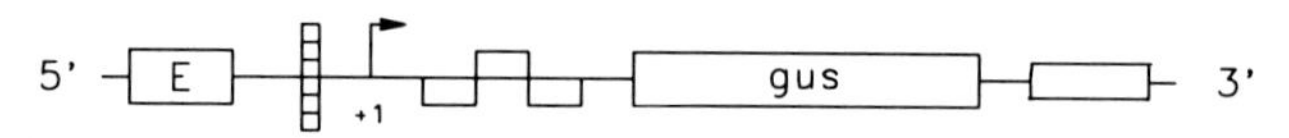

Symbols

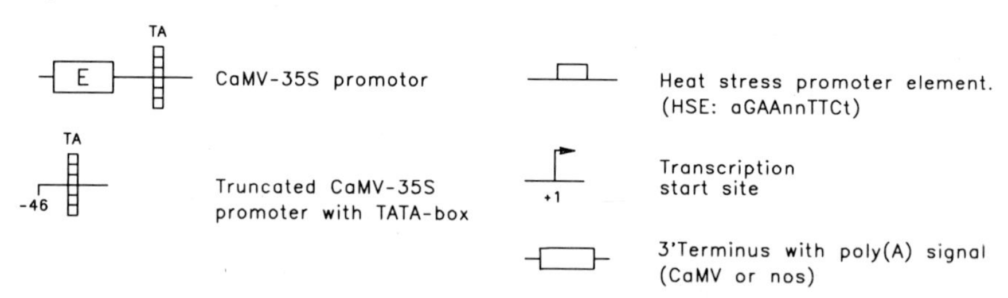

Fig. 7. Expression constructs for hsf cDNA clones and gus reporter constructs. The hsf cDNA (**A**) or the bacterial β-glucuronidase gene (gus, **B–D**) are ligated up to the consitutively active cauliflower mosaic virus 35S promoter (**A, B, D**) and to a heat stress-inducible promoter, respectively (**C**). In the repressor test construct both elements, the CaMV 35S enhancer upstream and heat stress elements downstream of the TATA box, are combined. For details of construction, see Töpfer et al. (1988) and Treuter et al. (1993)

pression plasmids and appropriate reporter constructs containing the bacterial β-glucuronidase gene (Fig. 7). High, constitutive expression of the *hsf* cDNA is achieved by combination with the cauliflower mosaic virus 35S promoter (Fig. 7A). The hs-inducible reporter construct contains a truncated 35S-promoter without its enhancer element but with heat stress elements inserted

Table 3. *Trans*-activation and *trans*-repression assay with the three tomato HSF clones. (Data from Treuter et al. 1993)

Reporter constructs		HSF expression plasmids (endog.)	HSF24	HSF8	HSF30
Activator test		A	B	C	D
A1 (gus)	25°C	110	3400	4100	1380
	40°C	1800	9200	12000	2050
A2 (gus)	25°C	350	1060	260	450
	40°C	390	1030	270	480
Repressor test					
R1 (E, gus)	25°C	22400	10750	2100	1300
	40°C	18600	3360	1600	400
R2 (E, gus)	25°C	30000 ± 3000			
	40°C				

Pictographs on the left show the basic structures of the gus reporter constructs with HSE insertions upstream (activator) and downstream (repressor) of the TATA box, respectively. For construction of the inactive control plasmids A2, R2 mutant HSE oligonucleotides (nTAAnnTTAn) were used for insertion instead of the wild-type HSE (nGAAnnTTCn). Data represent the relative GUS activities of protoplast preparations transformed with the indicated reporter plasmids only (endog.) or in presence of HSF24, HSF8 or HSF30. For heat stress induction protoplast samples were treated two-times 2 h at 40 °C during the 24-h cultivation and gene expression period (for details, see Treuter et al. 1993).

upstream of the TATA box (activator test construct, Fig. 7C). In addition, a repressor test construct (Fig. 7D) was created by introducing a HSE oligonucleotide downstream of the constitutively expressed CaMV-35S × *gus* reporter construct (Fig. 7B).

The transient expression assay with this tobacco protoplast system fulfills two important requirements. (1) The PEG-mediated, direct transformation (Krens et al. 1982) leads to a transient high copy situation with respect to the reporter gene (about 10^7 plasmids/cell in the transformation mix). Even if only a fraction of this is taken up and engaged in *gus* gene transcription, the endogenous level of HSF will not be sufficient to provide full activity. Hence, *trans*-activation can be achieved by cotransformation with exogenous HSF expression plasmids. (2) The hs-inducibility is maintained after protoplast preparation, PEG treatment and cultivation of the protoplasts. This is true for transformation with the reporter construct alone and also for cotransformation with reporter plus HSF expression constructs (Table 3).

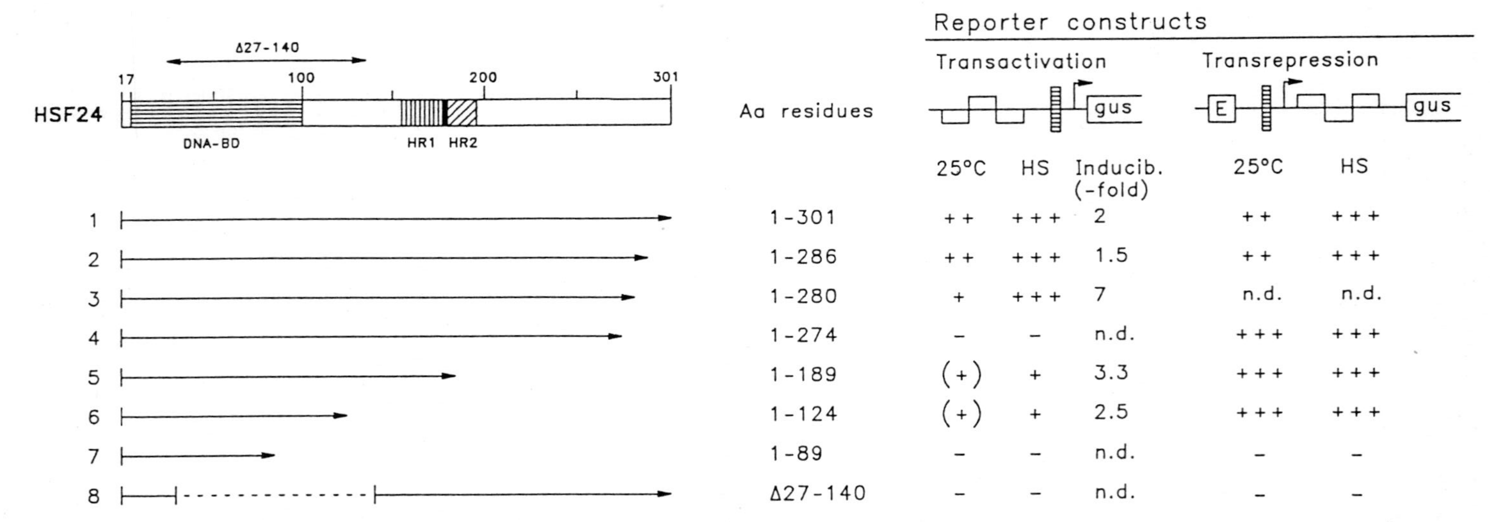

Fig. 8. Deletion analysis of HSF24 (for details, see Treuter et al. 1993). The basic structure of the full length HSF24 is indicated by the block diagram (*above*). All deletions are referred to in it. On the *right*, test results of the transactivation and transrepression assays are summarized: +++, ++, +,– symbolize strong, medium, low and lacking activity, respectively; *n.d.* not determined

The reduction in the *gus* gene expression in the repressor assay results from HSF binding downstream of the TATA box. Initiation complex formation by RNA polymerase is blocked. The reasons for creating these seemingly artificial constructs are considerations of the three functional parts of HSF described in Section 4. Deletion clones or mutants generated for a detailed characterization of the C-terminal domain may be inactive in the *trans*-activation assay. However, because the DNA-binding domain is not affected, they should be active as a repressor. Hence, the repressor assay can be used as a valuable control for the proper expression and/or stability of truncated or mutant HSF (see Fig. 8).

However, in addition to this function in the transient expression assay, the particular regulatory situation of the repressor construct may have biological significance for an important aspect of the hs-induced reprogramming of gene expression, i.e., the shutoff of transcription of housekeeping genes. In support of this, Westwood et al. (1991) reported on the localization of *Drosophila* HSF in polytene chromosomes of heat-shocked larvae. Not only the well-known hs-inducible puffs were stained by HSF antibodies, but many other chromosomal bands were stained as well. Some of them are well-known loci of hs-repressed housekeeping and developmental genes. Though detailed sequence data are still lacking for definitive proof, it is tempting to speculate that, depending on the position in the promoter, the HSF/HSE complexes act as activators or repressors of gene expression. A comparable dual role was described for the papillomavirus transcription factor BPV1E2 (Dostatni et al. 1991).

Together with a considerable number of additional modifications of the prototypes, the four basic types of expression and reporter constructs (Fig. 7) were used for a detailed analysis of the tomato *hsf* clones (Treuter et al. 1993). Results are exemplified in Table 3. The data represent relative activities of β-glucuronidase of protoplast preparations transformed with the *gus* reporter plasmids indicated on the left and with the HSF expression plasmids indicated at the top. For control of the specificity each type of reporter construct having a functional heat stress element (A1, R1) is complemented with an inactive mutant construct with a nonfunctional HSE (A2, R2, see legend to Table 3). There is a low basal level expression of *gus* with the hs-inducible activator test construct A1 alone cultivated at room temperature (25 °C), but this construct can be induced about 18-fold by heat stress treatment (40 °C). The GUS activity increases from 110 to 1800 relative units. This effect is due to the endogenous HSF of the tobacco protoplasts. Cotransformation with any one of the three HSF expression plasmids leads to a marked increase in GUS activities both under control and hs conditions. A maximum of 12 000 relative units of GUS is observed with the HSF8/A1 combination. In contrast, transformation with the mutant activator test construct A2 gives a low level of GUS expression, which is not influenced by heat stress nor by cotransformation with HSF expression plasmids.

Evaluation of the repressor situation (lower part of Table 3) starts best with the very high levels of GUS activities, i.e., with the 20–30 000 relative units found with the mutant construct R2 or with the repressor construct R1 without additional HSF. Cotransformation with or without subsequent hs treatments leads to a considerable reduction. Maximum repression to <5% is observed with the HSF30/R1 combination. Similar to the results with A2, the mutant repressor construct R2 does not respond to hs treatments nor to contransformation with HSF expression plasmids.

6.2 Deletion Analysis of *hsf* Clones

Sequence comparison, as discussed in Section 4, is useful to obtain first hints on conserved parts which might have functional significance. The subsequent creation of truncated and/or mutated HSF and the analysis of their activity in the transient expression assay are required to test these predictions. Results with all three tomato *hsf* clones are summarized in Figs. 8 and 9 (for details of the experimental data, see Treuter et al. 1993). The procedure with the dual testing of deletion constructs in the *trans*-activation and *trans*-repression situation is exemplified for a collection of *hsf24* deletions (Fig. 8). Transition between the fully active and inactive factor is found in a short C-terminal region between amino acid residues 274 and 286. All shorter versions are inactive as *trans*-activator but function perfectly as repressor unless truncations extend into the DNA-binding domain. These results confirm the functional separation of DNA-binding (N-terminal) and *trans*-activation domains (C-terminal). In addition, the repressor assay serves as control for the proper expression and stability of the truncated HSF.

The results with HSF30 are basically similar. Deletions within a small region (residues 304 to 292) lead from fully active to inactive HSF. In contrast, a much more extended part of HSF8 comprising 100 amino acid residues (aa 491 to 396) influences the activator function. If the data for all three HSF are combined (Fig. 9), short peptide motifs with 13 amino acid residues and a central tryptophan residue (Trp-1 and Trp-2 elements) are recognized. In HSF8 these motifs are accompanied by two repeats (R-1, R-2). The Trp elements of HSF8 and HSF30 are negatively charged and probably form short helical regions, whereas the single, somewhat distorted Trp element of HSF24 is embedded in a positively charged region. This summary of the deletion analysis of tomato HSF clones leads to a unifying concept on the role of short activator motifs centered around Trp residues. The functional anatomy is currently being tested with appropriate mutants.

7 Survey of the HSF World 1993

7.1 Multiplicity and Selectivity of Heat Stress Transcription Factors

Though the induction of HSP synthesis usually appears as a coordinate response to the elevation of temperature or the addition of chemical stressors,

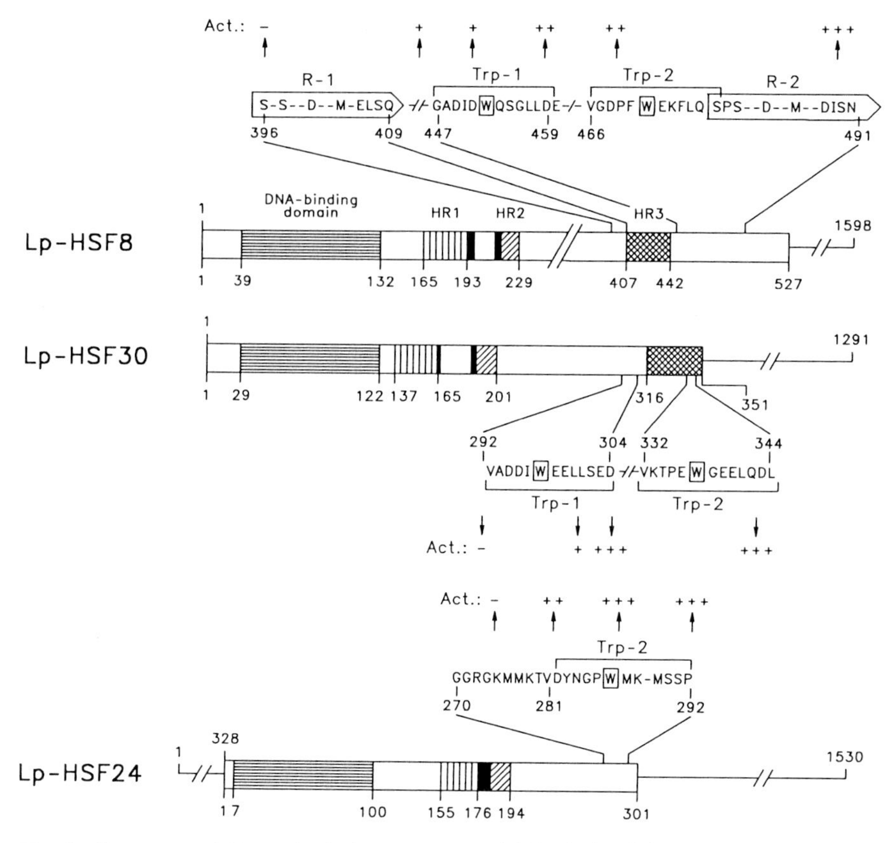

Fig. 9. Summary of C-terminal elements essential for *hsf* activity. Results are derived from experimental data with corresponding deletion clones (see example in Fig. 9; Treuter et al. 1993). The short sequence elements marking the transition between clones fully active (+++) and inactive (−) in the transactivation assays are positioned with *arrows* on the corresponding block diagrams. *Trp-1*, *Trp-2*, acidic elements with a central tryptophan residue (*W*); *R-1*, *R-2* terminal repeats flanking Trp-1 and Trp-2 in HSF8

detailed analysis revealed multifold evidence for variation between individual hs genes with respect to the time scale (early vs. medium vs. late HSP), the temperature optimum, tissue specificity of the HSP patterns and the differential response to chemical stressors (for summary, see Nover 1991). The regulatory basis for this gene-specific fine tuning reported for all types of organisms is largely unclear.

The general switch system for all eukaryotic hs genes is the combination of the conserved promoter element (HSE) with the corresponding HSF. Variations of this main system of hs control result from three elements:

1. The particular HSE configuration of a given promoter has a considerable influence. Thus, in *Drosophila* the low temperature *hsp82* gene is

characterized by a condensed HSE trimer (Blackman and Meselson 1986), whereas the five *hsp70* genes with dominant expression under high stress conditions have two essential 1.5 HSE separated by a 10-bp spacer (Fig. 1). Cooperativity of HSF binding contributes to the high effectivity of the *hsp70* promoter (Shuey and Parker 1986; Xiao et al. 1991). The third group are the *Drosophila hsp20* genes with HSE motifs scattered over a 2-kb region upstream of the TATA box. In one case, bending of the promoter region around a nucleosome is assumed to create the necessary proximity of two HSF/HSE complexes spaced by about 200 bp (Thomas and Elgin 1988).

2. Another frequent variation is the combination of HSE with other promoter elements (see Fig. 1). In these cases binding of HSF under stress conditions is just one additional regulatory factor of a highly complex "transcriptosome". These types of hs genes are usually expressed under stress and nonstress conditions. A similar "patchwork of different *cis*-regulatory elements" (Lamb and McKnight 1991) characterizes many, if not all eukaryotic promoters. The interaction of corresponding binding proteins with the high molecular weight TFIID complex (Sharp 1992) and RNAPII "can be viewed as a three dimensional jigsaw puzzle" (Lamb and McKnight 1991). It is essential to be aware of this complexity and its putative consequences for the qualitative and quantitative aspects of transcription control. Step by step we gain a more precise understanding of DNA protein interactions. However, studies on the nature and multiplicity of protein-protein contacts in the "transcriptosome" are only beginning.
3. Similar to other families of transcription factors in animal and plant systems, which are related by their conserved DNA-binding domains (Gehring 1987; Graham et al. 1989; Ruvkun and Finney 1991; Goff et al. 1992; Schena and Davis 1992; Tanaka et al. 1992; Williams et al. 1992), the multiplicity of HSFs may contribute to the regulatory fine tuning and selectivity of the stress response. The results on three HSFs in plants with distinct differences in structure, promoter affinity or temperature optimum of activation (see summary in Table 4) are to be compared to the findings of three HSFs in chicken (Nakai and Morimoto 1993) and of two HSFs in both human and mouse cells (Rabindran et al. 1991; Sarge et al. 1991; Schuetz et al. 1991). In this respect, it is very intriguing that in human erythroleukemia cells hs-induction affects mainly HSF1, whereas hemin stimulation involves HSF2 activation only. Using genomic footprinting with in vivo methylation by dimethyl sulfate, Sistonen et al. (1992) demonstrated that HSF1 and HSF2 recognize similar HSE-binding sites of the *hsp70* gene but with different footprints. This correlates with considerable differences in *hsp70* gene transcription, which is strong and transient for hs but moderate and delayed for the hemin induction. These striking differences between HSF1 and HSF2 are also valid for mouse cells. Only HSF1 but not HSF2 is activated by hs or chemical stressors (Sarge et al. 1991, 1993).

Table 4. Summary of properties of tomato HSFs

	HSF24	HSF8	HSF30
Map position (chromosome)[a]	2 (LA)	8 (SA)	8 (SA)
mRNA (kb)	1.5	2.0	1.4
ORF (codons)	301	527	351
Protein (kDa)	33.3	57.5	40.2
Expression (mRNA levels)			
25 °C	+	+	−
35 °C	++	(+)	+++
39 °C	(+)	+	+++
Trans-*activation assay with gus reporter constructs*[b]			
pGm-hsp17gus			
25 °C	−	+	+
2 h 40 °C[c]	++	++	+
8 h 35 °C	(+)	+++	++
8 h 40 °C	+++	++++	+
pHSE9gus			
25 °C	+	++	+
2 h 40 °C[b]	+++	++++	+
8 h 35 °C	++	++++	++
8 h 40 °C	+++	++++	+
Trans-*repression assay*			
pRT/HSE3gus	Strong	Medium	Weak[d]
pRT/HSE9gus	Weak[d]	Medium	Strong

[a] For details, see Scharf et al. (1990, 1993); LA, long arm; SA, short arm of the indicated chromosome (see Tanksley et al. 1992).
[b] Maximum GUS activities found after transformation with the soybean hsp17 gus construct are about 30% of those found with the pHSE9 gus (Treuter et al. 1993).
[c] Standard hs treatment with two-times 2 h at 40 °C interrupted by 16 h at 25 °C (see Treuter et al. 1993).
[d] Repression measured at room temperature is enhanced by hs (Treuter et al. 1993).

Regarding the complexity of the transcriptosome discussed above, our transient expression assays with tobacco protoplasts overproducing a single HSF appear to be a rather crude method. Usually, the high effectivity of GUS expression under these conditions is combined with an increase in the basal level expression, i.e., a significant loss of hs control. Further characterization must include HSF mixtures, the use of chemical stressors for activation (arsenite, heavy metal ions, salicylate), the use of natural promoters with a well-defined "patchwork" of *cis*-regulatory elements and, above all, a program with appropriate modifications of HSF expression in transgenic plants.

7.2 HSF Activation and the Role of the Oligomerization State

The results from deletion analysis of tomato HSF clones (see Treuter et al. 1993 and example given in Fig. 8) led to the identification of short sequence motifs, which are in the center of activity control (Fig. 9). In addition, other parts of the C-terminal domain evidently modulate the regulatory behavior with respect to hs-inducibility or overall activity. For comparison, it may be useful to briefly review results with the *Saccharomyces* and *Kluyveromyces* HSF, which are the only systems studied so far in comparable detail (Sorger and Pelham 1988; Sorger 1990; Sorger and Nelson 1989; Nieto-Sotelo et al. 1990; Jakobsen and Pelham 1991; Bonner et al. 1992):

1. The only yeast *hsf-1* gene is indispensible under all temperature and growth conditions, i.e., in addition to its role as stress activator, HSF1 has an undefined vital function for yeast (Sorger and Pelham 1988; Wiederrecht et al. 1988). Interestingly, it can be replaced in both functions by HSF8, HSF30 or the *Drosophila* HSF (Scharf et al. unpubl.). Hence, it is very likely that the function under nonstress conditions is also mediated by HSE/HSF interactions, because the DNA-binding domain is the only functional part common to all HSFs.
2. If parts of the C-terminus of the *Saccharomyces* HSF1 are combined with heterologous DNA-binding domains, elements with activating and repressing properties can be identified. Together with parts of the DNA-binding domain and the N-terminus, they exhibit the typical regulatory behavior of the yeast HSF1 (Nieto-Sotelo et al. 1990; Sorger 1990; Bonner et al. 1992). By sequence comparison between the *Kluyveromyces* and *Saccharomyces* HSF, Jakobsen and Pelham (1991) defined two short sequence elements (P1 and P2, see Fig. 2), which are indispensible for the repression of the HSF activity at 25 °C.
3. Essential for the control of HSF activity is the formation of trimers, which are the prevailing form of the yeast HSF independent of the binding state to DNA (Sorger and Nelson 1989). Deletion of the oligomerization domain HR1/HR2 of the *Kluyveromyces* HSF creates an unregulated superactivator (Jakobsen and Pelham 1991).
4. Inactivity of HSF under control conditions requires masking of the activator domain as a result of a particular conformation of the trimeric HSF. Participation of a HSF repressor in formation and/or maintenance of this inactive state is very likely (Nieto-Sotelo et al. 1990; Jakobsen and Pelham 1991).

Despite many similarities between structure and function of eukaryotic HSFs, there is an important difference between yeast and all other systems so far characterized. In the former, HSF binding to DNA is constitutive irrespective of the function as transcription activator, which is only observed under stress conditions (Sorger et al. 1987). In other systems, induced HSF binding and gene activation are coupled (Sorger et al. 1987; Zimarino and Wu 1987; Scharf et al. 1990; Gallo et al. 1991). One possible explanation for this difference was elaborated recently for the *Drosophila* and human HSFs (Rabindran et al. 1993; Westwood and Wu 1993). Together with most other HSFs they contain an additional heptad hydrophobic repeat (HR3) in the C-terminal domain (Figs. 2 and 4). The inactive state is a HSF monomer probably generated by intramolecular coiled-coil interactions of HR1/2 with

HR3. Stress-induced activation could result from release of this backfolded conformation and trimerization (Rabindran et al. 1993; see review by Morimoto 1993). These conclusions were drawn from careful size characterization of wild-type and C-terminally truncated HSF coupled with gel retardation assays. Unfortunately, activity tests using a transient expression assay are lacking. Corresponding results with the two tomato HSF with a C-terminal HR3 (HSF8, HSF30, see Figs. 5 and 9) do not support such a dominant role of HR3. In both cases deletion of HR3 has no marked influence, neither on the overall activity nor on the hs-inducibility (Treuter et al. 1993).

7.3 The Missing Link(s) – Model for Control of HSF Activity

The questions to the cell-specific thermometer or sensor of the stress treatment and to subsequent elements of the signal transduction mechanism have been the focus of interest for several years. The simplest concept that HSF itself is the stress sensor was supported by results of several groups demonstrating in vitro activation of human, mouse and *Drosophila* HSF. HSF activation evidently involves changes in protein conformation, achieved either by mild heating, treatment with protein denaturants (low pH, urea, nonionic detergent), oxidation (H_2O_2) or binding of HSF-specific antibodies (Larson et al. 1988; Mosser et al. 1990; Zimarino et al. 1990; for a summary, see Lis and Wu 1992). It is inhibited in the presence of 20% glycerol. Remarkably, temperature activation of human HSF in vitro usually needs 43 °C, but proceeds already at 37 °C in the presence of 2% Nonidet. The results based on gel retardation assays are incomplete because tests with in vitro transcription systems are lacking. However, it is evident that at least three independent parts influence the capability of HSF to promote hs gene transcription (Baler et al. 1993): (1) protein conformation, (2) the oligomerization state especially in connection with HSE binding (Abravaya et al. 1991b; Westwood et al. 1991; Rabindran et al. 1993; Westwood and Wu 1993) and (3) the phosphorylation state. Though changing HSF phosphorylation was reported for yeast, *Tetrahymena*, *Drosophila* and human cells (Larson et al. 1988; Sorger and Pelham 1988; Wiederrecht et al. 1988; Carmo-Avides et al. 1990; Gallo et al. 1991), its role in the activity control is unclear. Differences in the stability and reversibility of HSF binding to DNA observed after mild vs. severe heat stress (Abravaya et al. 1991b) may reflect differences in the phosphorylation state. Supportive evidence for the role of phosphorylation stems from in vitro (in situ) activation of HSF in digitonin-permeabilized NIH 3T3 cells. The two-step activation requires heat treatment and Ca^{2+} for the first step and ATP for the second step, which can be inhibited by the protein kinase inhibitor genistein (Price and Calderwood 1991). It is tempting to speculate that a DNA-dependent protein kinase phosphorylates HSF in situ, i.e., after binding to the hs promoter (Gottlieb and Jackson 1993).

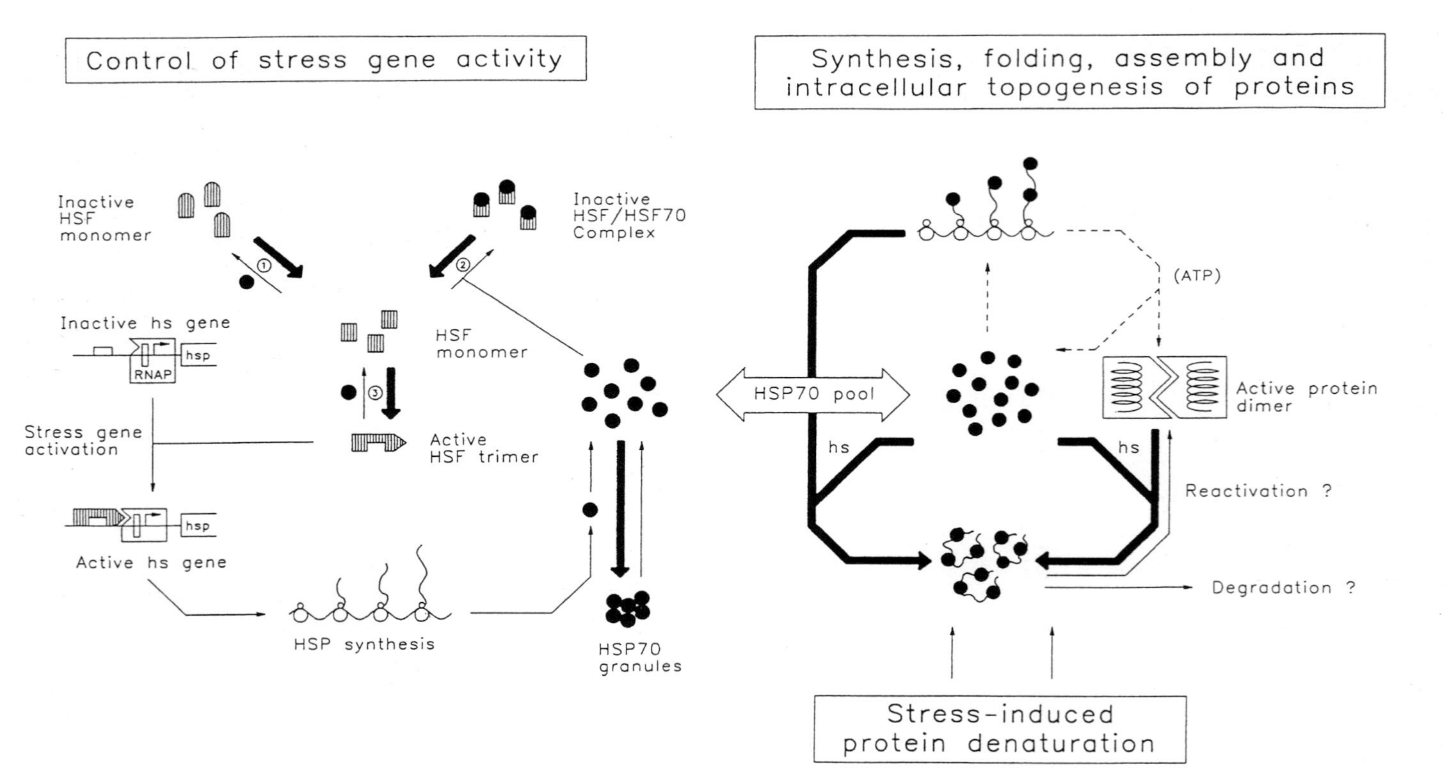

Fig. 10. Model for stress regulation of HSF activity. In the center of the model is the cellular pool of chaperones of the HSP70 family. They participate in the processing and intracellular transport of newly synthesized proteins (*right*) and in the activity control of HSF (*left*) either by forming HSF/HSP70 complexes (*2*) or by catalyzing the formation of the inactive state (reversion of reactions *1,3*). Other possibilities are discussed in the text

A more complex model of the stress-induced signal transduction mechanism is presented in Fig. 10. Central to it is the transient accumulation of denatured proteins. Direct evidence for the regulatory significance of this deviation from protein homeostasis stems from three types of observations. (1) A large number of chemical stressors characterized as inducers of HSP synthesis are well known for their negative effects on native protein structures (for a summary, see Nover 1991). (2) Activation of hs genes can be triggered by injection of denatured serum albumin into *Xenopus* oocytes (Ananthan et al. 1986) or the production of aberrant or recombinant proteins in bacteria (Goff and Goldberg 1985; Parsell and Sauer 1989; Allen et al. 1992). (3) Treatment of mammalian cells with hs and/or chemical stressors causes a massive aggregation of newly synthesized proteins. The salt- and detergent-insoluble aggregates are connected with chaperones of the HSP70 family (Palleros et al. 1991; Beckmann et al. 1992).

In contrast to the speculations that HSF itself is a direct temperature-sensitive protein are our results on the expression and proper function of heterologous HSFs in the tobacco protoplasts (Treuter et al. 1993). In appropriate expression constructs, both the human HSF1 and the *Drosophila* HSF are strong hs-inducible activators of *gus* gene expression in this plant system. Comparable to the tomato HSF30, optimum activity is achieved by a long-term treatment at 35 °C. At least for the human HSF1, this temperature is even below its normothermic condition. The simplest explanation is that the temperature threshold for induction is defined by intrinsic properties of the expression system and not by the HSF. This leads to considerations about heat stress proteins, expecially HSP70, playing a direct role in the cellular thermometer system (Fig. 10).

In support of this concept (Craig and Gross 1991; Sorger 1991; Morimoto 1993), early investigations with *Drosophila* cell cultures provided evidence that the expression of hs genes is controlled by autorepression at the transcription and translation levels and that HSP70 accumulation may be responsible for this (DiDomenico et al. 1982a,b). Genetic evidence for a special role of HSP70 in this respect came from Stone and Craig (1990). Deficiency of the *SSA1* gene encoding HSP70 caused overexpression of other HSPs. In the model presented in Fig. 10 the depletion of cellular pools of free HSP70 at the onset of the stress period is a consequence of denaturation and aggregation of proteins (Palleros et al. 1991; Beckmann et al. 1992) and results in HSF activation. The subsequent overproduction of HSP70 is assumed to restore the normal, inactive HSF state. Similar to findings with other transcription factors, e.g., the HSP90 × steroid hormone receptor complex (Picard et al. 1990; Muller and Renkawitz 1991; Rehberger et al. 1992), HSP70 may function as an HSF inhibitor shielding the DNA-binding domain. Alternatively, it may act as a cytoplasmic retention factor in the unstressed situation or as a chaperone which helps to create the inactive conformation of the free HSF monomer. The demonstration of HSP70 as a catalytic factor of cytoplasmic → nuclear transport of proteins

(Dingwall and Laskey 1992; Shi and Thomas 1992) points to an additional aspect in the autoregulation of HSP synthesis. Only in the case of the human and mouse HSF1 do we have sufficient experimental data supporting the tight coupling of hs- or Cd^{2+}-induced activation of DNA binding, oligomerization from monomer to trimer, phosphorylation and recompartmentation from the cytoplasm to the nucleus (Baler et al. 1993; Sarge et al. 1993). Interestingly, overexpression of HSF1 in mouse 3T3 cells is connected with constitutive DNA binding as well as nuclear localization. To explain these findings the existence of a titratable negative control factor not permanently associated with HSF1 is discussed (Sarge et al. 1993).

It is tempting to speculate that the C-terminal HR3 of HSFs serves the interaction with a putative hs repressor. Interestingly, the C-terminal peptide-binding domain of HSP70 contains highly conserved positions of hydrophobic amino acids forming two adjacent heptad repeats. This is exemplified for the human (Hs), *Drosophila* (Dm) and the tomato (Le) HSP70. Numbers in parentheses indicate the sequence position of the first amino acid (for a compilation of sequence data, see Nover 1991; Nover et al. 1990).

Hs-HSP70	(392)L(6X)L(6X)A(6X)L(6X)I(6X)I and (428)F(6X)Q(6X)V(6X)M(6X)L
Le-HSC70	(399)L(6X)L(6X)A(6X)L(6X)I(6X)V and (435)F(6X)Q(6X)V(6X)A(6X)L
Dm-HSP70	(392)L(6X)L(6X)A(6X)L(6X)I and (425)F(6X)Q(6X)V(6X)M(6X)L

Some interesting observations on HSP70 and on temperature-dependent changes in its properties may be relevant for the regulatory network depicted in Fig. 10:

1. The in vitro autophosphorylation reaction of the *E. coli* HSP70 (DnaK) is enhanced up to 400-fold by increasing the temperature from 37 to 52 °C (McCarty and Walker 1991). The phosphorylation site is a threonine residue (position 199) situated in a highly conserved loop between two β-sheet structures found in all prokaryotic and eukaryotic members of the HSP70 family (Nover 1991).
2. DnaK and its eukaryotic homolog HSP70 undergo conformation changes at 41–42 °C, leading to molten globules and aggregation (Palleros et al. 1992). Thus, not only the changes in the cellular levels of free HSP70 but also of its phosphorylation state and self-aggregation may influence the stability of the postulated HSP70/HSF complex.
3. Direct evidence for an interaction of HSF and HSP70 was reported for human cells (Baler et al. 1992; Abravaya et al. 1992). HSP70 blocks in vitro activation of HSF. Moreover, increased levels of free cellular HSP70, e.g., after application of cycloheximide, inhibit HSF activation.

8 Conclusions and Perspectives

The remarkable conservation of essential parts of the eukaryotic hs response extends also to the characteristic promoter element (HSE) and the corresponding heat stress transcription factor (HSF). All *hsf* genes cloned from plant, yeast and animals code for leucine zipper proteins with a novel type of DNA-binding domain close to the N-terminus, whereas the activity control elements are part of the C-terminal domain. In contrast to the other organisms, we detected three *hsf* genes in tomato, two of them being hs-inducible. Functional characterization of the proteins (HSF8, HSF30 and HSF24) in a tobacco transient expression assay revealed differences with respect to promoter recognition and the temperature profiles of activation. Despite the relatively low sequence conservation of the C-terminal activator domains, short peptide motifs with a central tryptophan residue were identified as essential parts of the activity control regions.

The rapidly advancing knowledge about HSF structure and function provides the basis for more detailed studies of stress-induced signal transduction, including the intriguing activation of HSF by hs and different types of chemical stressors resulting in binding to HSE, oligomerization, phosphorylation and relocalization to the nucleus. Other important problems concern the role of multiple HSFs in mammals and plants and the question of HSP70 or other chaperones acting as coregulator of HSF activity as summarized in the model in Fig. 10.

Acknowledgments. Work in the author's laboratories was supported by the Körber Foundation Hamburg (grant to L. Nover), the Volkwagenstiftung Hannover (grant to L. Nover), the Fonds der Chemischen Industrie Frankfurt (grants to L. Nover and K.D. Scharf) and the Bundesministerium für Forschung und Technologie Bonn (grant to K.D. Scharf). We gratefully appreciate the critical reading of the manuscript by Elizabeth Vierling (Tucson, Texas, USA).

References

Abravaya K, Phillips B, Morimoto RI (1991a) Heat shock-induced interactions of heat shock transcription factor and the human hsp70 promoter examined by in vivo footprinting. Mol Cell Biol 11:586–592

Abravaya K, Phillips B, Morimoto RI (1991b) Attenuation of the heat shock response in HeLa cells is mediated by the release of bound heat shock transcription factor and is modulated by changes in growth and in heat shock temperatures. Genes Dev 5:2117–2127

Abravaya K, Myers MP, Murphy SP, Morimoto RI (1992) The human heat shock protein hsp70 interacts with HSF, the transcription factor that regulates heat shock gene expression. Genes Dev 6:1153–1164

Allen SP, Polazzi JO, Gierse JK, Easton AM (1992) Two novel heat shock genes encoding proteins produced in response to heterologous protein expression in *Escherichia coli*. J Bacteriol 174:6938–6947

Amin J, Mestril R, Schiller P, Dreano M, Voellmy R (1987) Organization of the *Drosophila melanogaster* hsp70 heat shock regulation unit. Mol Cell Biol 7:1055–1062

Amin J, Ananthan J, Voellmy R (1988) Key features of heat shock regulatory elements. Mol Cell Biol 8:3761–3769

Ananthan J, Goldberg AL, Voellmy R (1986) Abnormal proteins serve as eukaryotic stress signals and trigger the activation of heat shock genes. Science 232:522–524

Ashburner M, Bonner JJ (1979) The induction of gene activity in *Drosophila* by heat shock. Cell 17:241–254

Baler R, Welch WJ, Voellmy R (1992) Heat shock gene regulation by nascent polypeptides and denaturated proteins – hsp70 as a potential autoregulatory factor. J Cell Biol 117:1151–1159

Baler R, Dahl G, Voellmy R (1993) Activation of human heat shock genes is accompanied by oligomerization, modification, and rapid translocation of heat shock transcription factor HSF1. Mol Cell Biol 13:2486–2496

Baumann G, Raschke E, Bevan M, Schöffl F (1987) Functional analysis of sequences required for transcriptional activation of a soybean heat shock gene in transgenic tobacco plants. EMBO J 6:1161–1166

Beckmann RP, Lovett M, Welch WJ (1992) Examining the function and regulation of hsp70 in cells subjected to metabolic stress. J Cell Biol 117:1137–1150

Bienz M (1984) *Xenopus* hsp70 genes are constitutively expressed in injected oocytes. EMBO J 3:2477–2483

Bienz M (1986) A CCAAT box confers cell-type-specific regulation on the *Xenopus* hsp70 gene in oocytes. Cell 46:1037–1042

Bienz M, Pelham HRB (1982) Expression of a *Drosophila* heat shock protein in *Xenopus* oocytes: conserved and divergent regulatory signals. EMBO J 1:1583–1588

Bienz M, Pelham HRB (1986) Heat shock regulatory elements function as an inducible enhancer in the *Xenopus hsp70* gene and when linked to a heterologous promoter. Cell 45:753–760

Blackman RK, Meselson M (1986) Interspecific nucleotide sequence comparisons used to identify regulatory and structural features of the *Drosophila* hsp82 gene. J Mol Biol 188:499–516

Bonner JJ, Heyward S, Fackenthal DL (1992) Temperature-dependent regulation of a heterologous transcriptional activation domain fused to yeast heat shock transcription factor. Mol Cell Biol 12:1021–1030

Burley S, Petsko GA (1985) Aromatic-aromatic interaction: a mechanism of protein structure stabilization. Science 229:23–28

Caplan AJ, Cyr DM, Douglas MG (1992) YIJ1p facilitates polypeptide translocation across different intracellular membranes by a conserved mechanism. Cell 71: 1143–1155

Carmo-Avides MD, Sunkel CE, Moradas-Ferreira P, Rodrigues-Pousada C (1990) Properties and partial characterization of the heat shock factor from *Tetrahymena pyriformis*. Eur J Biochem 194:331–336

Clos J, Westwood JT, Becker PB, Wilson S, Lambert K, Wu C (1990) Molecular cloning and expression of a hexameric *Drosophila* heat shock factor subject to negative regulation. Cell 63:1085–1097

Craig EA, Gross CA (1991) Is hsp70 the cellular thermometer? Trends Biochem Sci 16:135–140

Cunniff NFA, Wagner J, Morgan WD (1991) Modular recognition of 5-base-pair DNA sequence motifs by human heat shock transcription factor. Mol Cell Biol 11:3504–3514

Czarnecka E, Key JL, Gurley WB (1989) Regulatory domains of the Gmhsp17.5-E heat shock promoter of soybean. Mol Cell Biol 9:3457–3463

Dang CV, Barrett J, Villa-Garcia M, Resar LMS, Kato GJ, Fearon ER (1991) Intracellular leucine zipper interactions suggest c-Myc hetero oligomerization. Mol Cell Biol 11:954–962

DiDomenico BJ, Bugaisky GE, Lindquist S (1982a) The heat shock response is self-regulated at both the transcriptional and posttranscriptional levels. Cell 31:593–603
DiDomenico BJ, Bugaisky GE, Lindquist SS (1982b) Heat shock and recovery are mediated by different translational mechanisms. Proc Natl Acad Sci USA 79: 6181–6185
Dingwall C, Laskey R (1992) The nuclear membrane. Science 258:942–947
Dostatni N, Lambert PE, Sousa R, Ham J, Howley PM, Janiv M (1991) The functional BPV-1E2 *trans*-activating protein can act as a repressor by preventing formation of the initiation complex. Genes Dev 5:1657–1671
Foulkes NS, Sassone-Corsi P (1992) More is better: activators and repressors from the same gene. Cell 68:411–414
Frydman J, Nimmesgern E, Erdjument-Bromage H, Wall JS, Tempst P, Hartl F-U (1992) Function in protein folding of TRiC, a cystosolic ring complex containing TCP-1 and structurally related subunits. EMBO J 11:4767–4778
Fujita A, Kikuchi Y, Kuhara S, Misumi Y, Matsumoto S, Kobayashi H (1989) Domains of the Sf11 protein of yeast are homologous to myc oncoproteins or heat-shock transcription factor. Gene 85:321–328
Gallo GJ, Schuetz TJ, Kingston RE (1991) Regulation of heat shock factor in *Schizosaccharomyces pombe* more closely resembles regulation in mammals than in *Saccharomyces cerevisiae*. Mol Cell Biol 11:281–288
Gallo GJ, Prentice H, Kingston RE (1993) Heat shock factor is required for growth at normal temperatures in the fission yeast *Schizosaccharomyces pombe*. Mol Cell Biol 13:749–761
Gehring WJ (1987) Homeo boxes in the study of development. Science 236:1245–1252
Georgopoulos C (1992) The emergence of the chaperone machines. Trends Biochem Sci 17:295–299
Gething M-J, Sambrook J (1992) Protein folding in the cell. Nature 355:33–45
Glass DJ, Polvere RJ, van der Ploeg LHT (1986) Conserved sequences and transcription of hsp70 gene family in *Trypanosoma brucei*. Mol Cell Biol 6:4657–4666
Goff SA, Goldberg AL (1985) Production of abnormal proteins in *E. coli* stimulates transcription of lon and other heat shock genes. Cell 41:587–595
Goff SA, Cone VC, Chandler VL (1992) Functional analysis of the transcriptional activator encoded by the maize B gene: evidence for a direct functional interaction between two classes of regulatory proteins. Genes Dev 6:864–875
Gottlieb TM, Jackson SP (1993) The DNA-dependent protein kinase: requirement for DNA ends and association with Ku antigen. Cell 72:131–142
Graham A, Papalopulu N, Krumlauf R (1989) The murine and *Drosophila* homeobox gene complexes have common features of organization and expression. Cell 57:357–378
Grossniklaus U, Pearson RK, Gehring WJ (1992) The *Drosophila* sloppy paired locus encodes two proteins involved in segmentation that show homology to mammalian transcription factors. Genes Dev 6:1030–1052
Gurley WB, Key JL (1991) Transcriptional regulation of the heat-shock response: a plant perspective. Biochemistry 30:1–12
Häcker U, Grossniklaus U, Gehring WJ, Jäckle H (1992) Developmentally regulated *Drosophila* gene family encoding the fork head domain. Proc Natl Acad Sci USA 89:8754–8758
Härd T, Kellenbach E, Boelens R, Maler BA, Dahlman K, Freedman LP, Carlstedt-Duke J, Yamamoto KR, Gustafsson J-A, Kaptein R (1990) Solution structure of the glucocorticoid receptor DNA-binding domain. Science 249:157–160
Hartl FU, Martin J, Neupert W (1992) Protein folding in the cell – the role of molecular chaperones Hsp70 and Hsp60. Annu Rev Biophys Biomol Struct 21:293–322
Horwitz J (1992) α-Crystallin can function as a molecular chaperone. Proc Natl Acad Sci USA 89:10449–10453

Hu JC, O'Shea EK, Kim PS, Sauer RT (1990) Sequence requirements for coiled-coils. Analysis with lambda repressor-GCN4 leucine zipper fusions. Science 250:1400–1403

Jakobsen BK, Pelham HRB (1991) A conserved heptapeptide restrains the activity of the yeast heat shock transcription factor. EMBO J 10:369–375

Jentsch St (1992) The ubiquitin-conjugation system. Annu Rev Genet 26:177–205

Katagiri F, Chua NH (1992) Plant transcription factors – present knowledge and future challenges. Trends Genet 8:22–27

Kay RJ, Boissy RJ, Russnak RH, Candido EPM (1986) Efficient transcription of a *Caenorhabditis elegans* heat shock gene pair in mouse fibroblasts is dependent on multiple promoter elements which can function bidirectionally. Mol Cell Biol 6:3134–3143

Kingston RE, Schuetz TJ, Larin Z (1987) Heat-inducible human factor that binds to a human hsp70 promoter. Mol Cell Biol 7:1530–1534

Kobayashi N, McEntee K (1990) Evidence for a heat shock transcription factor-independent mechanism for heat shock induction of transcription in *Saccharomyces cerevisiae*. Proc Natl Acad Sci USA 87:6550–6554

Kobayashi N, McEntee K (1993) Identification of *cis* and *trans* components of a novel heat shock stress regulatory pathway in *Saccharomyces cerevisiae*. Mol Cell Biol 13:248–256

Krens FA, Molendijk L, Wullems GI, Schilperpoort RA (1982) In vitro transformation of plant protoplasts with Ti-plasmid DNA. Nature 296:72–74

Lamb P, McKnight SL (1991) Diversity and specificity in transcriptional regulation – the benefits of heterotypic dimerization. Trends Biochem Sci 16:417–422

Landschulz WH, Johnson PF, McKnight StL (1988) The leucine zipper: a hypothetical structure common to a new class of DNA binding proteins. Science 240:1759–1764

Larson JS, Schuetz TJ, Kingston RE (1988) Activation in vitro of sequence-specific DNA binding by a human regulatory factor. Nature 335:372–375

Lindquist S, Craig EA (1988) The heat-shock proteins. Annu Rev Genet 22:631–677

Lis J, Wu C (1992) Heat shock factor. In: McKnight SL and Yamamoto KR (eds) Transcriptional regulation, Vol 2. Cold Spring Harbor Press, Cold Spring Harbor, pp 907–930

Lovejoy B, Choe S, Cascio D, McRorie DK, DeGrado WF, Eisenberg D (1993) Crystal structure of a synthetic triple-stranded α-helical bundle. Science 259: 1288–1293

Luisi BF, Xu WX, Otwinowski Z, Freedman LP, Yamamoto KR, Sigler PB (1991) Crystallographic analysis of the interaction of the glucocorticoid receptor with DNA. Nature 352:497–505

McCarty JS, Walker GC (1991) Dnak as a thermometer: threonine-199 is the site of autophosphorylation and is critical for ATPase activity. Proc Natl Acad Sci USA 88:9513–9517

Mori M, Murata K, Kubota H, Yamamoto A, Matsushiro A, Morita T (1992) Cloning of a cDNA encoding the Tcp-1 (t complex polypeptide-1) homologue of *Arabidopsis thaliana*. Gene 122:381–382

Morimoto RI (1993) Cells in stress-transcriptional activation of heat shock genes. Science 259:1409–1410

Morimoto RI, Tissieres A, Georgopoulos C (eds) (1990) Stress proteins in biology and medicine. Cold Spring Harbor. Cold Spring Harbor Press

Mosser DD, Kotzbauer PT, Sarge KD, Morimoto RI (1990) In vitro activation of heat shock transcription factor DNA-binding by calcium and biochemical conditions that effect protein conformation. Proc Natl Acad Sci USA 87: 3748–3752

Muller M, Renkawitz R (1991) The glucocorticoid receptor. Biochim Biophys Acta 1088:171–182

Nakai A, Morimoto RI (1993) Characterization of a novel chicken heat shock transcription factor, heat shock factor 3, suggests a new regulatory pathway. Mol Cell Biol 13:1983–1997

Nieto-Sotelo J, Wiederrecht G, Okuda A, Parker CS (1990) The yeast heat shock transcription factor contains a transcriptional activation domain whose activity is repressed under nonshock conditions. Cell 62:807–817
Nover L (1987) Expression of heat shock genes in homologous and heterologous systems. Enzyme Microb Technol 9:130–144
Nover L (ed) (1991) Heat shock response. CRC Press, Boca Raton
Nover L, Munsche D, Ohme K, Scharf K-D (1986) Ribosome biosynthesis in heat shocked tomato cell cultures I. Ribosomal RNA. Eur J Biochem 160:297–304
Nover L, Scharf KD, Neumann D (1989) Cytoplasmic heat shock granules are formed from precursor particles and are associated with a specific set of mRNAs. Mol Cell Biol 9:1298–1308
Nover L, Neumann D, Scharf KD (eds) (1990) Heat shock and other stress response systems of plants. Results and problems of cell differentiation. Springer, Berlin Heidelberg New York
O'Shea EK, Klemm JD, Kim PS, Alber T (1991) X-ray structure of the GCN4 leucine zipper, a two-stranded, parallel coiled coil. Science 254:539–544
Palleros DR, Welch WJ, Fink AL (1991) Interaction of hsp70 with unfolded proteins: effects of temperature and nucleotides on the kinetics of binding. Proc Natl Acad Sci USA 88:5719–5723
Palleros DR, Reid KL, McCarty JS, Walker GC, Fink AL (1992) DnaK, hsp73, and their molten globules – two different ways heat shock proteins respond to heat. J Biol Chem 267:5279–5285
Paluh JL, Yanofsky Ch (1991) Characterization of *Neurospora* CPC1, a bZIP DNA-binding protein that does not require aligned heptad leucines for dimerization. Mol Cell Biol 11:935–944
Parker CS, Topol J (1984a) A *Drosophila* RNA polymerase II transcription factor binds to the regulatory site of an hsp70 gene. Cell 37:273–283
Parker CS, Topol J (1984b) A *Drosophila* RNA polymerase II transcription factor contains a promoter-region-specific DNA-binding activity. Cell 36:357–369
Parsell DA, Sauer RT (1989) Induction of a heat shock-like response by unfolded protein in *Escherichia coli*: dependence on protein level not protein degradation. Genes Dev 3:1226–1232
Parsell DA, Sanchez Y, Stitzel JD, Lindquist S (1991) Hsp104 is a highly conserved protein with 2 essential nucleotide-binding sites. Nature 353:270–273
Pelham HRB (1982) A regulatory upstream promoter element in the *Drosophila* hsp70 heat-shock gene. Cell 30:517–528
Peteranderl R, Nelson HCM (1992) Trimerization of heat shock transcription factor by a triple-stranded-helical coiled-coil. Biochemistry 31:12272–12276
Picard D, Khursheed B, Garabedian MJ, Fortin MG, Lindquist S, Yamamoto KR (1990) Reduced levels of hsp90 compromise steroid receptor action in vivo. Nature 348:166–168
Price BD, Calderwood SK (1991) Ca^{2+} is essential for multistep activation of the heat shock factor in permeabilized cells. Mol Cell Biol 11:3365–3368
Raabe T, Manley JL (1991) A human homologue of the *Escherichia-coli* DnaJ heat shock protein. Nucleic Acids Res 19:6645
Rabindran SK, Giorgi G, Clos J, Wu C (1991) Molecular cloning and expression of human heat shock factor. Proc Natl Acad Sci USA 88:6906–6910
Rabindran SK, Haroun RI, Clos J, Wisniewski J, Wu C (1993) Regulation of heat shock factor trimerization: role of a conserved leucine zipper. Science 259:230–234
Rasmussen R, Benvegnu D, O'Shea EK, Kim PS, Alber T (1991) X-ray scattering indicates that the leucine zipper is a coiled coil. Proc Natl Acad Sci USA 88:561–564
Rehberger P, Rexin M, Gehring U (1992) Heterotetrameric structure of the human progesterone receptor. Proc Natl Acad Sci USA 89:8001–8005
Rieping M, Schöffl F (1992) Synergistic effect of upstream sequences, CCAAT box elements, and HSE sequences for enhanced expression of chimaeric heat shock genes in transgenic tobacco. Mol Gen Genet 231:226–232

Ritossa F (1962) A new puffing pattern induced by heat shock and DNP in *Drosophila*. Experientia 18:571–573
Rochester DE, Winer JA, Shah DM (1986) The structure and expression of maize genes encoding the major heat shock protein, hsp70. EMBO J 5:451–458
Ruvkun G, Finney M (1991) Regulation of transcription and cell identity by POU domain proteins. Cell 64:457–478
Sanchez ER (1990) Hsp56 – a novel heat shock protein associated with untransformed steroid receptor complexes. J Biol Chem 265:22067–22070
Sarge KD, Zimarino V, Holm K, Wu C, Morimoto RI (1991) Cloning and characterization of two mouse heat shock factors with distinct inducible and constitutive DNA binding ability. Genes Dev 5:1902–1911
Sarge KD, Murphy SP, Morimoto RI (1993) Activation of heat shock gene transcription by heat shock factor 1 involves oligomerization, acquisition of DNA-binding activity, and nuclear localization and can occur in the absence of stress. Mol Cell Biol 13:1392–1407
Scharf K-D, Rose S, Zott W, Schöffl F, Nover L (1990) Three tomato genes code for heat stress transcription factors with a region of remarkable homology to the DNA-binding domain of the yeast HSF. EMBO J 9:4495–4501
Scharf K-D, Rose S, Thierfelder J, Nover L (1993) Nucleotide sequence of the cDNA clones encoding tomato heat stress transcription factors. Plant Physiol 102:1355–1356
Schena M, Davis RW (1992) HD-Zip proteins: members of an *Arabidopsis* homeodomain protein superfamily. Proc Natl Acad Sci USA 89:3894–3898
Schlesinger MJ, Ashburner M, Tissieres A (eds) (1982) Heat shock. From bacteria to man. Cold Spring Harbor Lab, Cold Spring Harbor, NY
Schöffl F, Raschke E, Nagao RT (1984) The DNA sequence analysis of soybean heat-shock genes and identification of possible regulatory promoter elements. EMBO J 3:2491–2497
Schuermann M, Hunter JB, Hennig G, Muller R (1991) Non-leucine residues in the leucine repeats of fos and jun contribute to the stability and determine the specificity of dimerization. Nucleic Acids Res 19:739–746
Schuetz TJ, Gallo GJ, Scheldon L, Tempst P, Kingston RE (1991) Isolation of a cDNA for HSF2: evidence for two heat shock factor genes in humans. Proc Natl Acad Sci USA 88:6911–6915
Schwabe JWR, Neuhaus D, Rhodes D (1990) Solution structure of the DNA-binding domain of the oestrogen receptor. Nature 348:458–461
Sharp PA (1992) TATA-binding protein is a class-less factor. Cell 68:819–821
Shi Y, Thomas JO (1992) The transport of proteins into the nucleus requires the 70-kilodalton heat shock protein or its cytosolic cognate. Mol Cell Biol 12:2186–2192
Shuey DJ, Parker CS (1986) Binding of *Drosophila* heat-shock gene transcription factor to the hsp70 promoter. J Biol Chem 261:7934–7940
Sistonen L, Sarge KD, Phillips B, Abravaya K, Morimoto RI (1992) Activation of heat shock factor 2 during hemin-induced differentiation of human erythroleukemia cells. Mol Cell Biol 12:4104–4111
Sorger PK (1990) Yeast heat shock factor contains separable transient and sustained response transcriptional activators. Cell 62:783–805
Sorger PK (1991) Heat shock factor and the heat shock response. Cell 65:363–366
Sorger PK, Nelson HCM (1989) Trimerization of a yeast transcriptional activator via a coiled-coil motif. Cell 59:807–813
Sorger PK, Pelham HRB (1987) Cloning and expression of a gene encoding hsc73, the major hsp70 – like protein in unstressed rat cells. EMBO J 6:993–998
Sorger PK, Pelham HRB (1988) Yeast heat shock factor is an essential DNA-binding protein that exhibits temperature-dependent phosphorylation. Cell 54:855–864
Sorger PK, Lewis MJ, Pelham HRB (1987) Heat shock factor is regulated differently in yeast and HeLa cells. Nature 329:81–84
Squires C, Squires CL (1992) The Clp proteins: proteolysis regulators or molecular chaperones? J Bacteriol 174:1081–1085

Stone DE, Craig EA (1990) Self-regulation of 70-kilodalton heat shock proteins in *Saccharomyces cerevisiae*. Mol Cell Biol 10:1622–1632

Stragier P, Parsol C, Bouvier J (1985) Two functional domains conserved in major and alternate bacterial sigma factors. FEBS Lett 187:11–15

Swindle J, Ajioka J, Eisen H, Sanwal B, Jacquemot C, Browder Z, Buck G (1988) The genomic organization and transcription of the ubiquitin genes of *Trypanosoma cruzi*. EMBO J 7:1121–1127

Tai P-K K, Albers MW, Chang H, Faber LE, Schreiber StL (1992) Association of a 59-kilodalton immunophilin with the glucocorticoid receptor complex. Science 256:1315–1318

Takahashi T, Komeda Y (1989) Characterization of two genes encoding small heat-shock proteins in *Arabidopsis thaliana*. Mol Gen Genet 219:365–372

Tanaka M, Lai J-S, Herr W (1992) Promoter-selective activation domains in Oct-1 and Oct-2 direct differential activation of an snRNA and mRNA promoter. Cell 68:755–767

Tanksley SD, Ganal MW, Prince JP, deVicente MC, Bonierbale MW, Broun P, Fulton TM, Giovanonni JJ, Grandillo S, Martin GB, Messeguer R, Miller JC, Miller L, Paterson AH, Pineda O, Röder MS, Wing RA, Wu W, Young ND (1992) High density molecular linkage maps of the tomato and potato genomes. Genetics 132:1141–1160

Thomas GH, Elgin SCR (1988) Protein-DNA architecture of the DNase I hypersensitive region of the *Drosophila* hsp26 promoter. EMBO J 7:2191–2201

Tissières A, Mitchell HK, Tracy UM (1974) Protein synthesis in salivary glands of *D. melanogaster*. Relation to chromosome puffs. J Mol Biol 84:389–398

Töpfer R, Schell J, Steinbiss HH (1988) Versatile cloning vectors for transient gene expression and direct gene transfer in plant cells. Nucleic Acids Res 16:8725

Topol J, Ruden DM, Parker CS (1985) Sequences required for in vitro transcriptional activation of a *Drosopila hsp70* gene. Cell 42:527–537

Treuter E, Nover L, Ohme K, Scharf K-D (1993) Promoter specificity and deletion analysis of three tomato heat stress transcription factors. Mol Gen Genet 240:113–125

Tropsha A, Bowen JP, Brown FK, Kizer JS (1991) Do interhelical side chain-backbone hydrogen bonds participate in formation of leucine zipper coiled coils? Proc Natl Acad Sci USA 88:9488–9492

Vierling E (1991) The roles of heat shock proteins in plants. Annu Rev Plant Physiol Plant Mol Biol 42:579–620

Weigel D, Jäckle H (1990) The fork head domain: a novel DNA binding motif of eukaryotic transcription factors. Cell 63:455–456

Westwood JT, Wu C (1993) Activation of *Drosophila* heat shock factor: conformational change associated with a monomer to trimer transition. Mol Cell Biol (in press)

Westwood JT, Clos J, Wu C (1991) Stress-induced oligomerization and chromosomal relocalization of heat shock factor. Nature 353:822–827

Wiederrecht G, Seto D, Parker CS (1988) Isolation of the gene encoding the *S. cerevisiae* heat shock transcription factor. Cell 54:841–853

Wieser R, Adam G, Wagner A, Schuller C, Marchler G, Ruis H, Krawiec Z, Bilinski T (1991) Heat shock factor-independent heat control of transcription of the CTT1 gene encoding the cytosolic catalase-T of *Saccharomyces cerevisiae*. J Biol Chem 266:12406–12411

Williams ME, Foster R, Chua NH (1992) Sequences flanking the hexameric G-box core CACGTG affect the specificity of protein binding. Plant Cell 4:485–496

Wu C (1984a) Activating protein factor binds in vitro to upstream control sequences in heat shock gene chromatin. Nature 311:81–84

Wu C (1984b) Two protein-binding sites in chromatin implicated in the activation of heat shock genes. Nature 309:229–234

Wu C (1985) An exonuclease protection assay reveals heat shock element and TATA box DNA-binding proteins in crude nuclear extracts. Nature 317:84–87

Xiao H, Perisic O, Lis JT (1991) Cooperative binding of *Drosophila* heat shock factor to arrays of a conserved 5 bp unit. Cell 64:585–593

Zhu JK, Shi J, Bressan RA, Hasegawa PM (1993) Expression of an *Atriplex nummularia* gene encoding a protein homologous to the bacterial molecular chaperone Dna J. Plant Cell 5:341–349

Zimarino V, Wu C (1987) Induction of sequence-specific binding of *Drosophila* heat shock activator protein without protein synthesis. Nature 327:727–730

Zimarino V, Wilson S, Wu C (1990) Antibody-mediated activation of *Drosophila* heat shock factor in vitro. Science 249:546–549

Notes added in proof:

Sect. 6.2: Functional tests with HSF30 mutants altered by W → A exchanges in the Trp-1 and/or Trp-2 element (see Fig. 9) demonstrated (i) that the central tryptophan residues are essential and (ii) that both Trp elements function alternatively (Treuter et al. in preparation).

Sect. 7.2: In contrast to the dominant role of the Trp elements for the activator function, hs-inducibility is evidently a result of additive effects of the HR1/HR2 and the HR3 regions. A mutants of HSF30 created by internal deletions of both regions is a hs-*in*dependent activator protein (Treuter et al. in preparation). In addition to its role for oligomerization after hs-activation, maintenance of the inactive monomeric state may require intramolecular interactions between HR1/2 and HR3 with a shielding effect on clusters of basic amino acid residues (K/R) serving as nuclear localization signals (L.E. Sheldon and R.E. Kingston, Genes Dev 7:1549–1558 (1993)).

Sect. 7.3: In support of HSP70 as negative control factor of HSF activity (Fig. 10) Mosser et al. (1993) (Mol Cell Biol 13:5427–5438) reported that the extent of HSF activation was reduced in human cells constitutively over expressing HSP70.

7 Regulatory Elements Governing Pathogenesis-Related (PR) Gene Expression

Imre E. Somssich

1 Introduction

Plants, being sedentary organisms, are inevitably confronted during their life cycle with numerous environmentally determined stress situations that can be detrimental to their survival. Apart from tolerating large fluctuations in temperature, changes in the nutrient composition and salinity of the soil, flooding, drought, etc., plants are continuously exposed to harmful air- and soilborne bacterial and fungal pathogens or insect pests. In order to meet such challenges, plants possess both preformed structural physical barriers, such as the cell wall, and have evolved inducible mechanisms which allow them to respond rapidly to external stimuli. In the case of pathogen attack, early perception of the invading agent and the rapidity with which the active defense responses are initiated can strongly influence the outcome of the interaction. Therefore, only by understanding in detail the molecular mechanisms underlying this multifacet plant defense response, can we devise efficient and safe strategies to prevent plant infections by means of genetic engineering techniques.

Numerous excellent reviews exist that deal with the various biochemical reactions frequently found associated with both host and nonhost plant/pathogen interactions (Collinge and Slusarenko 1987; Hahlbrock and Scheel 1987; Bowles 1990; Dixon and Harrison 1990). In nearly all of the systems studied to date, transcriptional activation of specific plant genes play a key role (Dixon and Lamb 1990). In some instances, the functions encoded by these genes are known and include lytic enzymes such as chitinases and β-1,3-glucanases, or key enzymes involved in the biosynthesis of lignin or antifungal phytoalexins, such as phenylalanine ammonia-lyase and 4-coumarate: CoA ligase. For others, no biological function as yet has been found.

Considerable efforts have been made to isolate and characterize defense-related genes and their promoters from a variety of plants. The availability of these genes is a prerequisite for the application of gene transfer methods in modern plant breeding programs aimed at improving plant tolerance to pathogens. In the scope of this review I will try to summarize our present knowledge concerning the transcriptional regulation of a somewhat arbitrarily

Max-Planck-Institut für Züchtungsforschung, Abteilung Biochemie, Carl-von-Linné-Weg 10, D-50829 Köln, FRG

Results and Problems in Cell Differentiation 20
L. Nover (Ed.)
Plant Promoters and Transcription Factors

defined subset of pathogen-induced genes termed pathogenesis-related (PR) protein genes. After scanning the rather large PR protein literature and finding only quite a limited number of reports available on the regulation of PR protein genes, it became obvious that this chapter can only provide a compilation of the current state of a rapidly expanding field of research. More information must be obtained before general conclusions on possible common regulatory mechanisms can be drawn.

2 Pathogenesis-Related (PR) Proteins

Pathogenesis-related (PR) proteins have been defined as proteins coded for by the host plant but induced by various pathogens as well as under stress situations similar to those provoked by pathogens (van Loon 1990). PR proteins were first described over two decades ago in the leaves of Samsun NN and *Xanthi-nc* tobacco plants infected by tobacco mosaic virus (TMV) (Gianinazzi et al. 1970; van Loon and van Kammen 1970). Originally, all of them were identified as additional bands present in native or SDS polyacrylamide gels in TMV-infected leaves. Initial characterization revealed them to be low molecular weight compounds that can be extracted at low pH and appear rather resistant to various plant and commercially available proteases. Proteins with similar characteristics were subsequently detected in a large variety of plants and are now generally assumed to be ubiquitous (van Loon 1985). In the absence of any known functions, the tobacco PR proteins were grouped, PR-1 to PR-5, solely on the basis of their electrophoretic mobilities and immunological cross-reactivities. The advent of recombinant DNA technology opened the way for the cloning of PR protein cDNAs and genes from various plant species, thereby revealing, based on sequence relationships, that the original criteria for classification are unsatisfactory (Linthorst 1991; Cutt and Klessig 1992). In addition, application of molecular techniques have led to the isolation of numerous additional pathogen-inducible plant genes whose products have not been detected by protein gel analysis, but were merely deduced from their nucleic acid sequences. Many of the deduced products have a low molecular weight, appear to range from highly acidic to highly basic, and may or may not be sequence related to the classical PR proteins of tobacco. As numerous such proteins have been included into the class of PR proteins, PR nomenclature has become rather confusing. Table 1 summarizes some features of the various PR protein families. Designation of the five classical tobacco families, PR-1 to -5, has been retained and comprises all proteins identified in several plant species showing serological or sequence relationships (Linthorst 1991). Proteins of the classes PR-2 and PR-3 have been identified as being β-1,3-glucanases and chitinases, respectively (Kauffmann et al. 1987; Legrand et al. 1987; Kombrink et al. 1988).

Table 1. Classification and some characteristics of PR proteins

Family	M_r ($\times 10^{-3}$)	Features	Mode of activation	Reference[a]
PR-1	15–17	Unknown	Gene activation	1
PR-2	31–36	β-1,3-Glucanase	Gene activation	1
PR-3	27–34	Chitinase	Gene activation	1
PR-4	13–15	Unknown	Gene activation	1
PR-5	24–26	Thaumatin-like	Gene activation	1
(PR-6) PI (tomato, potato)	7–13	Proteinase inhibitor	Gene activation	1
(PR-7) P69 (tomato)	69–70	Protease	?	1
(PR-10)	15–17	Unknown	Gene activation	
PcPR1; PcPR2 (parsley)				1
PvPR1; PvPR2 (bean)				1
pSTH-1; pSTH-2 (potato)				1
pI49 (pea)				1
AoPR1 (asparagus)				2
BetvI (birch)				1
Win1/2 (potato)	20–24	Unknown	Gene activation	1
WunI (potato)	12	Unknown	Gene activation	1
PRP1 (potato)	25	HSP26-like	Gene activation	3

[a] 1, See review (Linthorst 1991); 2, Warner et al. (1992); 3, Taylor et al. (1990).

It has been proposed by a committee organized to standardize PR nomenclature, after the 2nd Workshop on PR proteins (Valencia, Spain, October 1989), that PR-6 be alloted to the well-characterized family of proteinase inhibitors (PI) first described by Green and Ryan (1972), and PR-7 to proteases of the P69 type detected in tomato (Vera and Conejero 1988; van Loon 1990). The list of still unclassified PR proteins contained in Table 1 is by no means complete. For the sake of this review, I have considered only those proteins for which corresponding genes have been isolated and functional promoter studies have been initiated. In addition, no attempt is made to incorporate well-characterized proteins such as amylases, peroxidases or thionins into this group (Linthorst 1991). The designation, PR-10, has been arbitrarily given to a number of sequence-related proteins found in numerous dicot and monocot plants (Fristensky et al. 1988; Somssich et al. 1988; Breiteneder et al. 1989; Matton and Brisson 1989; van de Löcht et al. 1990; Walter et al. 1990; Warner et al. 1992).

Many of the PR proteins accumulate extracellularly or in the cell vacuole. The majority of the acidic type PR-1 to PR-5 proteins are located extracellularly, whereas , in general, their basic counterparts are found in the vacuole. Exceptions do exist, as has been shown for the basic tomato PR-4 class protein, P2, which is extracellular (Joosten and de Wit 1989). In contrast, the localization of the PR-6, PR-10, Win1/2, WunI and PRP1 proteins is very likely in the cytosol (Taylor et al. 1990; Linthorst 1991).

3 Stimulation of PR Protein Biosynthesis

PR protein biosynthesis is observed in response to various environmental signals. Infection by many different pathogens, including bacteria, fungi, viruses and viroids, has been shown to result in PR protein accumulation (van Loon 1985; Carr and Klessig 1989). Similarly, treatment of plants or plant tissue with compounds of microbial origin (elicitors) induces PR protein synthesis as do numerous chemical compounds (abiotic elicitors) such as salicyclic acid (SA), benzoic acid, arachidonic acid, 2-thiouracil, Ag^{+}, Cd^{2+} and some plant hormones such as ethylene and cytokinin (Cutt and Klessig 1992). PR protein accumulation is also observed in healthy plants in an organ- and tissue-specific manner during development and in some instances upon tissue wounding (Eyal and Fluhr 1991; Cutt and Klessig 1992). Induced accumulation of PR proteins by pathogens is often not locally restricted to infected areas but also occurs, with some delay, in uninfected parts of the plant (van Loon 1985). This implies that PR proteins may play an important role in enhancing the levels of resistance to secondary challenges by pathogens, a phenomenon nicely demonstrated by Ross (1966) and referred to as systemic acquired resistance or SAR.

As indicated in Table 1, substantial direct or circumstantial evidence exists suggesting that regulation of PR protein biosynthesis occurs primarily at the level of gene transcription. This implies that molecular analyses of the PR genes should reveal important regulatory elements governing their expression. Considering the different environmental signals and developmental cues that activate these genes, it will be of particular interest, once sufficient data have been obtained, to see if common *cis*-regulatory motifs, or varying combinations thereof, are used to differentially regulate PR protein gene expression in diverse plant species. Such functional elements may be present in the regulatory regions of other classes of defense-related genes that show similar expression upon pathogen infection as the PR genes, and are often also regulated in a developmental or tissue-specific mannar (Bowles 1990; Dixon and Harrison 1990; Scheel 1992). This kind of knowledge could possibly open the way for the identification and isolation of regulatory genes controlling the coordinate expression of a whole subset of interdependent defense genes.

4 Regulatory Elements Involved in PR Gene Expression

Delineations of the regulatory elements involved in PR protein gene expression have only recently been initiated. In most studies, rather large fragments of the respective promoters fused to appropriate reporter genes have been analyzed utilizing transgenic plants or plant protoplasts. Table 2 summerizes the current results obtained from such experiments at the time this chapter was written (December 1992).

Table 2. Important *cis*-regulatory regions identified in plant PR promoters

Gene	Inducer	Regulatory regions	References
Tobacco PR-1a	TMV; SA	−643 to −689 of promoter	van de Rhee et al. (1990)
	TMV; SA	Two protein-binding sites within −300 promoter; gene expression negatively controlled	Ohshima et al. (1990); Ohashi et al. (1992); Ohashi and Ohshima (1992)
	TMV; SA	−691 to −902 (element 1) −643 to −689 (element 2) −287 to −643 (element 3) +23 to −287 (element 4)	van de Rhee and Bol (1993)
Tobacco PRb-1b	Ethylene	−141 to −213 of promoter	R. Fluhr (unpubl.)
Bean CH5B (chitinase)	Ethylene	−195 to −422	Broglie et al. (1989)
	Ethylene; elicitor	−236 to −305	Roby et al. (1991)
Tobacco CHN50 (chitinase)		−345 to −788; silencer between −47 and −68	Fukuda et al. (1991)
Arabidopsis pMON8815 (chitinase)	SA	−192 promoter sufficient; −384 to −590 contains negative regulatory element; −590 to −1129 contains positive regulatory element	Samac et al. (1990); Samac and Shah (1991)
Tobacco GLB (glucanase)		Enhancer element −1305 to −1365 (ACG box)	Hart et al. (1993)
Tobacco PR-2b (glucanase)	TMV; SA	−300 promoter sufficient	van de Rhee et al. (1993)
Tobacco gglb50 (glucanase)	TMV; SA; ethephon	−1476 promoter inducible; −446 promoter not inducible	van de Rhee et al. (1993)
Barley EI (glucanase I)	Gibberellic acid (GA_3)	1.9-kb promoter GA_3 responsive only upon deletion of region between −402 and −552	Wolf (1992)
Barley EII (glucanase II)	Gibberellic acid (GA_3)	253-bp promoter sequence upstream of the TATA box sufficient	Wolf (1992)

Table 2. *Continued*

Gene	Inducer	Regulatory regions	References
Tobacco PR-5 (thaumatin-like)	TMV	−718 to −967 of promoter	Albrecht et al. (1992)
Tobacco osmotin	ABA; NaCl; ethylene	−1630-bp promoter	Nelson et al. (1992)
Potato E 32 (proteinase inhibitor II)	Wounding	−210 to −514 of promoter	Keil et al. (1990)
Potato proteinase inhibitor IIK	Wounding	−136 to −557 of promoter	Palm et al. (1990)
Potato proteinase inhibitor IIK	Wounding; methyl jasmonate (MJ); sucrose	Internal promoter fragment from −520 to −625 sufficient	Kim et al. (1992)
Parsley PR1-1	Elicitor	Two independent responsive elements located at −187 to −273 (iF) and −707 to −734	Meier et al. (1991); Tovar Torres et al. (unpubl.)
Parsley PR2	Elicitor	Internal promoter fragment from −52 to −168 sufficient	van de Löcht et al. (1990)
Asparagus AoPR1	Wounding	−982-bp promoter	Warner et al. (1993)
Potato win2	Wounding	Sequences between −558 and −2177 of promoter	Stanford et al. (1990)
Potato wun1	Wounding	1.2-kb promoter	Logemann et al. (1989)
Potato prp1	fungal infection	Internal promoter fragment from −130 to −402	Martini et al. (1993)

4.1 PR-1 Genes

Conflicting results have been obtained when various upstream regions of the tobacco PR-1a promoter fused to the coding region of the reporter gene, β-glucuronidase (GUS), were assayed for TMV and salicylic acid (SA) induced expression in transgenic tobacco plants. Ohshima et al. (1990) reported that the first 300 bp of the PR-1a promoter was sufficient for regulated gene expression. Histochemical staining of leaves for GUS activity following TMV infection confirmed that the 300-bp promoter fragment yielded qualitatively the same response as a 2.4-kb PR-1a promoter. In contrast, both Van de Rhee et al. (1990) and Beilmann et al. (1991) failed to observe TMV or SA inducibility in experiments using PR-1a promoter fragments of similar size. Instead, van de Rhee et al. (1990) identified a region between −643 and −689 to be important for TMV and SA inducibility. Possibly, *cis* activation by the 35S enhancer positioned immediately upstream of the PR1-a promoter/GUS construct in the transformation cassette used by Ohshima et al. (1990) may have influenced expression (Beilmann et al. 1992). However, the situation remains unresolved, as recent results with an enhancer that is not responsive to SA confirmed the sufficiency of the small 300-bp PR-1a promoter to drive TMV- and SA-dependent reporter gene expression (Ohashi and Ohshima 1992). Additionally, gel mobility-shift assays revealed two protein binding sites within the 300-bp PR-1a promoter. Binding was found using nuclear protein extracts from healthy Samsun NN tobacco leaves not expressing PR-1 protein, but not with nuclear extracts derived from an interspecific hybrid, *Nicotiana glutinosa* × *Nicotina debneyi*, in which the PR-1 protein is expressed constitutively (Ohashi and Oshima 1992; Ohashi et al. 1992). These experiments suggest that the interaction of a repressor with this region negatively regulates PR-1 gene expression. Indeed, multiple regulatory elements may modulate the tobacco PR-1a gene as has recently been implied by van de Rhee and Bol (1993).

Promoter deletion analysis of the tobacco basic PR-1 protein gene, prb-1b, identified sequences between −141 and −213 to be required for ethylene inducibility (R. Fluhr unpubl.). This region contains a G-box consensus core sequence, CACGTG, at position −187, which is also commonly found in numerous plant promoters and which has been shown to bind a subclass of bZIP DNA-binding proteins (Giuliano et al. 1988; Katagiri et al. 1989; Schulze-Lefert et al. 1989; Eyal et al. 1992; Schindler et al. 1992).

4.2 PR-2, PR-3 Genes

Gene expression of β-1,3-glucanase (PR-2) and chitinase (PR-3) appears coordinately regulated in various plant species upon induction by bacteria, fungi, fungal elicitors, ethylene, SA, wounding, cytokinin and during development (Meins et al. 1992). One common motif, AGCCGCC (AGC box), has been found to be present in nearly all chitinase and glucanase promoters so far analyzed (Ohme-Takagi and Shinshi 1990; Hart et al. 1993). A 61-bp fragment derived from the tobacco β-1,3-glucanase GLB promoter, containing two copies of the AGC box and one sequence identical at 6 of 7 bp, was shown to have enhancer activity in *Nicotiana plumbaginifolia* protoplasts (Hart et al. 1993). Point mutations in all three of these sequences eliminated enhancer activity. In addition, a slightly shorter version of this fragment (49 bp), still containing the two AGC boxes, showed specific binding to proteins derived from leaf extracts. Binding was more pronounced with extracts from old lower leaves than with those from young upper leaves, consistent with the endogenous expression behavior of the GLB gene (Felix and Meins 1986). Binding could be competed with diverse DNA fragments containing AGC box sequences, but not with the 49-bp fragment in which both AGC boxes were mutated, implying protein binding to the GLB enhancer element. Whether the AGC elements also function as enhancers when positioned at their original GLB promoter sites (−1305 to −1365), and whether they are alone capable of protein binding, were not addressed.

Functional analysis of the bean chitinase CH5B promoter in transgenic tobacco plants revealed a region between −195 to −422 important for ethylene responsiveness (Broglie et al. 1989). Although this region also contains one AGC box at position −205 to −211, subsequent transient expression assays using bean protoplasts more precisely defined the ethylene responsive element to reside between −236 and −305 (Roby et al. 1991). As deletion of the AGC box sequence also had no effect on overall expression, it does not appear to function as an enhancer element in the bean system. No other obvious similarities to known plant *cis*-acting sequence elements (Weising and Kahl 1991; Katagiri and Chua 1992; Kuhlemeier 1992), nor to the ethylene responsive segment of the tobacco PRb-1b, were found within the ethylene responsive promoter region of the CH5b gene. Therefore, a more precise analysis will be required to delimit the functional element(s).

Transient expression assays of various tobacco chitinase CHN50 promoter/GUS constructs in tobacco protoplasts identified a distal region between −345 to −788 containing sequences required for high level expression, whereas the region between −47 and −68 contains a silencer (Fukuda et al. 1991). Two 12-bp direct repeat sequences, which include one AGC box motif, are located at −698 and −637. The silencer region contains an inverted repeat located between −49 to −70.

Functional comparison of the barley glucanase isozyme I and II genes, using aleurone protoplasts and chloramphenicol acetyltransferase (CAT) as the reporter gene, revealed some interesting differences (Wolf 1992). In transient assays, the −1900 promoter of the isozyme I gene showed no gibberellic acid (GA_3) responsiveness. However, upon removal of a unique region between −402 and −552, not present in the isozyme II gene promoter, low GA_3 responsiveness was restored suggesting the presence of a silencer within this 150-bp DNA stretch. In contrast, the −2900 isozyme II promoter gave high GA_3 responsive expression. Truncation of this promoter down to within 253 bp of the TATA box was still sufficient for GA_3 stimulation. It contains a sequence, CAAACTATTTTTCTCAAAA, at about −240, which is similar to an element, CTCTTTTAAATGAG, designated the pyrimidin box, found upstream of the rice α-amylase gene, pOSamy-a-L, and which has been shown to interact with a GA_3-induced protein factor (Ou-Lee et al. 1988). A second sequence, TAACAAC, at position −170, is similar to a GA_3 responsive motif (GARE) found in the promoter of the barley α-amylase gene, Amy 1/6–4 (Skriver et al. 1991). However, many amylase genes contain the pyrimidin box sequence (C/T CTTTT C/T) within their promoters, yet the few functionally studied promoters imply that this box is not required for GA_3 inducibility (Huang et al. 1990; Skriver et al. 1991). Therefore, the importance of these findings must await experimental verification.

Developmental and pathogen-inducible expression of the *Arabidopsis* acidic chitinase gene, pMON8815, was addressed by stable integration of pMON8815 promoter/GUS fusion constructs into *Arabidopsis* and tomato (Samac et al. 1990; Samac and Shah 1991). Histochemical analysis of GUS expression in *Rhizoctonia solani*-inoculated *Arabidopsis* and *Alternaria solani* or *Phytophthora infestans*-inoculated tomato plants bearing the 1129-bp promoter constructs revealed staining of mesophyll cells surrounding the infection sites. In contrast, wounding had no effect. Experiments with four additional promoter deletion constructs, in transgenic *Arabidopsis*, delineate the presence of a negative regulatory element between −384 and −590 and the sufficiency of the first 192 bp of the promoter to drive tissue-specific and pathogen-inducible expression. Inspection of the promoter sequence failed to reveal the presence of any previously identified *cis*-acting motifs.

4.3 PR-5 Genes

Some limited studies dealing with the regulation of the tobacco acidic (thaumatin-like) and basic (osmotin) PR-5 protein genes have also been reported. Analysis of transgenic tobacco plants containing acidic PR-5/E2 promoter/GUS constructs of various lengths identified the region between −718 and −1364 to be involved in TMV induction of gene expression (Albrecht et al. 1992). GUS activity was found both locally within TMV-infected leaves, and systemically within virus-free leaf material. Promoter sequence comparison of the PR-5, gene E2 (van Kan et al. 1989), with upstream regions of other TMV-inducible tobacco genes, failed to detect obvious similarities.

Transcriptional activation by the hormones abscisic acid (ABA) and ethylene of a chimeric construct bearing 1630 bp of the osmotin promoter fused to the reporter gene, GUS, was observed in transgenic tobacco plants (Nelson et al. 1992). NaCl (100–200 mM) also activated this promoter. The promoter sequence lacks any previously identified, functionally important, ABA-responsive motifs (Guiltinan et al. 1990; Mundy et al. 1990; Skriver et al. 1991). Interestingly, recent histochemical data also demonstrate specific osmotin promoter activity during pollen and fruit desiccation (Kononowicz et al. 1992).

4.4 PR-6 Genes

Promoter studies of the potato proteinase inhibitor IIK gene in transgenic tobacco uncovered a 421-bp sequence between −136 to −557, required for wound-induced expression (Palm et al. 1990). A 10-bp sequence, AAGCGTAAGT, from within this region (−165 to −155), binds a wound-inducible nuclear protein(s) derived from tomato leaves. This fragment lies directly adjacent to an 8-bp motif, ACCTTGCC, which is present in several elicitor- and light-responsive promoters (Lois et al. 1989; Palm et al. 1990).

These results conflict somewhat with those obtained from Kim et al. (1991, 1992) who reported that an internal fragment between bp −520 and −625 was sufficient to confer a response to methyl jasmonate (MJ), wounding and sucrose when placed upstream of the nos −101 promoter. Low wound inducibility was still observed with 3′ promoter deletions lacking sequences downstream of −574, suggesting that wound responsive elements are located 5′ thereof. The presence of a number of wound-responsive elements within the proteinase inhibitor gene II promoter, as has also been resported and discussed by Keil et al. (1990), may explain this discrepency. In the case of MJ, the response was eliminated upon 5′ deletion down to −573. This, together with additional results using 3′ promoter deletions, indicate that the MJ responsive element coincides with the position of a G-box (CACGTG) motif at −574/573 (Giuliano et al. 1988; Kim et al. 1992).

4.5 PR-10 Genes

In parsley, three genes encoding two isoforms of a small (Mr 16500) acidic protein designated pcPR1 have been cloned and shown to be transcriptionally activated by fungal elicitor treatment of cultured cells (Somssich et al. 1988; Meier et al. 1991). By means of in vivo DNA footprinting, two putative sites of protein-DNA interaction were identified within the first 500bp of the pcPR1-1 promoter (Meier et al. 1991). The TATA distal footprint, designated iF and located between −220 and −255, was elicitor-dependent and correlated with the onset of PR1 transcription. The second footprinted region, cF, located around −130, was observed irrespective of elicitor treatment. A series of PR1-1 promoter deletion/GUS reporter gene constructs, tested transiently in parsley protoplasts, confirmed the functional importance of the iF region for elicitor-mediated expression (J. Tovar Torres and I.E. Somssich unpubl.). Block mutation of the cF region increased basal expression levels but had no influence on elicitor inducibility. By mutational analysis, a second elicitor responsive element, uE, that can function independently from the iF region, was detected further upstream between −707 and −734. No obvious sequence similarities exist between the iF and the uE regions. Interestingly, the iF region contains a 9-mer, GAGTAACAT, also present at position −678 to −686 in the tobacco PR-1a gene (Payne et al. 1988). It resides within that region of the PR-1a promoter which was demonstrated to be functionally important for TMV and SA inducibility (van de Rhee et al. 1990).

Functional analysis of parsley PR2 promoter/GUS constructs in parsley protoplasts identified a region between −52 and −168 to be both necessary and sufficient for strong elicitor-mediated gene expression (van de Löcht et al. 1990). We have recently demonstrated specific binding of a nuclear protein to this region and have isolated cDNA clones encoding putative transcription factors (U. van de Löcht, I. Meier, G. Trezzini and I.E. Somssich, unpubl.). We can find no sequence relationships between the functionally identified iF and uE regions of the PR1-1 promoter and the relevant region of the PR2 gene promoter.

A wound-responsive gene, AoPR1, sequence related to the parsley PR1 and PR2 protein genes, has recently been isolated from asparagus (Warner et al. 1992, 1993). Histochemical analysis of transgenic tobacco plants bearing a 982-bp AoPR1 promoter/GUS construct showed strong GUS expression localized to wounded sites and in the vicinity of necrotic lesions caused by infection with *Botrytis cinerea* (Warner et al. 1993). GUS activity was also observed in mature pollen grains. The AoPR1 promoter contains sequence motifs very similar to the iF and uE regions found in the parsley PR1 promoter, to the ACCTGACC motif first described by Lois et al. (1989) in the parsley phenylalanine ammonia-lyase gene, and to the H-box sequence $CCTACC(N)_7CT$ described by Loake et al. (1992) in the bean chalcone synthase promoter.

4.6 PRP1 Gene of Potato

Infection of potato, cultivar Datura, with the fungus *Phytophthora infestans* results in rapid transcriptional activation of a gene encoding a small (Mr = 25 054) acidic polypeptide, PRP1, sharing similarity to the 26-kDa heat shock protein of soybean (Taylor et al. 1990). Functional analysis of the prp1-1 promoter in transgenic potato plants identified a 273-bp sequence from −130 to −402 to be sufficient for rapid fungus-mediated transcription of the GUS reporter gene (Martini et al. 1993). Surprisingly, fusion of a longer promoter fragment (from +26 to −698) did not confer fungus inducibility, suggesting that sequences upstream of −403 contain negative regulatory *cis*-acting elements. No obvious similarities to other known functional elements were found.

5 Signal Transduction Leading to PR Gene Expression

Studies dealing with the identification of *cis*-acting elements and transcription factors controlling gene expression are always intimately linked to the questions concerning the endogenous signal molecules and the signal transduction pathways that are utilized. This becomes especially obvious e.g., for the systemic induction of the tobacco PR proteins upon TMV infection (Brederode et al. 1991). Local induction of the PR genes is shortly followed by the transcriptional activation of these genes in neighboring, uninfected plant tissues. A number of chemical compounds such as SA, ethylene, MJ and systemin have been shown to induce PR gene expression and are present, in vivo, at concentrations sufficient to act as inducers (Pearce et al. 1991; Farmer and Ryan 1992; Malamy and Klessig 1992; Raskin 1992; Staswick 1992). In addition, recent exciting results promote the somewhat neglected idea that propagation of electric activity can also be a systemic signal (Scott 1967; Wildon et al. 1992).

Very little is known about the signal transduction pathway(s) required for PR gene expression. Evidence obtained in tobacco suggests the involvement of multiple, independent transducing pathways. For instance, infection by TMV results in the concerted expression of all five PR gene classes (van Loon 1985). In contrast, differential expression of subsets of these genes is observed during normal development in an organ- and tissue-specific fashion (Lotan et al. 1989; Memelink et al. 1990; Neale et al. 1990; Ori et al. 1990; Cote et al. 1991). Cytokinin, xylanase and SA also differentially activate the expression of specific PR genes (Hooft van Huijsduijnen et al. 1986; Memelink et al. 1990; Eyal et al. 1992). Independent regulation of distinct members of the same gene family, as exemplified by the β-1,3-glucanases and chitinases, has also been demonstrated (see review by Meins et al. 1992). In cultured parsley cells, changes in the intensity of specific ion fluxes lead to differential activation of the PR1 and PR2 genes (Scheel et al. 1991). Differential accumulation of mRNAs for intracellular and extracellular PR proteins during compatible and incompatible plant/pathogen interactions has also been reported in tomato (van Kan et al. 1992). Therefore, the limited information available implies a rather complex regulatory network involving distinct or only partially overlapping pathways.

6 Conclusions

In plants, as in other higher eukaryotes, cell- and tissue-specific patterns of gene expression can be achieved by using common or diverse signal transduction pathways, by differential expression of separate genes of multigene families, and by the interaction of diverse transcription factors with defined *cis*-acting DNA sequences. The presence of multigene families is often observed in plants, as exemplified by the PR genes of tobacco (Eyal and Fluhr 1991; Linthorst 1991; Kuhlemeier 1992). Individual gene family members may respond to diverse stimuli through multiple nonoverlapping or partially overlapping *cis*-acting elements that can discriminate between distinct transcription factors. Alternatively, this is managed by separate family members containing only single or unique combinations of such elements. For this reason, functional analyses of the regulatory regions of entire PR gene families may be required before a more concise and general picture of PR gene expression can evolve. Considering the heterogeneity of genes classified as PR protein genes in diverse plant species, it is foreseeable that numerous distinct regulatory elements will be found. Other levels of control, such as post-transcriptional regulation, should also not be excluded. Recently, posttranscriptional regulation has been reported for the osmotin genes of tobacco (LaRosa et al. 1992).

Assessment of our current understanding on the regulatory elements involved in the expression of individual PR protein genes, as depicted in this review, reveals how limited our knowledge in this area of research still is. We have begun to delineate some important regulatory regions but, in most cases, are far from pinpointing specific, functionally defined, *cis*-acting elements. In some instances, PR promoter elements sharing homology to known *cis*-acting sequences have been identified, but their functional relevance has yet to be demonstrated. With one exception, no putative transcription factors have been cloned. Nevertheless, considering the inducible nature of the PR genes, their continued study will add considerably to a deeper insight into the general molecular mechanisms involved in plant gene regulation.

Acknowledgements. I would like to thank the many colleagues who, by providing me with preprints or unpublished data of their work, have helped make this review as up-to-date as possible. I am especially grateful to Drs. Klaus Hahlbrock, Dierk Scheel and Erich Kombrink (all Cologne) for discussions and critical reading of the manuscript.

References

Albrecht H, van de Rhee MD, Bol JF (1992) Analysis of *cis*-regulatory elements involved in induction of a tobacco PR-5 gene by virus infection. Plant Mol Biol 18:155–158

Beilmann A, Pfitzner AJP, Goodman HM, Pfitzner UM (1991) Functional analysis of the pathogenesis-related 1a protein gene minimal promoter region. Comparison of reporter gene expression in transient and in stable transfections. Eur J Biochem 196:415–421

Beilmann A, Albrecht K, Schultze S, Wanner G, Pfitzner UM (1992) Activation of a truncated PR-1 promoter by endogenous enhancers in transgenic plants. Plant Mol Biol 18:65–78

Bowles DJ (1990) Defense-related proteins in higher plants. Annu Rev Biochem 59:873–907

Brederode FT, Linthorst HJM, Bol JF (1991) Differential induction of acquired resistance and PR gene expression in tobacco by virus infection, ethephon treatment, UV light and wounding. Plant Mol Biol 17:1117–1125

Breiteneder H, Pettenburger K, Bito A, Valenta R, Kraft D, Rumpold H, Schreiner O, Breitenbach M (1989) The gene coding for the major birch pollen allergen, *BetvI*, is highly homologous to a pea disease resistance response gene. EMBO J 8:1935–1938

Broglie K, Biddle P, Cressman R, Broglie R (1989) Functional analysis of DNA sequences responsible for ethylene regulation of a bean chitinase gene in transgenic tobacco. Plant Cell 1:599–607

Carr JP, Klessig DF (1989) The pathogenesis-related proteins in plants. In: Setlow JK (ed) Genetic engineering, principles and methods. Plenum Press, New York, pp 65–109

Collinge DB, Slusarenko AJ (1987) Plant gene expression in response to pathogens. Plant Mol Biol 9:389–410

Cote F, Cutt JR, Asselin A, Klessig DF (1991) Pathogenesis-related acidic β-1,3-glucanase genes of tobacco are regulated by both stress and developmental signals. Mol Plant Microbe Interact 4:173–181

Cutt JR, Klessig DF (1992) Pathogenesis-related proteins. In: Boller T, Meins F (eds) Genes involved in plant defence. Springer, Berlin Heidelberg New York, pp 209–243

Dixon RA, Harrison MJ (1990) Activation, structure, and organization of genes involved in microbial defense in plants. Adv Genet 28:165–234

Dixon RA, Lamb CJ (1990) Molecular communication in interactions between plants and microbial pathogens. Annu Rev Plant Physiol Plant Mol Biol 41:339–367

Eyal Y, Fluhr R (1991) Cellular and molecular biology of pathogenesis related proteins. In: Miflin BJ (ed) Oxford surveys of plant molecular and cell biology. Oxford University Press, Oxford, pp 223–254

Eyal Y, Sagee O, Fluhr R (1992) Dark-induced accumulation of a basic pathogenesis-related (PR-1) transcript and a light requirement for its induction by ethylene. Plant Mol Biol 19:589–599

Farmer EE, Ryan CA (1992) Octadecanoid precursors of jasmonic acid activate the synthesis of wound-inducible proteinase inhibitors. Plant Cell 4:129–134

Felix G, Meins FJ (1986) Developmental and hormonal regulation of β-1,3-glucanase in tobacco. Planta 167:206–211

Fristensky B, Horovitz D, Hadwiger LA (1988) cDNA sequences for pea disease resistance response genes. Plant Mol Biol 11:713–715

Fukuda Y, Ohme M, Shinshi H (1991) Gene structure and expression of a tobacco endochitinase gene in suspension-cultured tobacco cells. Plant Mol Biol 16:1–10

Gianinazzi S, Martin C, Vallee JC (1970) Hypersensibilité aux virus, température et protéines soluble chez le *Nicotiana* Xanthi n.c. Apparition de nouvelles macromolécules lors de la répression de la synthèse virale. CR Acad Sci Paris Ser D 270:2383

Giuliano G, Pichersky E, Malik VS, Timko MP, Scolnic PA, Cashmore AR (1988) An evolutionary conserved protein binding sequence upstream of a plant light-regulated gene. Proc Natl Acad Sci USA 85:7089–7093

Green TR, Ryan CA (1972) Wound-induced proteinase inhibitor in plant leaves: a possible defense mechanism against insects. Science 175:776–777

Guiltinan MJ, Marcotte WR, Quatrano RS (1990) A plant leucine zipper protein that recognizes an abscisic acid response element. Science 250:267–271

Hahlbrock K, Scheel D (1987) Biochemical responses of plants to pathogens. In: Chet I (ed) Innovative approaches to plant disease control. Wiley, New York, pp 229–254

Hart CM, Nagy F, Meins FJ (1993) A 61 bp enhancer element of the tobacco β-1,3-glucanase B gene interacts with a regulated nuclear protein(s). Plant Mol Biol 21:121–131

Hooft van Huijsduijnen RAM, van Loon LC, Bol JF (1986) cDNA cloning of six mRNAs induced by TMV infection of tobacco and a characterization of their translation products. EMBO J 5:2057–2061

Huang N, Sutliff TD, Litts JC, Rodriguez RL (1990) Classification and characterization of the rice α-amylase multigene family. Plant Mol Biol 14:655–668

Joosten MHAJ, de Wit PJGM (1989) Identification of several pathogenesis-related proteins in tomato leaves inoculated with *Cladosporium fulvum* (syn. *Fulvia fulva*) as 1,3-b-glucanases and chitinases. Plant Physiol 89:945–951

Katagiri F, Chua N-H (1992) Plant transcription factors: present knowledge and future challenges. Trends Genet 8:22–27

Katagiri F, Lam E, Chua N-H (1989) Two tobacco DNA-binding proteins with homology to the nuclear factor CREB. Nature 340:727–730

Kauffmann S, Legrand M, Geoffroy P, Fritig B (1987) Biological function of "pathogenesis-related" proteins: four PR proteins of tobacco have 1,3-b-glucanase activity. EMBO J 6:3209–3212

Keil M, Sanchez-Serrano J, Schell J, Willmitzer L (1990) Localization of elements important for the wound-inducible expression of a chimeric potato proteinase inhibitor II-CAT gene in transgenic tobacco plants. Plant Cell 2:61–70

Kim S-R, Costa MA, An G (1991) Sugar response element enhances wound response of potato proteinase inhibitor II promoter in transgenic tobacco. Plant Mol Biol 17:973–983

Kim S-R, Choi J-L, Costa MA, An G (1992) Identification of G-box sequence as an essential element for methyl jasmonate response of potato proteinase inhibitor II promoter. Plant Physiol 99:627–631

Kombrink E, Schröder M, Hahlbrock K (1988) Several "pathogenesis-related" proteins in potato are 1,3-b-glucanases and chitinases. Proc Natl Acad Sci USA 85:782–786

Kononowicz AK, Nelson DE, Singh NK, Hasegawa PM, Bressan RA (1992) Regulation of the osmotin gene promoter. Plant Cell 4:513–524

Kuhlemeier C (1992) Transcriptional and post-transcriptional regulation of gene expression in plants. Plant Mol Biol 19:1–14

LaRosa PC, Chen Z, Nelson DE, Singh NK, Hasegawa PM, Bressan RA (1992) Osmotin gene expression is posttranscriptionally regulated. Plant Physiol 100:409–415

Legrand M, Kauffmann S, Geoffroy P, Fritig B (1987) Biological function of pathogenesis-related proteins: four tobacco pathogenesis-related proteins are chitinases. Proc Natl Acad Sci USA 84:6750–6754

Linthorst HJM (1991) Pathogenesis-related proteins of plants. Crit Rev Plant Sci 10:123–150

Loake GJ, Faktor O, Lamb CJ, Dixon RA (1992) Combination of H-box [CCTACC(N)$_7$CT] and G-box (CACGTG) *cis* elements is necessary for feed-forward stimulation of a chalcone synthase promoter by the phenylpropanoid-pathway intermediate *p*-coumaric acid. Proc Natl Acad Sci USA 89:9230–9234

Logemann J, Lipphardt S, Lörz H, Häuser I, Willmitzer L, Schell J (1989) 5′ Upstream sequences from the wunl gene are responsible for gene activation by wounding in transgenic plants. Plant Cell 1:151–158

Lois R, Dietrich A, Halbrock K, Schulz W (1989) A phenylalanine ammonia-lyase gene from parsley: structure, regulation and identification of elicitor and light responsive *cis*-acting elements. EMBO J 8:1641–1648

Lotan T, Ori N, Fluhr R (1989) Pathogenesis-related proteins are developmentally regulated in tobacco flowers. Plant Cell 1:881–887

Malamy J, Klessig DF (1992) Salicylic acid and plant disease resistance. Plant J 2:643–654

Martini N, Egen M, Rüntz I, Strittmatter G (1993) Promoter sequences of a potato pathogenesis-related gene mediate transcriptional activation selectively upon fungal infection. Mol Gen Genet 236:179–186

Matton DP, Brisson N (1989) Cloning, expression, and sequence conservation of pathogenesis-related gene transcripts of potato. Mol Plant Microbe Interact 2:325–331

Meier I, Hahlbrock K, Somssich IE (1991) Elicitor-inducible and constitutive in vivo DNA footprints indicate novel *cis*-acting elements in the promoter of a parsley gene encoding pathogenesis-related protein 1. Plant Cell 3:309–315

Meins FJ, Neuhaus J-M, Sperisen C, Ryals J (1992) The primary structure of plant pathogenesis-related glucanohydrolases and their genes. In: Boller T, Meins F (eds) Genes involved in plant defense. Springer, Berlin Heidelberg New York, pp 245–282

Memelink J, Linthorst HJM, Schilperoort RA, Hoge JHC (1990) Tobacco genes encoding acidic and basic isoforms of pathogenesis-related proteins display different expression patterns. Plant Mol Biol 14:119–126

Mundy J, Yamaguchi-Shinozaki K, Chua N-H (1990) Nuclear proteins bind conserved elements in the abscisic acid-responsive promoter of a rice *rab* gene. Proc Natl Acad Sci USA 87:1406–1410

Neale AD, Wahleithner JA, Lund M, Bonnett HT, Kelly A, Meeks-Wagner DR, Peacock WJ, Dennis ES (1990) Chitinase, β-1,3-glucanase, osmotin, and extensine are expressed in tobacco explants during flower formation. Plant Cell 2:673–684

Nelson DE, Raghothama KG, Singh NK, Hasegawa PM, Bressan RA (1992) Analysis of structure and transcriptional activation of an osmotin gene. Plant Mol Biol 19:577–588

Ohashi Y, Ohshima M (1992) Stress-induced expression of genes for pathogenesis-related proteins in plants. Plant Cell Physiol 33:819–826

Ohashi Y, Ohshima M, Itoh H, Matsuoka M, Watanabe S, Murakami T, Hosokawa D (1992) Constitutive expression of stress-inducible genes, including pathogenesis-related 1 protein gene in a transgenic interspecific hybrid of *Nicotiana glutinosa* × *Nicotiana debneyi*. Plant Cell Physiol 33:177–187

Ohme-Takagi M, Shinshi H (1990) Structure and expression of a tobacco β-1,3-glucanase gene. Plant Mol Biol 15:941–946

Ohshima M, Itoh H, Matsuoka M, Murakami T, Ohashi Y (1990) Analysis of stress-induced or salicylic acid-induced expression of the pathogenesis-related 1a protein gene in transgenic tobacco. Plant Cell 2:95–106

Ori N, Sessa G, Lotan T, Himmelhoch S, Fluhr R (1990) A major stylar matrix polypeptide (sp41) is a member of the pathogenesis-related proteins superclass. EMBO J 9:3429–3436

Ou-Lee T-M, Turgeon R, Wu R (1988) Interaction of a gibberellin-induced factor with the upstream region of an α-amylase gene in rice aleurone tissue. Proc Natl Acad Sci USA 85:6366–6369

Palm CJ, Costa MA, An G, Ryan CA (1990) Wound-inducible nuclear protein binds DNA fragments that regulate a proteinase inhibitor II gene from potato. Proc Natl Acad Sci USA 87:603–607

Payne G, Parks TD, Burkhart W, Dincher S, Ahl P, Metraux JP, Ryals J (1988) Isolation of the genomic clone for pathogenesis-related protein 1a from *Nicotiana tabacum* cv. *Xanthi-nc*. Plant Mol Biol 11:89–94

Pearce G, Strydom D, Johnson S, Ryan CA (1991) A polypeptide from tomato leaves induces wound-inducible proteinase inhibitor proteins. Science 253: 895–898

Raskin I (1992) Salicylate, a new plant hormone. Plant Physiol 99:799–803

Roby D, Broglie K, Gaynor J, Broglie R (1991) Regulation of a chitinase gene promoter by ethylene and elicitors in bean protoplasts. Plant Physiol 97:433–439

Ross AF (1966) Systemic effects of local lesion formation. In: Beemster ABR, Dijkstra J (eds) Viruses of plants. North-Holland, Amsterdam, pp 127–150

Samac DA, Shah DM (1991) Developmental and pathogen-induced activation of the *Arabidopsis* acidic chitinase promoter. Plant Cell 3:1063–1072

Samac DA, Hironaka CM, Yallaly PE, Shah DM (1990) Isolation and characterization of the genes encoding basic and acidic chitinases in *Arabidopsis thaliana*. Plant Physiol 93:907–914

Scheel D (1992) Molecular aspects of host defense responses after infection by pathogenic fungi: an overview. In: Stahl U, Tudzynski P (eds) Molecular biology of filamentous fungi. VCH, Weinheim, pp 125–138

Scheel D, Colling C, Hedrich R, Kawalleck P, Parker JE, Sacks WR, Somssich IE, Hahlbrock K (1991) Signals in plant defense gene activation. In: Hennecke H, Verma DPS (eds) Advances in molecular genetics of plant-microbe interactions, vol 1. Kluwer, Dordrecht, pp 373–380

Schindler U, Beckmann H, Cashmore AR (1992) TGA1 and G-box binding factors: two distinct classes of *Arabidopsis* leucine zipper proteins compete for the G-box-like element TGACGTGG. Plant Cell 4:1309–1319

Schulze-Lefert P, Dangl JL, Becker-André M, Hahlbrock K, Schulz W (1989) Inducible in vivo DNA footprints define sequences necessary for UV light activation of the parsley chalcone synthase gene. EMBO J 8:651–656

Scott BIH (1967) Electrical fields in plants. Annu Rev Plant Physiol 18:409–418

Skriver K, Olsen FL, Rogers JC, Mundy J (1991) *Cis*-acting DNA elements responsive to gibberellin and its anagonist abscisic acid. Proc Natl Acad Sci USA 88:7266–7270

Somssich IE, Schmelzer E, Kawalleck P, Hahlbrock K (1988) Gene structure and in situ transcript localization of pathogenesis-related protein 1 in parsley. Mol Gen Genet 213:93–98

Stanford AC, Northcote DH, Bevan MW (1990) Spatial and temporal patterns of transcription of a wound-induced gene in potato. EMBO J 9:593–603

Staswick PE (1992) Jasmonate, genes, and fragrant signals. Plant Physiol 99:804–807
Taylor JL, Fritzemeier K-H, Häuser I, Kombrink E, Rohwer F, Schröder M, Strittmatter G, Hahlbrock K (1990) Structural analysis and activation by fungal infection of a gene encoding a pathogenesis-related protein in potato. Mol Plant Microbe Interact 3:72–77
van de Löcht U, Meier I, Hahlbrock K, Somssich IE (1990) A 125 bp promoter fragment is sufficient for strong elicitor-mediated gene activation in parsley. EMBO J 9:2945–2950
van de Rhee MD, Bol JF (1993) Multiple regulatory elements are involved in the induction of the tobacco PR-1a gene by virus infection and salicylate treatment. Plant J 3:71–82
van de Rhee MD, van Kan JAL, Gonzales-Jaen MT, Bol JF (1990) Analysis of regulatory elements involved in the induction of two tobacco genes by salicylate treatment and virus infection. Plant Cell 2:357–366
van de Rhee MD, Lemmers R, Bol JF (1993) Analysis of regulatory elements involved in stress-induced and organ-specific expression of tobacco acidic and basic β-1,3-glucanase genes. Plant Mol Biol 21:451–461
van Kan JAL, Joosten MHAJ, Wagemakers CAM, van den Berg-Velthuis GCM, De Wit PJGM (1992) Differential accumulation of mRNAs encoding extracellular and intracellular PR proteins in tomato induced by virulent and avirulent races of *Cladosporium fulvum*. Plant Mol Biol 20:513–527
van Kan JAL, van de Rhee MD, Zuidema D, Cornelissen BJC, Bol JF (1989) Structure of the tobacco genes encoding thaumatin-like proteins. Plant Mol Biol 12:153–155
van Loon LC (1985) Pathogenesis-related proteins. Plant Mol Biol 4:111–116
van Loon LC (1990) The nomenclature of pathogenesis-related proteins. Physiol Mol Plant Pathol 37:229–230
van Loon LC, van Kammen A (1970) Polyacrylamide disc electrophoresis of the soluable leaf proteins from *Nicotiana tabacum* var. "Samsun" and "Samsun NN". II. Changes in protein constitution after infection with TMV. Virology 40:199
Vera P, Conejero V (1988) Pathogenesis-related proteins of tomato. P-69 as an alkaline endoproteinase. Plant Physiol 87:58–63
Walter MH, Liu J-W, Grand C, Lamb CJ, Hess D (1990) Bean pathogenesis-related (PR) proteins deduced from elicitor-induced transcripts are members of a ubiquitous new class of conserved PR proteins including pollen allergens. Mol Gen Genet 222:353–360
Warner SAJ, Scott R, Draper J (1992) Characterization of a wound-induced transcript from the monocot asparagus that shares similarity with a class of intracellular pathogenesis-related (PR) proteins. Plant Mol Biol 19:555–561
Warner SAJ, Scott R, Draper J (1993) Isolation of an asparagus intracellular PR gene (AoPR1) wound-responsive promoter by the inverse polymerase chain reaction and its characterization in transgenic tobacco. Plant J 3:191–201
Weising K, Kahl G (1991) Towards an understanding of plant gene regulation: the action of nuclear factors. Z Naturforsch 46c:1–11
Wildon DC, Thain JF, Minchin PEH, Gubb IR, Reilly AJ, Skipper YD, Doherty HM, O'Donnell PJ, Bowles DJ (1992) Electrical signalling and systemic proteinase inhibitor induction in the wounded plant. Nature 360:62–65
Wolf N (1992) Structure of the genes encoding *Hordeum vulgare* (1→3,1→4)-β-glucanase isoenzymes I and II and functional analysis of their promoters in barley aleurone protoplasts. Mol Gen Genet 234:33–42

8 Analysis of Tissue-Specific Elements in the CaMV 35S Promoter

Eric Lam

1 Introduction

Development of a multicellular organism is critically dependent on the proper temporal and spatial cues that regulate gene expression. Thus, in any one cell at a particular time, only a subset of the total genetic information is expressed. Manifestation of a gene's function is a multistep process that starts at the chromosome of a cell in the form of transcription and ends at the production of a functional protein in the proper place and time. Thus, regulation can be applied at numerous points in this pathway. The early step of transcription initiation represents one of such points of control. Typically, the regulatory elements that are responsible for directing specific transcription initiation reside within the 5′ promoter region of the gene. The elucidation of the mechanisms how different promoter sequences generate tissue-specific gene expression will be invaluable to our understanding of how transcription may be modulated during development.

The 35S promoter of cauliflower mosaic virus (CaMV) has been a useful paradigm for the study of promoter elements responsible for tissue- and developmental stage-specific gene expression in higher plants (Benfey and Chua 1990). In the host, this promoter is responsible for the production of a 35S transcript that covers the entire length of the viral genome. Promoter fragments of about 1000 to 800 base pairs (bp) upstream from the transcription start site have been shown to be active in cells of monocots and dicots (Fromm et al. 1985; Odell et al. 1985; Harpster et al. 1988). Recently, it has also been found to be active in *E. coli* (Assaad and Signer 1990) as well as the fission yeast *S. pombe* (Pobjecky et al. 1990). In *E. coli*, multiple transcription start sites that are distinct from the unique start site found in plant cells were found. Sequences upstream from the strongest transcript start site, at about −315 from the start site found in plant cells, show good homology to prokaryotic promoter −10/−35 elements. In fission yeast, the transcript start site is identical to that observed in plant cells, suggesting that the basal transcription machinery for these eukaryotes may be similar and that some common sequence motifs for transcription factors are likely to be conserved throughout evolution. In higher plants, the strength of the 35S

AgBiotech Center, Waksman Institute of Microbiology, Rutgers State University of New Jersey, Piscataway, New Jersey 08854, USA

Results and Problems in Cell Differentiation 20
L. Nover (Ed.)
Plant Promoters and Transcription Factors

promoter has been compared with the nopaline synthase (*nos*) promoter in transgenic petunia (Sanders et al. 1987). On the average, the 35S promoter is about 30 to 110 times stronger than that of the *nos* promoter. The 35S promoter is also a stronger promoter in transformed callus tissues of tobacco, sugar beet, and oilseed rape when compared with the divergent promoters from the 1′, 2′ genes of *Agrobacterium tumefaciens* in addition to the *nos* promoter (Harpster et al. 1988). Initial studies with transformed tobacco revealed that the 35S promoter appears to be active in most, if not all, of the tissues in the plant (Odell et al. 1985). These observations suggest that the 35S promoter is a strong constitutive promoter and can be used as a general promoter element in plants. Indeed, it has been one of the most commonly used promoters for expression of desired gene products in plant cells.

Since the activity of the 35S promoter does not depend on polypeptides encoded by the CaMV genome, host transcription factors must be responsible for its activity in plant cells. In addition to its use in plant gene expression vectors, the 35S promoter thus offers a convenient system to dissect promoter organization and to identify functional *cis*-acting elements. The strong activity of the 35S promoter allows one to easily quantitate the contribution of various regions of the promoter. Also, by analogy to the elegant studies carried out with the SV40 promoter in mammalian cells (Ondek et al. 1987; Schirm et al. 1987; Fromental et al. 1988), it is likely that the 35S promoter will provide a rich resource for functional sequence motifs. In this chapter, I will summarize our present knowledge concerning the organization of the 35S promoter with emphasis on tissue-specific elements. Implications on general promoter architecture will also be discussed.

2 Functional Analysis of the CaMV 35S Promoter

Three types of approaches have been carried out to functionally dissect the 35S promoter: (1) 5′, 3′ and internal deletions. (2) Gain-of-function analyses with different promoter regions. (3) Site-specific mutations of known factor-binding sites. These studies have been carried out by transient assays with protoplasts as well as by stable transformants obtained through *A. tumefaciens*-mediated DNA transfer. Each of these functional assays contribute in their own way to our understanding of the 35S promoter, with the site-specific mutation approach being more informative about the possible role of each individual factor-binding site in the context of the intact promoter.

2.1 Promoter Regions Required for in vivo Activity

The early 5′ deletion analysis of the 35S promoter by Odell et al. (1985) was carried out predominantly in transformed tobacco calli with the human growth hormone coding sequence as a reporter gene. This study demonstrated that a promoter fragment containing sequences 343 bp upstream from the transcription start site is sufficient to activate transcription to similar levels as a "full" promoter with 941 bp. Furthermore, a truncation to −105 brings the activity of the promoter about threefold lower, whereas further deletion to −46 results in a 20-fold drop in activity. 5′ S1-nuclease mapping demonstrated that transcription of the chimeric genes are initiated at the cognate start site of the 35S promoter, even with the very low expression of the −46 truncated promoter. In this study, a −168 truncated promoter has essentially the same activity as the full promoter. Ow et al. (1987) has subsequently carried out a more detailed promoter deletion analysis with the firefly luciferase gene as a reporter. Transient assays with carrot cell suspension protoplasts were used to quantitate expression levels of various truncated 35S promoter constructs. Their results place the 5′ limit of the promoter at the region between −148 and −134. Further deletions to −108 or −104 result in a fivefold drop in activity. The next drastic lost of activity occurred between −89 and −73 with a 20-fold drop in activity. These results are thus consistent with those of Odell et al. (1985) and suggest that similar sequence elements may be used by nuclear factors of tobacco and carrot to drive expression of the 35S promoter. One interesting observation by Ow et al. (1987) is that the region between −89 and −73 is absolutely required for the ability of the upstream sequence (−148 to −89) to activate transcription in the protoplast assay. The upstream region is also unable to drive significant levels of expression when fused upstream of the −110 truncated CaMV 19S promoter. These results point to the possibility that a critical element exists between −89 and −73 of the 35S promoter, the presence of which is required for the upstream sequences to activate transcription. More recent transient assays with tobacco and soybean protoplasts support this conclusion (Odell et al. 1988).

Although the protoplast transient assay and transformed tobacco calli provide more rapid analyses of deletion constructs, the fact that they employ heterogeneous plant cells whose developmental state is ill-defined makes their relevance to *in planta* promoter activity uncertain. A detailed deletion analysis of the 35S promoter has been carried out by Fang et al. (1989) with transgenic tobacoo plants. In this study, only mature leaves of the transgenic plants were used for promoter activity quantitation. Results from a combination of 5′ and 3′ deletions suggest the following model of functional architecture for the 35S promoter: (1) the region from −343 to −46 contains at least three domains of functional importance for leaf expression. The domain between −343 and −208 is responsible for about 50% of the promoter activity. Deletion of the sequences between −208 and −90 further decrease the remaining activity about twofold. The third domain, from −90 to −46, is required for the synergistic activation observed with upstream sequences (−343 to −107). (2) The region from −107 to −78 can also interact with upstream sequences to drive high levels of expression. (3) Promoters truncated at −90 or −105 have little activity above background, unlike in the case of transient assays with protoplasts where these derivatives are significantly more active. These results of Fang et al. (1988) thus established that for maximal levels of expression in tobacco leaves, multiple sequence elements between −343 to −46 are required. Synergistic interactions between elements from different regions of the promoter are likely to be critical, as demonstrated by the interdependence of the upstream fragment (−343 to −208) and the sequences between −90 and −46 for activation. The apparent lack of requirement for the upstream sequences (from −343 to −148) in protoplast-based transient assays may reflect the lack or inactivation of *trans*-acting factors that interact with this region (Fang et al. 1989). On the other hand, the apparent higher activity of the −90 derivative of the 35S promoter in transient assays (Ow et al. 1987) suggests the factor(s) that interacts with this region may be activated or induced during protoplasts isolation. Interestingly, Poulsen and Chua (1988) reported that the −90 derivative of the 35S promoter is, in fact, active in the root tissues of transgenic tobacco. These data thus point to the fact that promoter interaction with nuclear factors is dynamic and the observed activity can be highly dependent on the developmental context of the tissue in question.

2.2 Tissue Specificity of Functional Elements

From the above deletion analyses, it became apparent that the 35S promoter is likely a composite of *cis*-acting elements that have distinct functional properties. To establish a qualitative model for the tissue specificity of subdomains within this promoter, various upstream regions were tested for activity in transgenic tobacco and petunia (Benfey et al. 1989; Benfey and Chua 1990). The coding region of bacterial β-glucuronidase (GUS) was used as reporter gene in order to detect gene expression in tissue sections by histochemistry. Using this reporter gene, the −90 truncated 35S promoter was found to be active in roots of transgenic tobacco, while the upstream sequences (−343 to −90) are preferentially expressed in leaves (Benfey et al. 1989). Histochemical assays revealed that the activity of the −90 truncated promoter is restricted mainly to meristematic regions of the root, thus extending the previous observation of Poulsen and Chua (1988). A −72 truncated promoter is inactive under all the conditions examined, thus implicating sequences in the −90 to −72 region as being critical for the observed activity in roots. The sequences between −343 and −90 were further divided into five different subdomains and their activities studied individually in transgenic tobacco plants (Benfey and Chua 1990). These were designated as: B1 (−105 to −90), B2 (−155 to −106), B3 (−208 to −156), B4 (−301 to −209), and B5 (−343 to −302). Individual monomers (B2 to B5) and tetramers of these five subdomains were fused upstream of the −46 or −90 truncated 35S promoters to test for their activities in transgenic plants. Multimerization of the individual elements is expected to enhance their activity under conditions where cooperativity between neighboring factors may be essential for transcriptional activation (Ondek et al. 1987, 1988; Schirm et al. 1987; Fromental et al. 1988). In the study of Benfey and Chua (1990), significant dependence of the expression pattern on multimerization of the element was observed only for the B2 subdomain. Although the expression of each construct was not quantitated, the observations suggest strongly that each subdomain of the 35S promoter is capable of activating gene expression in a distinctive manner. With the exception of the B3 subdomain, however, they all have very low or non-detectable levels of activity when placed upstream of the −46 truncated promoter. When combined with sequences between −90 and −46, these regions then activate highly specific patterns of expression in various tissues in transgenic plants. Of the five different subdomains, B1 is the most dramatic in this respect since essentially no activity is observed without the promoter proximal sequence. This interdependence of different *cis*-acting elements to drive cell-type specific expression has also been reported in transgenic mice (Crenshaw et al. 1989) and illustrates the combinatorial nature of promoter elements. The results of Benfey and Chua (1990) also indicate that the −90 to −46 region of the 35S promoter must contain element(s) that can interact with various types of upstream factors.

3 Promoter Interaction with Plant Nuclear Proteins

In order to gain an understanding of the 35S promoter at the molecular level, the interaction between nuclear factors and various sequence elements has to be clearly defined. We and others have carried out DNase I footprinting and gel mobility shift assays to identify sequences of the 35S promoter that are potential targets of *trans*-acting factors. This approach has two main limitations: (1) the apparent absence of protein binding to certain regions does not rule out the ability of these sequences to participate in productive factor interactions in vivo. The possibility that proteins binding to these sites are very low in abundance or are lost during nuclear extract preparation cannot be ruled out easily. (2). In vitro binding of a certain factor to a particular region of the promoter does not necessarily mean that the particular site is in fact bound in vivo by the same factor. Thus, under factor-limiting conditions, it is likely that many possible binding sites for a specific factor may remain unoccupied. Alternatively, multiple factors in the nucleus may be capable of interacting with the particular binding site. The relative affinity and abundance of the different factors toward the particular binding site sequence may determine the composition of the final DNA-protein complex. Thus, in order to test whether a particular factor binding site is functional in the promoter in question, one has to corroborate results from DNA-protein interaction studies in vitro with site-specific mutation analyses and gain-of-function assays in vivo. To date, two different factor-binding sites of the 35S promoter have been studied in detail.

3.1 Activating Sequence Factor 1 (ASF-1)

The first nuclear factor reported to bind to the 35S promoter is called *A*ctivating *S*equence *F*actor 1 (ASF-1). The sequence that this factor binds to (−82 to −62) is designated as *a*ctivating *s*equence 1 (*as-1*) since it coincided with the region known to be required for activating the function of upstream elements. DNase I footprinting with site-specific mutations of *as-1* and methylation interference assays demonstrated that two TGACG motifs within this site are critical for DNA-protein interaction (Lam et al. 1989a,b). Mutations of 2 bp within each of the TGACG motifs (*as-1c*) abolish ASF-1 binding in vitro and dramatically lowered the activity of the 35S promoter in

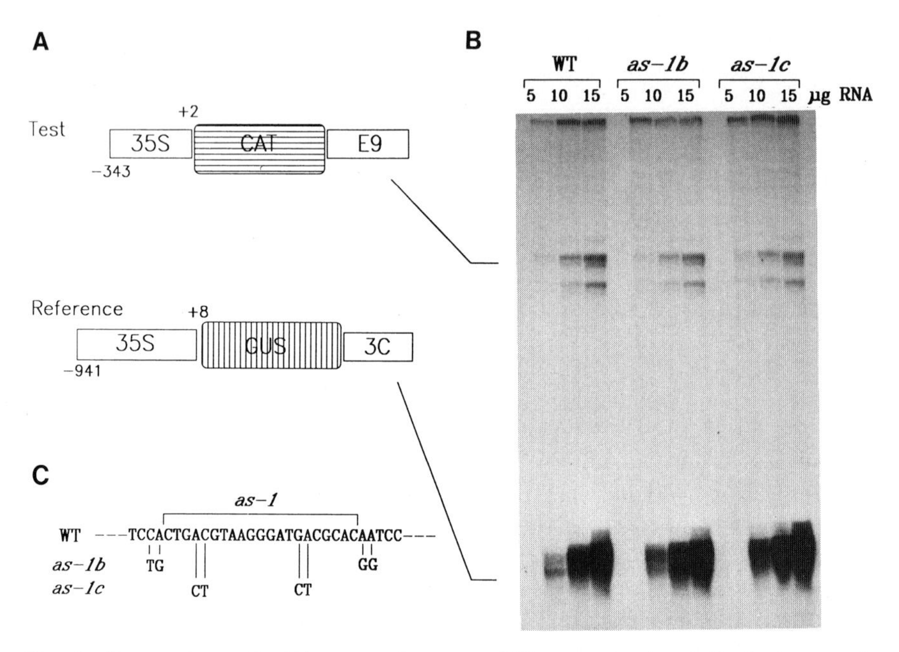

Fig. 1. Comparison of wild-type and mutant 35S promoters' activity in transgenic tobacco leaves. **A** Schematic depiction of test and reference genes in constructs used for transformation. More details of the constructs are described in Lam et al. (1989a). The test gene contains the coding region of bacterial chloramphenicol acetyl-transferase (*CAT*) fused downstream of the wild-type (*WT*) CaMV 35S promoter or derivatives with mutations in the *as-1* region (see **C**). The reference gene in all three constructs consists of the bacterial β-glucuronidase (*GUS*) coding sequence fused to the full 35S promoter. The 3′ regions of the pea *rbcS-E9* and *rbcS-3C* genes were used as polyadenylation signals for the test and reference genes, respectively. **B** Mature leaves from seven independent transformed lines of tobacco were collected from each construct and total RNA was prepared. 10 µg RNA from each of the seven plants was then pooled for each of the three constructs. 5, 10 and 15 µg of pooled RNA for each of the three constructs was then used for 3′ S1 nuclease protection analysis. The probe used is the 3′ region of pea *rbcS-E9* which produces a 230-nucleotide (nt) and an 89-nt product with transcripts from the test and reference genes, respectively. **C** Sequences of the *as-1* region of the WT 35S promoter and mutations introduced in the *as-1b* and *as-1c* derivatives

tobacco leaf protoplasts. Interestingly, mutations of 4 bp surrounding the *as-1* site fortuitously increased the binding affinity of ASF-1 toward this sequence (Lam et al. 1989a). This mutated 35S promoter (*as-1b*) showed an approximately twofold increase in activity in tobacco protoplast transient assays (Lam et al. 1989b). The sequence of these mutant promoters is shown with that of the wild-type *as-1* element in Fig. 1C. These observations thus indicate a tight correlation between binding affinity of ASF-1 and 35S promoter activity in tobacco protoplasts. In transgenic tobacco, mutations of *as-1c* caused a drastic decrease in promoter activity in root and stem, while their effects on leaf expression are comparatively minor (Lam et al. 1989a). The mutations of *as-1b* showed relatively little effect on promoter activities in these tissues. The activity of these constructs in mature leaves of light-grown transgenic tobacco is shown in Fig. 1B. RNA titration with 3′ S1 nuclease protection assays indicates that at most a twofold decrease in promoter activity is observed with both mutant promoters as compared to the wild-type 35S promoter.

To examine whether ASF-1 may play a role in leaf expression under other conditions, total RNA was prepared from light-grown and dark-adapted tobacco plants containing each of the three constructs. Figure 2 shows that the wild-type 35S promoter has a slightly higher level of

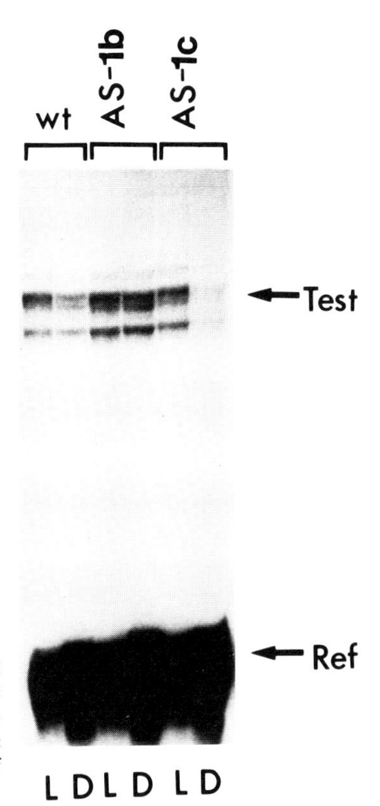

Fig. 2. Mutations in the *as-1* region alters light sensitivity of the 35S promoter. RNA samples were collected from mature leaves of light-grown (*L*) or 2-day dark-adapted transgenic tobacco plants. The constructs and methods of RNA analysis are as described in Fig. 1

expression in light versus dark, with the CAT-coding sequence as a reporter gene. Interestingly, the reference gene containing the GUS-coding sequence fused to a full 35S promoter show a higher transcript level in the dark sample versus that from the light. This is likely due to a higher stability of GUS mRNA in the dark. In the light-grown leaves, the two mutant 35S promoter show similar levels of expression when compared to the wild type. In the dark-adapted leaves, however, the *as-1b* promoter showed no decrease in the activity, whereas the *as-1c* promoter showed a drastic decrease in promoter activity. These results suggest that in the dark, a functional ASF-1 binding site in the 35S promoter is critical for maximal leaf expression. Analogous to the situation in leaf protoplasts (Lam et al. 1989b), the reciprocal effects of the *as-1b* and *as-1c* mutations suggest that the affinity of ASF-1 to the *as-1* site determines the level of promoter activity in the dark. These results also indicate that there is at least one element within the upstream region of the 35S promoter that preferentially functions in light-grown leaves.

Although we have shown previously that a functional ASF-1 binding site is important for activity in roots and stems (Lam et al. 1989a), it will also be interesting to examine the effects that site-specific mutations of *as-1* may have on promoter activity during development of the vegetative tissues. In Fig. 3, the activity of our promoter constructs is compared in 2-week-old transgenic tobacco seedlings and juvenile leaves. A single transformed line with each of these constructs was selected for comparison. It is clear that whereas the *as-1c* mutations have little effect on the expression in mature leaves, the activity in etiolated seedlings as well as in juvenile leaves is drastically reduced. Under these developmental conditions, the *as-1b* mutations did not cause any significant increase in activity as compared to that in mature leaves. The effects of these constructs in light-grown transgenic tobacco seedlings were examined as well. Essentially the same results were obtained, with the exception that in the hypocotyls + leaves samples, slightly higher activity than that of etiolated seedlings was observed for the *as-1c* construct (data not shown). These results indicate that ASF-1 is also likely to play an important role in maintaining high activity of the 35S promoter in seedlings as well as in leaves during early development. Benfey et al. (1989) have previously reported that the −90/−46 region is not required for expression of the 35S promoter in the cotyledons of tobacco seedlings. In that study, an internal truncation of the promoter, which places the upstream elements closer to the promoter, as well as the introduction of linker sequences may have caused the quantitative differences from those shown in Fig. 3.

A compelling approach to demonstrate the function of a putative *cis*-acting element is by gain-of-function assays. This can be done in at least two different ways: (1) fusion of the element upstream of a minimal promoter which interacts with only the basal factors required for transcription initiation. (2) Insertion of the element into a well-defined heterologous promoter and assay for alterations in expression pattern. Both of these approaches have

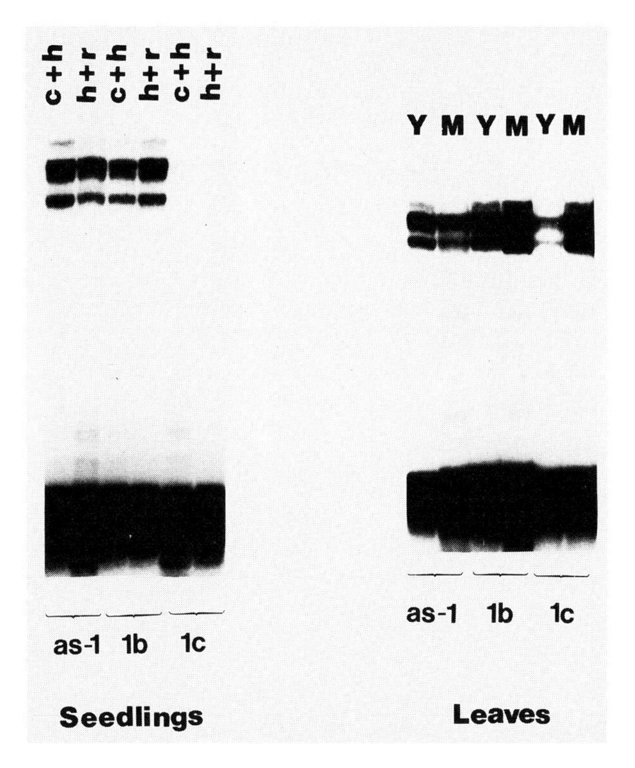

Fig. 3. Effects of mutations in the *as-1* region on 35S promoter activity in seedlings and young leaves. A single line of transgenic tobacco plant from each of the three constructs described in Fig. 1 was selected for analysis. *Left* RNA samples were prepared from 2-week-old etiolated tobacco seedlings. The seedlings were cut in the middle to produce two portions: cotyledons + hypocotyls (*c* + *h*) and hypocotyls + roots (*h* + *r*). 20 μg of total RNA for each sample was used for analysis as described in Fig. 1. *Right* Samples were collected from a single mature transgenic plant of the same lines that were used for the samples on the *left*. Mature leaves (*M*) from the middle of the plant or young, unopen leaves (*Y*) near the apices of the plants were used for RNA analysis

been applied toward the *as-1* element with the following results: (1) a single copy of the *as-1* site is sufficient to activate root expression (E. Lam, unpubl. data) when fused to the −46/+8 derivative of the 35S promoter. Multiple copies of this site (four or five copies) can activate expression in all the vegetative tissues examined (Lam and Chua 1990). (2) Insertion of a functional *as-1* site into the pea *rbcS-3A* promoter, which normally expresses only in green tissues (Kuhlemeier et al. 1988), activates root expression (Lam et al. 1989a). In both of these experiments, mutations that abolish ASF-1 binding in vitro also eliminated any detectable activity under our assay conditions. Since a −166 truncated *rbcS-3A* promoter is preferentially expressed in mature leaves (Kuhlemeier et al. 1988), we wanted to test also if insertion of a functional *as-1* site will activate expression in

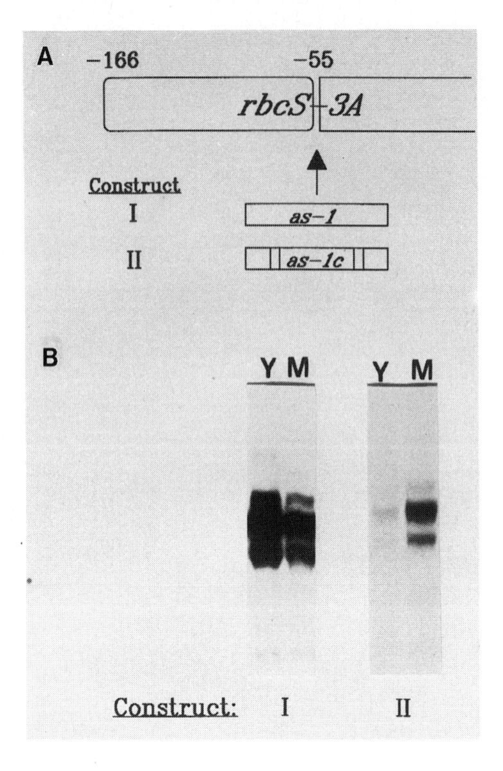

Fig. 4. Insertion of a functional *as-1* site augments expression of a truncated *rbcS-3A* promoter in young leaves. For *construct* I, a 21-bp *as-1* sequence was inserted into a unique *BstX* I restriction site at the −55 position of the *rbcS-3A* promoter. A mutant *as-1c* sequence (Fig. 1C) was inserted similarly for *construct II*. More details of these constructs can be found in Lam et al. (1989a). Mature (*M*) and young (*Y*) leaves from an individual transformant of each construct are compared by S1 nuclease protection as described in Fig. 1

juvenile leaves. Figure 4 shows that this is indeed the case. Insertion of a wild-type *as-1* element elevated the activity of the promoter in juvenile leaves to a level comparable to that observed in mature leaves. The developmental dependence of the truncated *rbcS-3A* promoter is observed when a mutant *as-1c* sequence was inserted instead. These results thus indicate that ASF-1 in young tobacco leaves can interact productively with factors that bind to the *rbcS-3A* promoter.

Based on our results so far, we have proposed the following model for the apparent tissue specificity of the *as-1* element in transgenic tobacco (Lam et al. 1989a): functional ASF-1 activity is hypothesized to be limiting in leaf tissues, whereas it may be more abundant in roots. Thus, a single binding site for this factor with a minimal promoter will confer activity predominantly in roots. In leaves, however, interaction with other adjacent *cis*-acting elements will allow *as-1* to synergistically activate transcription. Consistent with this model, tobacco mRNA encoding TGA1a, a family of transcription factors that binds to *as-1*, is found to express about ten times higher in roots versus that in leaves (Katagiri et al. 1989).

3.2 Activating Sequence Factor 2 (ASF-2)

The study of Benfey et al. (1989) has demonstrated that the −343 to −90 region of the 35S promoter contains at least one leaf-specific element. The best-characterized factor in this region is ASF-2 which binds to the −105 to −85 region (*as-2*) of the 35S promoter (Lam and Chua 1989). Methylation interference and competition analyses demonstrated that this factor is distinct from another nuclear factor GT-1, which is hypothesized to mediate light-responsive *rbcS-3A* expression (Green et al. 1988; Kuhlemeier et al. 1988). Four tandem copies of *as-2* are able to potentiate leaf expression when fused upstream of a −90 truncated 35S promoter but not to a minimal −46 promoter. In fact, its expression pattern is strikingly similar to that observed with the B1 fragment which contains sequences from −105 to −90 (Benfey and Chua 1990). ASF-2 apparently binds to the GAT(A/G) repeat within the core of the *as-2* site (Fig. 5A). A GATA repeat sequence was also found in a similar position of the promoters from 12 different *cab* (chlorophyll *a*/*b* binding protein) genes from different species (Lam and Chua 1989). Mutations within the GAT(A/G) repeats of *as-2* abolish ASF-2 binding in vitro and also the ability of the resultant mutant sequence to activate leaf expression. These results thus directly correlate the activity of the *as-2* element with its ability to bind the ASF-2 factor. Although the *as-2* element can potentiate expression in tobacco leaves, its activity is not dependent on light. Thus, our present results (see Fig. 2) indicate there may be additional elements in the −343 to −105 region which can potentiate leaf expression in a light-dependent manner. This is consistent with the result of Fang et al. (1989) who showed that an internal deletion of the −105/−90 region in the 35S promoter does not significantly reduce leaf expression, thus also implicating additional elements for leaf expression other than *as-2*.

3.3 Others

In addition to ASF-1 and ASF-2, functional studies with various upstream regions clearly indicate that there are likely to be more nuclear factors which may interact with the 35S promoter (Benfey and Chua 1990). Further molecular dissection of the 35S promoter will require characterization of additional factor-binding sites within the region between −343 and −105. Using internal restriction sites, one can isolate four different fragments that cover the region from −343 to −46. The position and size of these fragments are shown in Fig. 5A,B. Using each of these fragments as probes, sequence-specific DNA-protein complexes can be detected that bind to each one of the four fragments (indicated by asterisks in Fig. 5C). For fragment

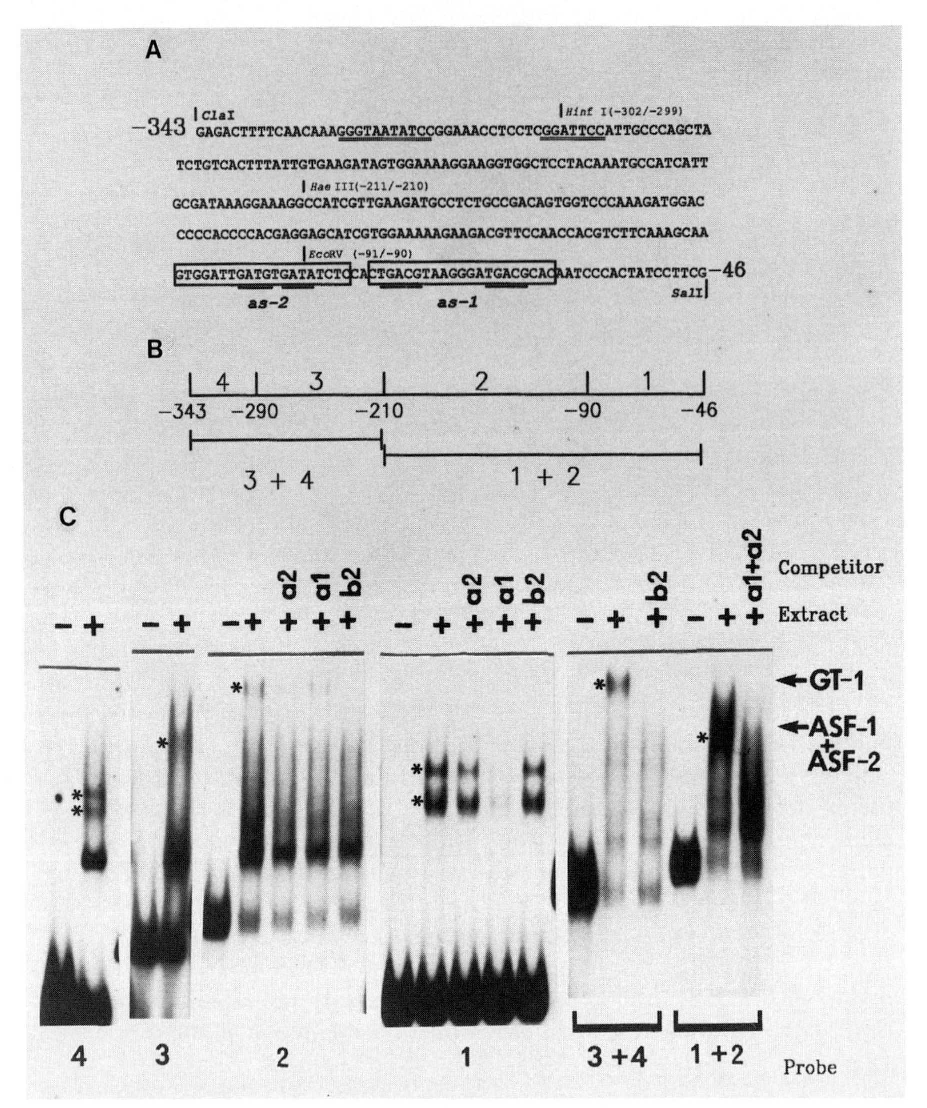

Fig. 5. Multiple factors interact with the CaMV 35S promoter. **A** Sequence of the CaMV 35S promoter from −343 to −46. The *Cla* I and *Sal* I sites are within the vector polylinker region in which the −343 to −46 fragment was subcloned (provided by R.X. Fang). The restriction sites used to prepare subfragments are indicated along with their location in the promoter. The *as-1* and *as-2* sequences used for tetramer construction (Lam and Chua 1989; Lam et al. 1989a) are *boxed*, with the repeat sequences *underlined*. The putative *box II*-like sequences in the −300 to −343 region are *double-underlined*. **B** The designation and location of the fragments used for gel mobility shift assays. The four subfragments and the two larger promoter fragments are shown. **C** Gel mobility shift assays with tobacco nuclear extracts. The fragments indicated in **B** were labeled with *E. coli* DNA polymerase and mixed with tobacco nuclear extracts (+) containing 2 μg total protein. Where indicated, 20 ng of

1, which covers the −90/−46 region, the two predominant complexes are competed specifically by *as-1* tetramers and are thus assigned as due to ASF-1. These two complexes are likely to result from occupancy of one or both TGACG sequence elements within *as-1* (Lam et al. 1989b). Interestingly, fragment 2 containing the −210/−91 region does not show any complex with mobility expected of ASF-2. Using a tetramer of *as-2* as probe, ASF-2 activity with mobility similar to the ASF-1/*as-1* complex can be readily detected with the same nuclear extract (data not shown). Previously, we have shown that ASF-2 can readily bind to a fragment of the 35S promoter consisted of the sequences from −130 to −46 (Lam and Chua 1989). Thus, the behavior of fragment 2 is not due to a lack of ASF-2 factors in the particular nuclear extract. However, a slow-migrating complex is observed with this fragment. Competition studies indicate that this complex can be competed by either a synthetic *as-2* tetramer or a *box II* sequence from the pea *rbcS-3A* promoter (Green et al. 1988), but not by a tetramer of *as-1*. The opposite is observed for the complexes observed with fragment 1. These results suggest that the complex observed with fragment 2 may involve both ASF-2 and GT-1 factors. Fragment 3, covering the −299 to −211 region, shows a relatively weak complex distinct from those observed with the other fragments. Two closely spaced complexes are observed with the most distal fragment (−343 to −302), but they are not found with the others. Since these binding activities may in fact be related to nuclear factors that are known to bind to other characterized promoter elements, they are not designated at present with a specific name.

It is important to bear in mind that the isolation of subfragments may cause alterations in factor binding since the restriction sites or deletion limits that were chosen to prepare the probes may reside within *cis*-acting elements. This is likely the case for fragment 2 in this study where the removal of about 5 bp from the 3′ end of the *as-2* sequence may have prevented the normal ASF-2/*as-2* complex. Interestingly, under this condition, a putative ASF-2/GT-1 complex became evident. Another example of the disruption of factor binding by the choice of restriction site is also shown in Fig. 5C. A slow-migrating complex is clearly observed with an upstream fragment that contains sequences in both fragments 3 and 4. In this case, competition analysis with a *box II* tetramer demonstrates that it is the nuclear factor GT-1. In this case, addition of *as-2* or *as-1* tetramers did not show any competition (data not shown). The finding that GT-1 may be able

---

the following competitor DNA was added: tetramer of *as-2* (*a2*); tetramer of *as-1* (*a1*); tetramer of *box II* (*b2*). After incubating at room temperature for 20 min, the samples (10 μl total) were analyzed by gel electrophoresis (3.5% acrylamide gel, 0.25 × TBE). The data shown were from a single gel mobility shift assay and the relative positions of the complexes are preserved. The data were separated in the final prints so that the exposure for each set of the gel shift data involving different probes can be individually optimized for clarity. DNA/protein complexes apparently specific for each of the probes are indicated (*). The positions for the DNA-protein complexes of GT-1 and ASF-1 + ASF-2 are indicated by *arrows* on the *right*

to interact weakly with the 35S promoter is interesting because this factor has been implicated to mediate light-responsive and leaf-specific transcription activation (Kuhlemeier et al. 1988; Lam and Chua 1990). The finding that the upstream sequences of the 35S promoter can indeed potentiate light-responsive expression (i.e., Fig. 2) is consistent with this role. Since disruption of the upstream fragment at −300 appears to abolish binding of GT-1, it is likely that the binding site for this factor may be close to or overlap the *Hinf* I restriction site located in this region. In fact, there are several sequences that have apparent homologies to the core sequence of GGTTAA defined as important for GT-1 binding (Green et al. 1988). These are underlined in Fig. 5A. The downstream fragment (1 + 2) shows the expected ASF-1 plus ASF-2 complex which can be competed by the appropriate competitors. Interestingly, the putative ASF-2/GT-1 complex is no longer visible, suggesting that this complex is not significant for the 35S promoter in the presence of a complete *as-2* site.

Two other DNA-binding activities that interact with the 35S promoter have been observed. CAF is a tobacco leaf activity that was reported to bind to the −208/−155 region (Benfey and Chua 1990) and MNF1 is a maize leaf factor that binds to the −281 to −235 region (Yanagisawa and Izui 1992). At present, the relationship of these factors to the DNA/protein complexes reported here is not clear and their precise binding sites or possible functions have not been defined.

4 Concluding Remarks

In this chapter, I sought to summarize our present knowledge concerning the functional properties of the CaMV 35S promoter with respect to *cis*-acting elements and *trans*-acting factors. It is apparent from the earlier studies and results presented here that the 35S promoter is composed of multiple elements which are likely to interact with distinct factors. The complex interplay between factor abundance (e.g., ASF-1) and factor interaction (e.g., ASF-2/GT-1), as well as synergistic interdependence (e.g., *as-1* and *as-2*) of *cis*-acting elements for activity, are all important parameters in considering the functional architecture of this promoter. These are probably important considerations for the analyses of other promoters in general. It should be obvious that nuclear factors recruited by the 35S promoter of CaMV must have normal cellular target genes. For ASF-1, it has been shown to bind to a conserved hexamer motif of the histone *H3* promoter (Katagiri et al. 1989) as well as of the promoters of the tobacco auxin-inducible gene *GNT35* (Liu and Lam, unpubli. results). For ASF-2, the conserved GATA repeat from the petunia *cab22L* promoter has been found to interact with this nuclear factor (Lam and Chua 1989). It is expected that further characterization of other nuclear factors that interact with the 35S promoter will enrich our understanding of plant transcription factor and promoter organization.

Acknowledgments. I would like to thank my colleagues at the Laboratory of Plant Molecular Biology (Rockefeller University, New York, NY) for their collaboration and stimulating discussions. I am grateful to Nam-Hai Chua for support and encouragement during my postdoctoral training in his laboratory. I would also like to thank John Tonkyn for providing some of the tobacco nuclear extracts used in the present work. Critical reading of this manuscript by Noureddine Hajeb is appreciated. This work is supported in part by funding from the National Institutes of Health.

References

Assaad FF, Signer ER (1990) Cauliflower mosaic virus P35S promoter activity in *Escherichia coli*. Mol Gen Genet 223:517–520

Benfey PN, Chua NH (1990) The cauliflower mosaic virus 35S promoter: Combinatorial regulation of transcription in plants. Science 250:959–966

Benfey PN, Ren L, Chua NH (1989) The CaMV 35S enhancer contains at least two domains which can confer different developmental and tissue-specific expression patterns. EMBO J 8:2195–2202

Crenshaw III EB, Kalla K, Simmons DM, Swanson LW, Rosenfeld MG (1989) Cell-specific expression of the prolactin gene in transgenic mice is controlled by synergistic interactions between promoter and enhancer elements. Genes Dev 3:959–972

Fang RX, Nagy F, Sivasubramaniam S, Chua NH (1989) Multiple *cis* regulatory elements for maximal expression of the cauliflower mosaic virus 35S promoter in transgenic plants. Plant Cell 1:141–150

Fromental C, Kanno M, Nomiyama H, Chambon P (1989) Cooperativity and hierarchical levels of functional organization in the SV40 enhancer. Cell 54:943–953

Fromm M, Taylor LP, Walbot V (1985) Expression of genes transferred into monocot and dicot plant cells by electroporation. Proc Natl Acad Sci USA 82:5824–5828

Green PJ, Yong MH, Cuozzo M, Kano-Murakami Y, Silverstein P, Chua NH (1988) Binding site requirements for pea nuclear protein factor GT-1 correlate with sequences required for light-dependent transcriptional activation of the *rbcS-3A* gene. EMBO J 7:4035–4044

Harpster MH, Townsend JA, Jones JDG, Bedbrook J, Dunsmuir P (1988) Relative strengths of the 35S califlower mosaic virus, 1′, 2′ and nopaline synthase promoters in transformed tobacco sugarbeet and oilseed rape callus tissue. Mol Gen Genet 212:182–190

Katagiri F, Lam E, Chua NH (1989) Two tobacco DNA-binding proteins with homology to the nuclear factor CREB. Nature 340:727–730

Kuhlemeier C, Cuozzo M, Green PJ, Goyvaerts E, Ward K, Chua NH (1988) Localization and conditional redundancy of regulatory elements in *rbcS-3A*, a pea gene encoding the small subunit of the ribulose-bisphosphate carboxylase. Proc Natl Acad Sci USA 85:4662–4666

Lam E, Chua NH (1989) ASF-2: A factor that binds to the cauliflower mosaic virus 35S promoter and a conserved GATA motif in *cab* promoters. Plant Cell 1:1147–1156

Lam E, Chua NH (1990) GT-1 binding site confers light responsive expression in transgenic tobacco. Science 248:471–474

Lam E, Benfey PN, Gilmartin PM, Fang RX, Chua NH (1989a) Site-specific mutations alter in vitro factor binding and change promoter expression pattern in transgenic plants. Proc Natl Acad Sci USA 86:7890–7894

Lam E, Benfey PN, Chua NH (1989b) Characterization of AS-1: a factor binding iste on the 35S promoter of cauliflower mosaic virus. In: Lamb C, Beachy R (eds) Plant gene transfer. UCLA Symp on Molecular and cellular biology, new series, vol 129. Academic Press, New York, pp 71–79

Odell JT, Nagy F, Chua NH (1985) Identification of DNA sequences reqired for activity of the cauliflower mosaic virus 35S promoter. Nature 313:810–812
Odell JT, Knowlton S, Lin W, Mauvais JC (1988) Properties of an isolated transcription stimulating sequence derived from the cauliflower mosaic virus 35S promoter. Plant Mol Biol 10:263–272
Ondek B, Shepard A, Herr W (1987) Discrete elements within the SV40 enhancer region display different cell-specific enhancer activities. EMBO J 6:1017–1025
Ondek B, Gloss L, Herr W (1988) The SV40 enhancer contains two distinct levels of organization. Nature 333:40–45
Ow DW, Jacobs JD, Howell SH (1987) Functional regions of the cauliflower mosaic virus 35S RNA promoter determined by use of the firefly luciferase gene as a reporter of promoter activity. Proc Natl Acad Sci USA 84:4870–4874
Pobjecky N, Rosenberg GH, Dinter-Gottlieb G, Kaufer NF (1990) Expression of the β-glucuronidase gene under the control of the CaMV 35S promoter in *Schizosaccharomyces pombe*. Mol Gen Genet 220:314–316
Poulsen C, Chua NH (1988) Dissection of 5′ upstream sequences for selective expression of *Nicotiana plumbaginifolia rbcS-8B* gene. Mol Gen Genet 214:16–23
Sanders PR, Winter JA, Barnason AR, Rogers SG, Fraley RT (1987) Comparison of cauliflower mosaic virus 35S and nopaline synthase promoters in transgenic plants. Nucleic Acids Res 15:1543–1558
Schirm S, Jiricny J, Schaffner W (1987) The SV40 enhancer can be dissected into multiple segments, each with a different cell type specificity. Genes Dev 1:65–74
Yanagisawa S, Izui K (1992) MNF1, a leaf tissue-specific DNA-binding protein of maize, interacts with the cauliflower mosaic virus 35S promoter as well as the C_4 photosynthetic phosphoenolpyruvate carboxylase gene promoter. Plant Mol Biol 19:545–553

9 Analysis of Ocs-Element Enhancer Sequences and Their Binding Factors

Karam B. Singh, Bei Zhang, Soma B. Narasimhulu, and Rhonda C. Foley

1 Introduction

A detailed analysis of the control of plant gene expression is necessary to understand both the molecular basis of plant growth and development, and how plants respond to external and internal changes in their environment. A powerful approach for studying gene regulation is to use pathogens as model systems. Such systems provide excellent windows into the control of gene expression in the host, since in many cases the pathogens have taken advantage of important, existing cellular transcription factors. Plants studies on bacterial pathogens such as *Agrobacterium tumefaciens* (DeGreve et al. 1983; Ellis et al. 1987; Leisner and Gelvin 1988; Bouchez et al. 1989) and viruses such as the caulimoviruses (Benfey et al. 1989; Bouchez et al. 1989) and maize streak virus (Fenoll et al. 1988) have helped elucidate the control of gene expression. Interestingly, the same regulatory element has been found to be utilized by both bacteria and viruses to express pathogen-encoded genes in the plant (Bouchez et al. 1989; Lam et al. 1989). The *cis*-acting DNA component of this plant transcription system has been called the ocs-element, as the sequence was first discovered in the promoter of the octopine synthase (*ocs*) gene of the plant pathogen *Agrobacterium tumefaciens* (Ellis et al. 1987). The ocs-element sequence in the CaMV promoter is also referred to as the as-1 site (Lam et al. 1989). In this review we briefly describe the initial identification and characterization of the ocs-element followed by a more detailed account of the isolation and characterization of proteins in maize and *Arabidopsis* which interact with the ocs-element.

2 The Ocs-Element

2.1 Characterization of the First Ocs-Element Enhancer Sequence

The initial ocs-element was identified through analysis of the promoter of the octopine synthase gene (*ocs*). The *ocs* gene, found on the tumor-

Department of Biology, University of California Los Angeles, 405 Hilgard Avenue, Los Angeles, California 90024, USA

Results and Problems in Cell Differentiation 20
L. Nover (Ed.)
Plant Promoters and Transcription Factors

inducing plasmid of the plant pathogen *Agrobacterium tumefaciens*, is not expressed in *Agrobacterium*, but is expressed after integration into the plant genome (DeGreve et al. 1983). A 16-base-pair palindrome, ACGTAAGCGCTTACGT, was identified in the promoter of the *ocs* gene and shown to function as an enhancer in plant protoplasts (Ellis et al. 1987). A saturation mutagenesis of the *ocs* enhancer was performed to identify the bases important for enhancer activity in the Black Mexican Sweetcorn (BMS) cell line (Singh et al. 1989). These studies found a significant degree of flexibility in the nucleotide requirements for the *ocs* enhancer and facilitated the identification of similar sequences in the promoters of other plant pathogens. A *trans*-acting protein factor, initially called OCSTF but now referred to as OTF for simplicity, was identified as the major ocs-element binding activity in nuclear extracts from maize and tobacco suspension cell lines (Singh et al. 1989; Tokuhisa et al. 1990). A number of the *ocs* enhancer mutants were used to characterize the binding properties of the protein factor OTF. These studies included the demonstration that OTF must bind to both halves of the ocs-element for transcriptional activation in vivo (Singh et al. 1989; Tokuhisa et al. 1990).

2.2 Identification and Characterization of Additional Ocs-Element Enhancer Sequences

Elements with sequence homology to the *ocs* enhancer have been identified in the promoter regions of other plant genes including other T-DNA genes and plant virus genes (Bouchez et al. 1989; Lam et al. 1989; Cooke 1990; Medbury et al. 1992). These sequences, which we have called ocs-element sequences, all bind OTF in vitro and enhance transcription in vivo. Comparison of ocs-element sequences gave rise to a 20-bp consensus sequence, which includes the 16-bp palindrome present in the *ocs* gene promoter (Bouchez et al. 1989). In transgenic tobacco plants ocs-element sequences, when linked to a minimal plant promoter and a reporter gene, direct tissue-specific expression patterns that are developmentally regulated. In young tobacco seedlings expression is primarily in the root tips, whereas in older seedlings a low level of expression also occurs in the shoot apex (Benfey et al. 1989; Fromm et al. 1989; Kononowicz et al. 1992). The root tip and the shoot apex contain meristematic regions which consist of rapidly dividing, nondifferentiated cells from which all the other cell types of the root and shoot are derived. Ocs-element sequences have also been reported to be active in the root and shoot meristem of older tobacco plants as well as in specific cell types in leaves (Kononowicz et al. 1992).

Sequences similar to an ocs-element half-site that contain the core motif ACGT have been found to be involved in the expression of a number of plant genes in response to diverse stimuli (Fig. 1). These include the G-box sequence required for the expression of a number of genes (Block et al. 1990; Donald and Cashmore 1990), the closely related abscisic acid response

Ocs-element consensus	-	TGACGTAAGC GCTTACGTCA
GNT35	-	TTAGCTAAGT GCTTACGTAT
Hex	-	TGACGTGG
dbp	-	TGACGTGG
G-box	-	CCACGTGG
ABRE	-	CACGTGGC

Fig. 1. The sequence that best fits the consensus ocs-element sequence, related plant regulatory sequences, and sequences in the promoters of auxin-regulated genes that resemble the ocs-element and Hex site. The ACGT core is *underlined* and nucleotides homologous to the ocs-element consensus sequence are in *boldface*

element (ABRE) required for ABA-induced expression of the wheat Em gene (Guiltinan et al. 1990), and the Hex site required for expression of the wheat histone H3 promoter (Nakayama et al. 1989). For some time it remained unclear if ocs-element sequences were involved in the expression of plant genes, although this seemed likely as sequences similar to ocs-element sequences and/or the Hex site are present in the promoters of a number of genes including, as shown in Fig. 1, some auxin-inducible genes such as the *Arabidopsis* dbp (Alliotte et al. 1989), and tobacco GNT35 genes (Van der Zaal et al. 1991). Recently, Ellis et al. (1993) have shown that the promoter of the soybean heat-shock gene, Gmhsp26-A (Czarnecka et al. 1988), contains a functional ocs-element sequence.

3 Maize Ocs-Element Binding Factors

3.1 Isolation of the Genes for Two Maize Ocs-Element Binding Proteins

The studies outlined above laid the framework for efforts to isolate genes encoding plant DNA-binding proteins that interact with ocs-element sequences. Two cDNA clones encoding specific ocs-element binding proteins were isolated by screening a maize root tip cDNA expression library with a 20-bp ocs-element sequence probe that matched the ocs-element consensus sequence (Singh et al. 1990). The cDNA clones, initially called OCSBF-1 and OCSBF-2 but now referred to as OBF1 and OBF2, have little DNA or protein sequence homology except for a small basic amino acid region which has homology to the DNA-binding domains of the bZIP (*b*asic domain-leucine *zip*per) family of transcription factors.

bZIP proteins occur in a range of eukaryotic organisms and bind to DNA as dimers (McKnight 1991). DNA binding results from two domains; the leucine zipper region which typically contains a leucine at every seventh amino acid and is required for dimerization of the protein as well as a basic region of about 20 amino acids, immediately N-terminal of the leucine zipper region which is involved in the recognition of DNA. In addition, there are one or more activation domains that are responsible, by mechanisms that are not well understood, for the increase in transcription once the factor is bound to DNA. A number of animal bZIP proteins such as CREB, Jun, and Fos appear to play major roles in cellular decision-making, interacting with various signal transduction pathways to determine whether a cell will undergo proliferation or differentiation. This is well illustrated by the members of the Jun/Fos family (reviewed in Ransone and Verma 1990; Vogt and Bos 1990; Herschman 1991). It remains to be determined if any of the plant bZIP proteins play analogous roles.

3.2 Properties of OBF1 and OBF2

A preliminary characterization of the structure of the gene, the RNA expression patterns during development, and the DNA-binding properties has been performed for OBF1. The OBF1 gene contains no introns, no obvious TATA box, and a large 5′-untranslated leader sequence (Singh et al. 1990). Interestingly, a gradient of OBF1 RNA levels was found in developing maize leaves with the basal portions of the leaf, which contains dividing and differentiating cells, having 40- to 50-fold higher levels of OBF1 transcripts than the apical protion of the leaves, where the cells are fully differentiated (Singh et al. 1990). Roots and shoots of young plants had levels of OBF1 mRNA similar to the basal portion of developing leaves. Studies with OBF2 reveals that OBF1 and OBF2 have overlapping but distinct DNA binding and RNA expression patterns. For example, OBF2, while able to bind to the consensus ocs-element sequence, was not able to bind to the ocs-element sequence present in the 35S promoter. The levels of OBF2 transcripts were found to be significantly lower in a number of tissues than OBF1 and in contrast to OBF1, OBF2 was expressed constitutively along the basipetal gradient of developing maize leaves.

3.3 Relationship of OBF1 and OBF2 to OTF

The DNA-binding properties of OBF1 and OBF2 were compared to OTF, the major ocs-element binding activity present in nuclear extracts from maize suspension cells (Tokuhisa et al. 1990). While OBF1 and OTF recognize similar DNA sequences, there are differences in the number and pattern of the DNA/protein complexes obtained in gel electrophoresis. Binding of OTF to ocs-element sequences gives rise to two bands in gel shift assays (Singh et al. 1989; Tokuhisa et al. 1990), while binding of OBF1 gives a single band of different mobility to either of the bands obtained with OTF (Singh et al. 1990). OBF2 does not bind to the ocs-element sequence in the CaMV 35S promoter. The levels of OBF1 and OBF2 transcripts in the maize BMS cell line were found to be low, although extracts from the BMS cells contain strong OTF binding activity. On the basis of both their DNA binding properties and RNA expression patterns neither OBF1 and OBF2 appeared to correspond to the OTF complex present in the BMS cell line.

3.4 Isolation of Additional OBF Clones Expressed in the BMS Cell Line

Since neither OBF1 nor OBF2 appeared to be OTF, an expression cDNA library prepared from the maize BMS cells was screened using the consensus ocs-element oligonucleotide probe (Foley et al. 1993). Two cDNA clones were isolated that were very similar in their coding sequences but distinct from either OBF1 or OBF2. The two clones form part of an OBF3 subfamily. The two OBF3 proteins, OBF3.1 and OBF3.2, are closely related to the OBF3.1 protein sharing 95.8% amino acid homology with the carboxyl two-thirds of the OBF3.2 protein. The OBF3.2 protein extends for an additional 137 amino acids at the N-terminal. The DNA binding domain of the OBF3 proteins was found to be most closely related to the tobacco TGA1a (Katagiri et al. 1989) and the wheat HBP-1b protein (Tabata et al. 1991), both of which were isolated as Hex motif binding proteins. In contrast to other plant bZIP proteins, the homology among the OBF3, TGA1a, and HBP1b proteins extended to regions outside the bZIP motif.

3.5 OBF3 Comprises a Small Multigene Family in Maize

Evidence for a small OBF3 gene family was derived from genomic Southern blot analysis using the recombinant inbred lines of Burr et al. (1988). Following digestion with HindIII, four cross-hybridizing bands were detected in all the inbred lines with the OBF3.1 probe. Two of these HindIII bands were found to be polymorphic among the parental lines. The positions of these bands were mapped to 3L105 and 8L075 on the maize genome (Foley et al. 1993). More recently, we have isolated an additional OBF3 cDNA clone that from sequence analysis and restriction mapping is similar to OBF3.1 and OBF3.2 in the carboxyl half including the bZIP motif but distinct in the N-terminal half, bringing the number of maize members of the OBF3 family to at least three (Narasimhulu and Singh, unpubl. results).

3.6 DNA Binding Properties of the OBF3 Proteins

A major goal was to determine if the OBF3 clones were candidates for encoding OTF. To examine the DNA binding properties of the OBF proteins the proteins were expressed by coupled in vitro transcription/translation reactions using T7 polymerase and rabbit reticulocyte lysates (Foley et al. 1993). In this review we focus on the OBF3.1 protein. The OBF3.1 protein was able to bind well to ocs-element probes but not to mutant versions of these probes. A titration of the OBF3.1 extract demonstrated that OBF3.1, like OTF, can form two distinct retarded bands with an ocs-element probe. The mobility of the retarded bands observed with the OTF activity present in nuclear extracts from the BMS cell line was very similar to the mobility of the retarded bands produced by the OBF3.1 protein. Based on the mobility and pattern of DNA binding, OBF3.1 appeared to be a good candidate for at least part of the OTF activity.

Competition studies with synthetic oligonucleotides demonstrated that OBF3.1 had high affinity for the Hex site and low affinity for the G-box and the ABRE sequence (Foley et al. 1993). The high affinity of OBF3.1 for the Hex site, a sequence required for the expression of a cell cycle-regulated, wheat histone 3 gene (Nakayama et al. 1989), is interesting. As already mentioned, the Hex sequence or a closely related sequence is also present in the promoters of a number of auxin-inducible genes; for example, the soybean GH3 gene promoter contains three such sequences, TGACGTGG, TGACGTAA, and TGACGCAG (Hagen et al. 1991).

3.7 OBF3.1 DNA Binding Differs from Both OBF1 and TGA1a

The binding of the OBF1, OBF3.1, and tobacco TGA1a proteins was compared (Foley et al. 1993). In these experiments equal molar amounts of protein produced by in vitro coupled transcription/translation reactions were used (Foley et al. 1993). Relative to an equal amount of OBF3.1, OBF1 was found to have a lower affinity for ocs-element sequences and to bind better to the G-box sequence than to the ocs-element sequences. In contrast, OBF3.1 was able to bind well to the ocs-element sequences but not to the G-box sequence. These results demonstrated that there are differences in the binding affinity of OBF1 and OBF3.1 for particular DNA sequences and that OBF3.1 is much more likely to be a major part of the OTF activity in the BMS cell line than OBF1.

Although OBF3.1 and TGA1a recognize similar DNA sequences, OBF3.1 appeared to recognize the ocs-element sequences more efficiently than TGA1a (Foley et al. 1993). A titration of both the TGA1a and OBF3.1 proteins was performed and revealed that there is about an eightfold difference in the ability of OBF3.1 to form the upper band compared to TGA1a. This difference in the ability of OBF3.1 and TGA1a to form the upper band is potentially significant since formation of the upper band appears to be a prerequisite for in vivo transcriptional activity for ocs-element sequences in plant cells (Bouchez et al. 1989; Singh et al. 1989). From our work on the *Arabidopsis* OBF proteins, described in the next section, we know that OBF3.1 and TGA1a are distinct members of an OBF3/TGA family.

4 *Arabidopsis* Ocs-Element Binding Factors

4.1 Isolation of *Arabidopsis* OBF Proteins

We have attempted to isolate *Arabidopsis* homologues for the maize OBF3 proteins thereby facilitating an analysis of their function through reverse genetic approaches (Zhang et al. 1993). To isolate *Arabidopsis* OBF proteins related to the maize OBF3 proteins we exploited the homology among the maize OBF3 proteins, the wheat HBP1b protein and the tobacco TGA1a protein, and the small size of the *Arabidopsis* genome. The only substantial homology among these proteins at the DNA level was in the basic domain of the bZIP motif. Two degenerate 20-mer oligonucleotides were synthesized corresponding to sequences in the basic domain. We screened duplicate filters from an *Arabidopsis* genomic library and isolated four phages that cross-hybridized with both sets of degenerate oligonucleotides. Upon detailed analysis two of these phages were found to contain basic domains essentially identical at the amino acid level to each other and to those in the maize OBF3, wheat HBP1b protein, and tobacco TGA1a protein. However, the two genomic clones shared less homology in the basic domain between each other at the DNA level, suggesting that they were distinct genes.

To isolate the corresponding cDNA clones, we used the genomic fragments containing the bZIP motif as probes to screen a cDNA library that we had constructed from *Arabidopsis* callus tissue; a tissue in which we knew the ocs-element sequence was active in *Arabidopsis* (Zhang and Singh, unpubl. results). We isolated and sequenced cDNA clones for both the genomic clones (Zhang et al. 1993). We called the cDNA clones OBF4 and OBF5. The deduced protein sequence revealed that both of the *Arabidopsis* cDNAs encoded proteins that were related to the maize OBF3, wheat HBP1b, and the tobacco TGA1a proteins (Katagiri et al. 1989; Fromm et al. 1991; Tabata et al. 1991; Foley et al. 1993). When we used the programs to compare DNA homologies among these various proteins, we found that the *Arabidopsis* OBF5 protein was similar to the maize OBF3.1 protein. This level of conservation over the entire coding sequence between a putative transcription factor from monocot and dicot plants is striking and makes it likely that the OBF5 protein is a homologue of one of the maize OBF3 proteins. Since this will be a difficult question to unequivocally answer, we have maintained the OBF5 name for now. In contrast, the *Arabidopsis* OBF4 protein showed little DNA homology with the maize OBF3.1 protein and was more similar to the tobacco TGA1a protein, although there were clear differences in the NH_2 terminus. Interestingly, when the DNA binding properties of the OBF4 and OBF5 proteins were compared, we found that OBF5 forms the upper retarded band with the ocs-element probes more efficiently than OBF4 (Zhang et al. 1993). As described earlier, similar results were obtained when the maize OBF3.1 protein was compared to the tobacco TGA1a protein (Foley et al. 1993).

4.2 Comparison of the *Arabidopsis* OBF Clones to Other *Arabidopsis* bZIP Proteins

Two groups have isolated *Arabidopsis* proteins similar to either the OBF4 or OBF5 proteins. Kawata et al. (1992) isolated a cDNA encoding a protein called aHBP1b and Schindler et al. (1992a) isolated a cDNA clone encoding a protein called TGA1. Both of these *Arabidopsis* proteins share close homology with either OBF4 or OBF5. The aHBP1b protein is similar to the OBF5 protein, while the TGA1 protein is similar to the OBF4 protein. For all four *Arabidopsis* genes the 3′-untranslated regions share no homology, implying distinct genes. The *Arabidopsis* OBF proteins share limited homology, confined to the bZIP motif, with the *Arabidopsis* GBF proteins which bind to G-box sequences (Schindler et al. 1992b) and with the PosF21 protein (Aeschbacher et al. 1991), the other *Arabidopsis* bZIP proteins isolated to date.

5 Conclusion

Ocs-elements are a group of related, bipartite promoter elements which have been exploited by two distinct groups of plant pathogens, *Agrobacterium*, and certain viruses to express genes in plants. The composition of the plant-derived, protein complex, OTF, that interacts with ocs-element sequences is potentially complex, since four identical or distinct bZIP proteins are involved in forming the functionally important, upper band binding complex with an ocs-element sequence. Moreover, the composition of OTF may vary in different tissues and stages in development. We have cloned maize bZIP proteins that may form part of the OTF complex (Foley et al. 1993). The OBF3.1 protein has a high binding affinity for various ocs-element sequences and binding properties very similar to OTF; it appears to be a good candidate for comprising at least part of the OTE complex in the BMS cell line. The other maize OBF3 proteins are also candidates for forming part of OTF.

In *Arabidopsis* the situation is also complex with a minimum of four different proteins, OBF4, OBF5, aHBP1b, and TGA1, having been identified that could form part of OTF. At this stage we are not sure if any of the *Arabidopsis* OBF, aHBP1b, or TGA1 proteins are the homologue of a particular bZIP protein isolated in other plant species. This will not become clear until the functions of these proteins have been determined in their respective plants. Why *Arabidopsis* contains so many closely related proteins is also unclear, although consistent with other eukaryotic bZIP families like the mammalian Jun/Fos family which also contains a number of closely related members (reviewed in Ransone and Verma 1990; Vogt and Bos 1990). We have found differences in the DNA binding properties of the OBF4 and OBF5 proteins with OBF5 able to bind simultaneously to both halves of the ocs-element more efficiently than OBF4 (Zhang et al. 1993). This difference in binding to the ocs-element between two closely related proteins from the same species is potentially significant, since binding to both halves of the ocs-elements is a prerequisite for in vivo transcriptional activity (Bouchez et al. 1989; Singh et al. 1989). Through detailed characterization of the *Arabidopsis* OBF/aHPB1b/TGA1 proteins, including a number of reverse genetic experiments, the precise functions of these proteins may be determined. Such an analysis should shed some light on why the different pathogens have selected the ocs-element sequence and OTF complex for expression of pathogen-encoded genes in the plant. Finally, these studies will also contribute to an understanding of the control of gene expression in plants and whether structurally related transcription factors interact during plant development and if so how?

Acknowledgments. Early parts of the work described in this review were performed in the laboratory of Drs. Jim Peacock and Liz Dennis. We gratefully acknowledge the contributions made by Drs. David Bouchez, Jeff Ellis, Danny Llewellyn, and Jim Tokuhisa and the technical expertise of Kay Faulkner and Janice Norman. The work described in this article was supported in part by USDA Grant No 91-37301-6369 to K.B.S.

References

Aeschbacher RA, Schrott M, Protrykus I, Saul MW (1991) Isolation and molecular characterization of PosF21, an *Arabidopsis thaliana* gene which shows characteristics of a b-Zip class transcription factor. Plant J 1:303–316

Alliotte T, Tire C, Engler G, Peleman J, Caplan A, Van Montagu M, Inze D (1989) An auxin-regulated gene of *Arabidopsis thaliana* encodes a DNA-binding protein. Plant Physiol 89:743–752

Benfey PN, Ren L, Chua N-H (1989) The CaMV 35S enhancer contains at least two domains which can confer different developmental and tissue-specific expression patterns. EMBO J 8:2195–2202

Block A, Dangl JL, Hahlbrock K, Schulze-Lefert P (1990) Functional borders, genetic fine structure, and distance, requirements of cis elements mediating light responsiveness of the parsley chalcone synthase promoter. Proc Natl Acad Sci USA 87:5387–5391

Bouchez D, Tokuhisa JG, Llewellyn DJ, Dennis ES, Ellis JG (1989) The ocs-element is a component of the promoters of several T-DNA and plant viral genes. EMBO J 8:4197–4204

Burr B, Burr FA, Thompson KH, Albertsen MC, Stuber CW (1988) Gene mapping with recombinant inbreds in maize. Genetics 118:519–526

Cooke R (1990) The figwort mosaic virus gene VI promoter region contains a sequence highly homologous to the octopine synthase (ocs) enhancer element. Plant Mol Biol 15:181–182

Czarnecka E, Nagao R, Key JL, Gurley WB (1988) Characterization of Gmhsp26-A, a stress gene encoding a divergent heat shock protein of soybean: heavy-metal-induced inhibition of intron processing. Mol Cell Biol 8:1113–1122

DeGreve H, Dhaese P, Seurinck J, Lemmers M, Van Montagu M, Schell J (1983) Nucleotide sequence and transcript map of the *Agrobacterium tumefaciens* Ti plasmid-encoded octopine synthase gene. J Mol Appl Genet 1:499–511

Donald RGK, Cashmore AR (1990) Mutation of either G box or I box sequences profoundly affects expression from the *Arabidopsis* rbc-1A promoter. EMBO J 9:1717–1726
Ellis JG, Llewellyn DJ, Walker JC, Dennis ES, Peacock WJ (1987) The ocs element: a 16 base pair palindrome essential for activity of the octopine synthase enhancer. EMBO J 6:3203–3208
Ellis JG, Tokuhisa JG, Llewellyn DJ, Bouchez D, Singh KB, Dennis ES, Peacock WJ (1993) Does the ocs-element occur as a functional component of the promoters of plant genes? Plant 4:433–443
Fenoll C, Black DM, Howell SH (1988) The intergenic region of maize streak virus contains promoter elements involved in rightward transcription of the viral genome. EMBO J 7:1589–1596
Foley RC, Grossman C, Ellis JG, Llewellyn DJ, Dennis ES, Peacock WJ, Singh KB (1993) Isolation of a maize bZIP subfamily; candidates for the ocs-element transcription factor. Plant J 3:669–679
Fromm H, Katagiri F, Chua N-H (1989) An octopine synthase enhancer element directs tissue-specific expression and binds ASF-1, a factor from tobacco nuclear extracts. Plant Cell 1:977–984
Guiltinan MJ, Marcotte WR, Quatrano RS (1990) A plant leucine zipper protein that recognises an abscisic acid response element. Science 250:267–271
Hagen G, Martin G, Li Y, Guilfoyle TJ (1991) Auxin-induced expression of the soybean GH3 promoter in transgenic tobacco plants. Plant Mol Biol 17:567–579
Herschman HR (1991) Primary response genes induced by growth factors and tumor promoters. Annu Rev Biochem 60:281–319
Katagiri F, Lam E, Chua N-H (1989) Two tobacco DNA-binding proteins with homology to the nuclear factor CREB. Nature 340:727–730
Kawata T, Imada T, Shiraishi H, Okada K, Shimura Y, Iwaduchi M (1992) A cDNA clone encoding HBP-1b homologue in *Arabidopsis thaliana*. Nucleic Acids Res 20:1141
Kononowicz H, Wang E, Habeck LL, Gelvin SB (1992) Subdomains of the octopine synthase upstream activation element direct cell-specific expression in transgenic tobacco plants. Plant Cell 4:17–27
Lam E, Benfey PN, Gilmartin PM, Rong-Xiang F, Chua N-H (1989) Site-specific mutations alter in vitro factor binding and change promoter expression pattern in transgenic plants. Proc Natl Acad Sci USA 86:7890–7894
Leisner SM, Gelvin SB (1988) Structure of the octopine synthase upstream activator sequence. Proc Natl Acad Sci USA 85:2553–2557
McKnight SL (1991) Molecular zippers in gene regulation. Sci Am 264:54–64
Medberry SL, Lochart BEL, Olszewski NE (1992) The Commelina Yellow Mottle Virus promoter is a strong promoter in vascular and reproductive tissues. Plant Cell 4:185–192
Nakayama T, Ohtsubo N, Mikami K, Kawata T, Tabata T, Kanazawa H, Iwabuchi M (1989) Cis-acting sequences that modulate transcription of wheat histone H3-gene and 3′ processing of H3 premature messenger RNA. Plant Cell Physiol 30:825–832
Ransone LJ, Verma IM (1990) Nuclear proto-oncogenes fos and jun. Annu Rev Cell Biol 6:539–557
Schindler U, Beckmann H, Cashmore AR (1992a) TGA1 and G-box binding factors: two distinct classes of *Arabidopsis* leucine zipper proteins compete for the G-box-like element TGACGTGG. Plant Cell 4:1309–1319
Schindler U, Terzaghi W, Beckmann H, Kadesch T, Cashmore AR (1992b) Heterodimerization between light-regulated and ubiquitously expressed *Arabidopsis* GBF bZIP protieins. EMBO J 11:1261–1273
Singh K, Tokuhisa JG, Dennis ES, Peacock WJ (1989) Saturation mutagenesis of the octopine synthase enhancer: correlation of mutant phenotypes with binding of a nuclear protein factor. Proc Natl Acad Sci USA 86:3733–3737
Singh K, Dennis ES, Ellis JG, Llewellyn DJ, Tokuhisa JG, Wahleithner JA, Peacock WJ (1990) OCSBF-1, a maize ocs enhancer binding factor: isolation and expression during development. Plant Cell 2:891–903

Tabata T, Nakayama T, Mikami K, Iwabuchi M (1991) HBP-1A and HBP-1b: leucine zipper-type transcription factors of wheat. EMBO J 10:1459–1467
Tokuhisa JG, Singh K, Dennis ES, Peacock WJ (1990) A DNA-binding protein factor recognises two binding domains within the octopine synthase enhancer element. Plant Cell 2:215–224
Van der Zaal EJ, Droog FNJ, Boot CJM, Hensgens LAM, Hoge JHC, Schilperoort RA, Libbenga KR (1991) Promoters of auxin-induced genes from tobacco can lead to auxin-inducible and root tip-specific expression. Plant Mol Biol 16:983–998
Vogt PK, Bos TJ (1990) Jun oncogene and transcription factor. Adv Cancer Res 55:1–35
Zhang B, Foley RC, Singh KB (1993) Isolation and characterization of two related *Arabidopsis* ocs-element bZIP binding proteins. Plant J 4:711–716

10 Regulation of α-Zein Gene Expression During Maize Endosperm Development

Milo J. Aukerman[1] and Robert J. Schmidt[1,2]

1 Introduction

Development in higher eukaryotes is a complex process involving the differentiation of various cell types that subsequently perform specific functions in the organism. In order to achieve this specification, certain genes must be expressed in some tissues but not in others. Many genes are also expressed only during well-defined developmental periods in order to synchronize the events leading to cell differentiation. One way in which tissue-specific and temporal gene expression can be established is through the action of specific regulatory factors that control the genes in *trans*. Although examples exist of gene control at posttranscriptional levels, most regulatory events during development are thought to occur at the level of transcription. The mechanisms by which regulatory factors control the transcription of specific target genes are currently being elucidated in many higher organisms. Through these studies, it is becoming increasingly clear that developmental gene regulation involves complex networks of regulatory factors that control multiple downstream targets. One example of this is the regulation of the zein storage protein genes during development of the maize seed. Investigation into the mechanisms responsible for the coordinated regulation of the zein multigene family may provide insight into the general regulatory strategies employed by higher organisms with regard to multigene families. In this chapter we will present a brief overview of maize seed storage protein gene expression. Emphasis will be given to recent contributions to this field of study, especially with regard to the role of *trans*-acting factors. For more detailed reviews relating to zein gene structure and organization, the reader is referred to Rubenstein and Geraghty (1986), Thompson and Larkins (1989) and Shewry and Tatham (1990).

2 Background

The seeds of higher plants are composed primarily of three different tissues which form within the ovary after pollination. The diploid embryo, which

[1] Department of Biology, 0116 and Center for Molecular Genetics, University of California at San Diego, La Jolla, California 92093, USA.
[2] To whom correspondence should be addressed.

Results and Problems in Cell Differentiation 20
L. Nover (Ed.)
Plant Promoters and Transcription Factors

will give rise to the plant seedling, derives from the fusion of a pollen sperm cell and the egg cell housed in the embryo sac. The endosperm is formed by another fertilization event characterized by the fusion of a second sperm cell with two polar nuclei. This produces the triploid primary endosperm nucleus. Both sperm cells are products of a single haploid mitotic division occurring in the pollen grain, and therefore contribute identical genetic information to the embryo and endosperm. Surrounding these two tissues in the mature seed is the maternally derived pericarp, which protects the seed from the environment. The main function of the endosperm is to store nutrients which will be utilized during later developmental events. In most dicots, the endosperm is absorbed during seed development, and is replaced by an embryo-derived storage tissue, the cotyledons. In monocots and some dicots, the endosperm continues to develop throughout seed maturation.

Development of the endosperm/cotyledon is characterized by the accumulation of two major components, starch and storage proteins. After synthesis, both are deposited in specialized organelles, starch in amyloplasts and storage proteins in protein bodies. In mature seed, starch can represent more than 80% of the dry weight, and storage proteins can account for more than half of the total seed protein (Nelson 1980; Shewry and Tatham 1990). The reserves of stored protein and starch are ultimately utilized by the developing embryo during germination. The abundance of storage proteins in the seed has facilitated their biochemical analysis for over a century. Osborne (1924) first described the fractionation of seed proteins into various classes depending on their solubility properties. The four classes he defined were the albumins (water-soluble), the globulins (salt-soluble), the prolamins (soluble in aqueous alcohol), and the glutelins (acid/alkali-soluble). Most storage proteins from legume species are globulins, whereas those of cereal species are primarily prolamins and glutelins.

Both the legume globulins and the cereal prolamins have been demonstrated to be encoded by multigene families, with individual members displaying high similarity to each other (Higgins 1984). Since storage proteins accumulate during a defined period of seed development and within a specific tissue, the multiple genes encoding these proteins are necessarily regulated in a coordinate manner. Investigation into the regulation of seed storage protein synthesis may therefore provide answers to a basic question in biology, i.e., what mechanisms are involved in the coordinate regulation of a multigene family? One paradigm which has been studied by investigators for several years is the regulation of the maize prolamins, also known as the zeins. Much of what we know about the regulation of storage protein synthesis in cereals has been derived from studies on zeins. Due to the ease of maize genetics, it has been possible to identify loci responsible for zein regulation, and to study their interactions with each other. With both regulators and gene targets identified, it is hoped that the regulatory network leading to the coordinate expression of the zeins will soon be dissected.

Maize is one of the most important crops with regard to worldwide nutrition. Nevertheless, studies on the amino acid composition of maize zein

(Nelson 1979) indicate that it is a less than perfect crop. Like the prolamins from other cereal species, zein is rich in proline and glutamine and severely deficient in lysine and tryptophan, two essential amino acids for humans and monogastric animals. Because the endosperm represents such a large percentage of the total seed protein content in maize, this deficiency is reflected in the whole seed and makes maize a relatively poor protein source. One of the major goals of zein research is to increase the lysine and tryptophan content of maize seed either through traditional breeding methods, or by generating seeds that express zein proteins genetically engineered to contain higher amounts of lysine and tryptophan. In order to achieve this goal, it is imperative that we understand the details concerning the regulation of zein expression in maize.

3 Zein Proteins

The synthesis of zein occurs during the development of the maize kernel, beginning at around 12 days after pollination (DAP) and concluding at seed maturity about 50 DAP (Jones et al. 1977b). At maturity, zein can represent as much as 60% of the total seed protein (Nelson 1979). Accumulation of zein mRNA during seed development follows the same time scale as zein protein (Langridge et al. 1982a; Marks et al. 1985b; Kriz et al. 1987), and occurs exclusively in the endosperm (Kriz et al. 1987; Bianchi and Viotti 1988; Dolfini et al. 1992). Zein protein is synthesized by polyribosomes present on rough endoplasmic reticulum and deposited into protein bodies, the vesicular storage organs of the endosperm (Burr and Burr 1976; Larkins and Hurkman 1978). Like the prolamin storage proteins of other cereal seeds, zein has no known enzymatic function and serves primarily as an amino acid source during seed germination.

Zein is a family of highly hydrophobic proteins which are all soluble in alcohol under reducing conditions. Their extensive heterogeneity allows them to be distinguished from one another by standard biochemical techniques. On one-dimensional SDS/polyacrylamide gels, zeins can be resolved into several different size classes (Lee et al. 1976). The most abundant zein polypeptides have apparent molecular weights of 19 and 22 kilodaltons (kDa), while other minor zein components migrate at 16, 14, 10 and 27 kDa. The 19- and 22-kDa size classes are similar in that they are soluble in alcohol alone, and are therefore grouped together as the α-zeins. The other zeins require reducing agents to be solubilized (Esen 1986), and are classified as β- (16- and 14-kDa), γ- (27-kDa), and δ- (10-kDa) zeins, due to structural differences. Both the 19- and 22-kDa classes, which together comprise about 70% of the total zein fraction, are composed of multiple polypeptides which differ in their overall charge (Vitale et al. 1980). In contrast, the other zein classes appear to be represented by single polypeptides. Individual zein species can be separated on the basis of both charge and size by electrophoresis on two-dimensional isoelectric focusing/SDS gels (Hagen and Rubenstein 1980; Burr and Burr 1981; Galili et al. 1987). These studies, plus others utilizing HPLC separatory techniques (Wilson 1991), indicate that most inbreds of maize express 10 to 15 different zein polypeptides.

Sequence inspection of cDNAs encoding the α-zeins reveals a common arrangement of primary structural motifs (reviewed in Shewry and Tatham 1990). Both the 19- and 22-kDa size classes have a 21 amino acid signal sequence which is involved in a cotranslational processing event (Burr and Burr 1981). One particularly striking motif present in the α-zeins is a sequence of 20 amino acids which is repeated several times in the C-terminal two-thirds of the polypeptide (Geraghty et al. 1981; Marks and Larkins 1982). The most straightforward analysis of this region reveals that the 19-kDa polypeptides contain nine of these repeats, whereas the 22-kDa polypeptides have ten, thus accounting for the size difference between the two α-zein groups (Messing 1987). It is thought that these repeats form antiparallel α-helices which facilitate intermolecular interactions between individual zein proteins (Argos et al. 1982). Although it has not been experimentally tested, this model might help explain the orderly packing of zein proteins in protein bodies (Lending and Larkins 1989). Furthermore, the ability of zein to self-aggregate into protein body-like structures when expressed in *Xenopus* oocytes (Wallace et al. 1988) is consistent with the idea that zein deposition is mediated by interactions between zeins.

4 Zein Genes

Due to their slight charge differences, as visualized by isoelectric focusing, individual zeins can be mapped to separate locations in the maize genome (Soave et al. 1981a). This suggests that the zeins are encoded by a multigene family. Hybridization analyses utilizing zein cDNAs have indicated that the zein multigene family contains approximately 50 to 100 individual members (Viotti et al. 1979; Hagen and Rubenstein 1981; Burr et al. 1982; Wilson and Larkins 1984; Heidecker and Messing 1986). The vast majority of zein genes are highly similar to those encoding polypeptides of the α-zein class, while the other classes (β, γ, and δ) are probably represented by only one or two genes (Hagen and Rubenstein 1981; Burr et al. 1982; Marks and Larkins 1982; Wilson and Larkins 1984). The discrepancy between the number of zein genes and the number of zein polypeptides observed to be expressed in a given inbred (only 10–15) suggests that the majority of zein genes are pseudogenes. In fact, several zein genes have been isolated which contain stop codons within the normal coding sequence, and thus may correspond to pseudogenes (Spena et al. 1982, 1983; Kridl et al. 1984; Wandelt and Feix 1989; Liu and Rubenstein 1992b; Thompson et al. 1992).

Studies have shown that the α-zein genes are clustered in the maize genome. Zein gene clusters can be mapped genetically to chromosomes 4 and 7, with a single gene found on chromosome 10 (Soave and Salamini 1984). Hybridization of zein cDNAs to maize chromosomes in situ is consistent with these assignments (Viotti et al. 1982). Genomic clones containing multiple zein genes separated by only a few kilobases (kb) have been isolated (Spena et al. 1983; Schmidt et al. 1990; Liu and Rubenstein 1992a), providing further molecular evidence for the clustering of zein genes. In another analysis, two clusters of approximately ten zein genes each were found to be entirely encompassed within two 100–200 kb fragments (Heidecker et al. 1991). The average distance between each gene in these two clusters was estimated to be about 15 kb. The significance of zein gene clustering is not entirely clear. It has been proposed that this arrangement has facilitated gene conversion and gene amplification events, resulting in high levels of zein gene expression in the endosperm (Messing 1987).

Although the 19- and 22-kDa zein polypeptides possess a similar overall architecture (see Sect. 3), the two classes differ significantly at the amino acid and nucleotide level. In one study, for example, three cDNAs encoding 22-kDa zeins displayed 92% nucleotide sequence identity to each other and only 60 to 65% sequence identity to five cDNAs encoding 19-kDa zeins (Marks et al. 1985a). Similarity between the five 19-kDa zein cDNAs ranged between 75 and 95%. The 19- and 22-kDa classes, together comprising the α-zein group, are not closely related to the other minor zein groups, displaying less than 40% similarity to the β-, γ-, and δ-zeins (Marks et al. 1985a; Pedersen et al. 1986; Rubenstein and Geraghty 1986; Shewry and Tatham 1990). In general, the classification of α-zein cDNAs by sequence homology parallels their classification by molecular weight. There are, however, exceptions to this general rule (see Schmidt 1992 for further discussion).

It has been proposed that the 22-kDa zein gene originated from the 19-kDa zein gene by duplication and divergence (Marks et al. 1985a). Recent evidence obtained from a related grass species, *Coix lacryma-jobi*, seems to indicate the opposite (Leite et al. 1990). Hybridization analyses performed with a 22-kDa zein gene probe identified *Coix* prolamin (coixin) genes homologous to the 22-kDa zein class. However, when a 19-kDa zein gene probe was used, no homologues to the 19-kDa class were found. Based on this result, the authors suggest that an ancestral 22-kDa gene duplicated to give rise to the 19-kDa gene, and that this occurred after the divergence of maize and *Coix* (Leite et al. 1990).

Comparison of zein cDNA and genomic clones indicates that the α-zein genes, unlike most eukaryotic genes, do not contain introns (Pedersen et al. 1982; Spena et al. 1982). A second unusual aspect of the zein genes is that two distinct promoters, P1 and P2, have been identified. P1 gives rise to an 1.8-kb transcript containing more than 1000 nucleotides of 5′ leader sequence (Langridge et al. 1982b). A smaller transcript of 0.9 kb originates from the P2 promoter. Although the utilization of these two promoters is highly variable in heterologous systems (Langridge and Feix 1983; Langridge et al. 1984; Brown et al. 1986), it appears that in vivo the P2 promoter is pre-

A

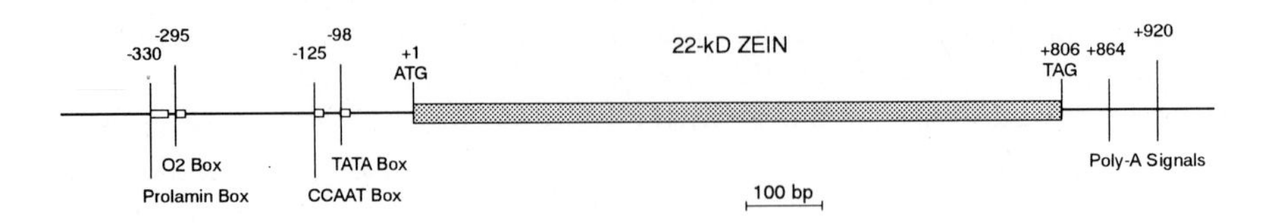

B

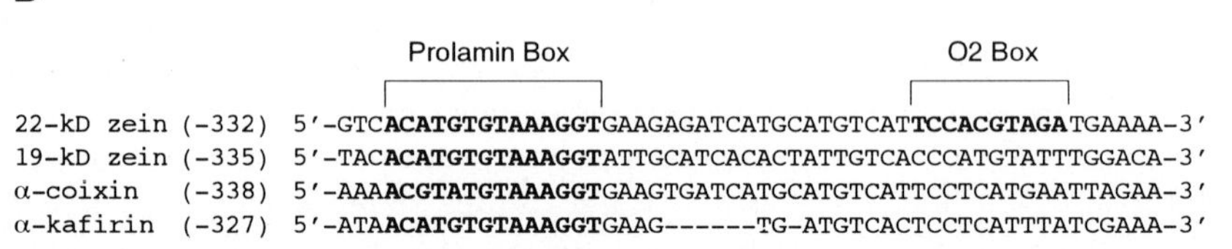

Fig. 1. **A** Diagram of 22-kDa zein gene. Zein genes are intron-less and contain canonical eukaryotic TATA (TATAAATA), CCAAT (CCAAAAAAT) and polyadenylation (AATAAT) signals. The positions of the highly conserved *prolamin box* (see **B**) and *O2 box* (TCCACGTAGA) are indicated. The closely related 19-kDa zein genes have these same features except for the O2 box. Nucleotide positions of *cis* elements are based on the sequence of 22 Z-4 (Schmidt et al. 1992). The distal P1 promoter is not shown. **B** Sequence comparison of *prolamin box* and *O2 box* in α-zeins, α-coixins, and α-kafirins. Both 19- and 22-kDa α-zein genes as well as the closely related 22-kDa-like α-coixin and α-kafirins have a conserved prolamin box sequence. Only 22-kDa zeins have the O2 box. Numbering is relative to the A of the initiation codon. The sequences are from Schmidt et al. (1992), Spena et al. (1983), Ottoboni et al. (1993), and DeRose et al. (1989)

ferentially utilized (Kriz et al. 1987; Quattrocchio et al. 1990). The small amount of larger transcript observed from P1 in vivo may have some function, as yet unknown, in regulating the zein genes. Aside from their lack of introns and unique promoter structure, the zein genes have features typical of eukaryotic genes (Fig. 1A), including a TATA and CCAAT consensus sequence in the 5′ flanking region and an AATAAA polyadenylation signal (Pedersen et al. 1982; Kridl et al. 1984; Liu and Rubenstein 1992a).

Both the 19- and 22-kDa families of zein genes have the aforementioned dual promoter structure (Langridge and Feix 1983; Kriz et al. 1987); however, the 5′ flanking regions of these classes are considerably diverged at the sequence level. The longest stretch of similarity between the 19- and 22-kDa gene families is a conserved sequence, ACATGTGTAAAGGT, located at around −330 relative to the ATG initiation codon (Brown et al. 1986). A portion of this sequence, TGTAAAG, is also found in the promoters of the β-, γ-, and δ-zein genes as well as in the promoters of prolamin genes isolated from other cereal species including wheat, barley, sorghum, and *Coix* (Forde et al. 1985; DeRose et al. 1989; Ottoboni et al. 1993). This 7-bp motif is consistently positioned between −300 and −350 in these promoters and is therefore referred to as the −300 element, or prolamin box. The fact that this element is exclusively shared by genes expressed in the endosperm suggests that it may be a *cis* determinant for endosperm-specific gene expression. Interestingly, the prolamin box shows similarity to the SV40 core transcriptional enhancer element, GTGGAAAG. By DNase footprint analysis, a nuclear protein from maize endosperm has been demonstrated to bind to the prolamin box (Maier et al. 1987). Although further work is needed to make sense of these correlations, it is possible that this sequence element plays some role in zein regulation (see below).

5 *Cis* Elements Involved in Zein Regulation

By studying the regulation of zein genes we may obtain insight regarding how a multigene family is coordinately controlled in a tissue-specific manner. One approach towards investigating zein regulation is to determine the *cis* sequences in the promoters of zein genes which are required for their proper regulation when they are reintroduced into a suitable expression system. Ideally, manipulated zein genes could be transformed into maize, and their expression detected by an attached reporter gene. Maize is not readily transformable at this point, however, and investigators have instead utilized heterologous species in which to express zein genes.

Unfortunately, heterologous systems have proven to be of limited value in the analysis of zein promoter regulation. In some cases, the relative utilization of the P1 and P2 promoters observed in vivo is not mirrored in these systems (Langridge et al. 1984; Brown et al. 1986; Quattrocchio et al. 1990). In general, only relatively weak expression from zein promoters has been obtained (Roussell et al. 1988; Schernthaner et al. 1988; Ueng et al. 1988), or there is considerable variation from transformant to transformant (Goldsbrough et al. 1986). Both of these drawbacks make it very difficult to compare the relative activities of different zein promoter fragments, a key step in identifying important *cis* regulatory elements in a promoter. If meaningful correlations are to be made, it is critical to demonstrate endosperm-specific expression of zein promoters. Unfortunately, this is not always observed (Ueng et al. 1988). In the case of heterologous tissue-culture systems derived from nonendosperm tissue, it is not even possible to address whether endosperm specificity has been retained (Roussell et al. 1988; Thompson et al. 1990).

Schernthaner et al. (1988) demonstrated that in tobacco a transgenic zein promoter/β-glucuronidase (GUS) fusion is expressed exclusively in the endosperm, as would be expected for correct zein regulation. This facilitated a deletion analysis of the zein promoter, which resulted in a puzzling observation: a 174-bp promoter which lacked the conserved prolamin box found in all zein genes nevertheless conferred normal endosperm-specific expression (Matzke et al. 1990). One possible explanation of these results is that another sequence in the 174-bp promoter similar to the prolamin box was determining endosperm specificity. However, one must be wary of drawing firm conclusions from these results, since the experiments were performed in a heterologous system. It is possible, for example, that tobacco seeds do not express nuclear proteins sufficiently similar to those which bind the prolamin box in maize and which have been postulated to be involved in zein gene regulation (Maier et al. 1987; Quayle and Feix 1992).

Clearly, zein expression should be studied in a homologous system in order to facilitate more confident interpretation of results. Recent strides in this direction, although preliminary, are encouraging. Zein promoter activity has been detected in transiently transformed protoplasts made from fresh endosperm (Schwall and Feix 1988) or endosperm tissue culture cells (Quayle et al. 1991; Ueda and Messing 1991). The regulation appears to be tissue-specific, since the same zein promoter constructs are not active when transformed into a nonendosperm-derived maize cell line (Schwall and Feix 1988; Quayle et al. 1991). As an initial step toward the identification of important *cis* elements in the zein promoter, Quayle and Feix (1992) have demonstrated that a 43-bp DNA fragment containing the prolamin box has a positive effect on transcription of a downstream reporter gene in an endosperm cell line. Endosperm nuclear proteins will bind to a DNA fragment containing the prolamin box sequence, further supporting the idea that this element is involved in endosperm-specific expression (Maier et al. 1987; Quayle and Feix 1992). Similarly, it has been shown that a sequence from a wheat gliadin gene promoter containing the prolamin box is required for high-level expression of a reporter gene in tobacco endosperm (Colot et al. 1987). It appears then that the prolamin box may indeed be a functional *cis* element involved in prolamin gene regulation.

Endosperm cell cultures, as well as the particle bombardment transformation technique (Klein et al. 1987), should prove useful in further delineating important *cis* elements in the promoters of the zein genes. Although most of the studies performed thus far have utilized promoters corresponding to the 19-kDa zein gene family, recent studies have focused on the promoters of the 22-kDa zein genes (Schmidt et al. 1992; Ueda et al. 1992; see below). The 22-kDa zein genes are significantly divergent relative to the 19-kDa family and are regulated in a significantly different manner than the 19-kDa family (see below); therefore, it will be important to determine which *cis* regulatory elements are shared between the two families and which are unique to each family.

6 Mutants Affecting Zein Synthesis

Identification of *cis* determinants of zein gene expression is a crucial step toward understanding zein gene regulation. Another approach is to identify the *trans*-acting regulators of the zein genes. Analysis of various maize mutants which affect zein synthesis has been invaluable in this type of approach. All of these mutants have a similar phenotype, that of an opaque, chalky endosperm (Mertz et al. 1964; Nelson et al. 1965). This is in contrast to the normal hard, translucent endosperm of the wild-type maize kernel. The opaque phenotype of these mutants is correlated with a decrease in zein accumulation in the endosperm relative to wild type (Lee et al. 1976), and it can be seen that the synthesis of different α-zein proteins are affected in the different mutants. For example, mutations in the *opaque-2* (*o2*) and *Defective endosperm B-30* (*DeB-30*) loci cause a preferential reduction in the 22-kDa class of zeins (Lee et al. 1976; Jones et al. 1977a; Salamini et al. 1979), whereas *opaque-7* (*o7*) mutants show a more dramatic reduction in the 19-kDa class of zeins (Di Fonzo et al. 1980). In three other mutants, *floury-2* (*fl2*), *opaque-6* (*o6*), and *Mucronate* (*Mc*), the synthesis of the 19- and 22-kDa classes appears to be roughly equally affected (Jones 1978; Salamini et al. 1983; Motto et al. 1989). It appears, then, that the genes corresponding to these mutants somehow play a positive role in zein synthesis, and evidence obtained from genetic mapping further suggests that these regulatory genes are *trans*-acting. For example, the *o2* locus maps to chromosome 7, whereas the 22-kDa genes, the predicted target of regulation by *o2*, map to chromosomes 4 and 10 (Soave and Salamini 1984). Likewise, the 19-kDa zeins are located on chromosomes 4 and 7, whereas their specific regulator *o7* is located on chromosome 10. Zein mRNA levels are reduced in the *o2*, *o7*, and *fl2* mutants (Pedersen et al. 1980; Burr and Burr 1982; Langridge et al. 1982a; Marks et al. 1985b), indicating that the products of these loci act either directly or indirectly to regulate zein gene transcription or message stability.

There are two other proteins which have an altered accumulation pattern in certain opaque mutants, one which migrates at 32 kDa and another at 70 kDa when analyzed by SDS/PAGE. The 32-kDa protein, called b-32, is an abundant protein found in the soluble cytoplasm of the endosperm. In both the *o2* and *o6* mutants, the expression of the b-32 protein is dramatically reduced relative to wild type (Soave et al. 1981b). A molecular clone of b-32 was isolated and used to demonstrate that the reduction in b-32 expression in *o2* and *o6* is due at least in part to reduced accumulation of b-32 message (Di Fonzo et al. 1988). The b-32 polypeptide is similar in sequence to a group of ribosomal inactivating proteins (RIPs). Like RIPs, the b-32 protein has been demonstrated to inactivate eukaryotic ribosomes in vitro, resulting in translational inhibition (Bass et al. 1992). Inactivation by b-32 appears to be specific for nonplant ribosomes, suggesting that b-32 may act to inhibit protein synthesis of invading pathogens. The implication from these studies is that the products of the *o2* and *o6* loci, in addition to their roles in regulating zein synthesis, also may play a role in plant defense by activating the expression of b-32 in the endosperm.

Overexpression of a 70-kDa protein, b-70, has been observed in the *fl2*, *Mc*, and *DeB-30* mutant backgrounds (Marocco et al. 1991). The b-70 polypeptide has been shown to be homologous to the mammalian immunoglobulin binding protein (BiP), a molecular chaperone (Fontes et al. 1991). Similar to BiP, b-70 binds with high affinity to ATP and is found localized to the ER lumen; in addition, b-70 is found in protein bodies. In the *fl2*, *Mc*, and *DeB-30* mutants, b-70 transcripts accumulate to higher levels, suggesting a transcriptional effect of these mutants on the gene encoding b-70 (Boston et al. 1991). It is likely, however, that the effect of these mutants on b-70 expression is indirect and may actually be due to the reduction in specific zein members or their misexpression. For example, one role of the b-70 chaperone may be to mediate the association of zeins into higher order structures in the ER and protein bodies (Argos et al. 1982). In this scenario, the reduced synthesis of zein in *fl2*, *Mc*, and *DeB-30* may affect the ability of the remaining zeins to orderly pack into protein bodies. In turn, this might result in an upregulation of b-70 expression in order to provide more chaperone to assist in zein folding. Consistent with this interpretation, recent investigations (Lending and Larkins 1992; Zhang and Boston 1992) have demonstrated that protein body morphology is dramatically affected by these mutations, and these alterations are correlated with increased levels of protein body-associated b-70.

Double mutants of the *o2*, *o7*, and *fl2* loci have been constructed in order to study their genetic interaction. The results of these studies indicate that *o2* and *o7* are epistatic to *fl2*. The preferential reduction in 22-kDa zeins observed in the *o2* mutant is also seen in the *o2/fl2* double mutant, and the pattern of zeins seen in the *o7* mutant is similar to that observed in the *o7/fl2* double mutant (Di Fonzo et al. 1980; Fornasari et al. 1982). The *o2/o7* combination appears to be additive, in that both the 19- and 22-kDa classes are severely reduced (Di Fonzo et al. 1980). The implication of these results is that there are separate pathways for regulating the synthesis of the 19- and 22-kDa zein classes. The gene product of *o2* is in the 22-kDa regulatory pathway, whereas the *o7* gene product is in the 19-kDa pathway. If both *o2* and *o7* are mutated, then both pathways are knocked out. The gene product of *fl2* appears to be involved in both pathways. Although *fl2* cannot be positioned accurately within these pathways, the epistatic relationship of *o2* and *o7* to *fl2* suggests that the action of the normal *fl2* gene product is mediated through direct or indirect interactions with the normal *o2* and *o7* gene products. In order to determine whether these interactions indeed exist, it will be necessary to isolate the gene products encoded by *o2*, *o7*, and *fl2*.

7 Opaque-2

The most extensively studied of the zein regulatory mutants is *o2*, first described in 1935 by R. A. Emerson and first characterized by Mertz et al. (1964). It was found in these early studies that the relative percentage of lysine in *o2* mutant seed is significantly higher than that found in wild-type seed. The likely explanation for this is that the lysine-poor zeins, being reduced in the *o2* background, no longer account for such a huge percentage of the total endosperm protein. The creation of a more balanced amino acid composition in this major commercial grain was hailed as a tremendous breakthrough for worldwide agriculture. Unfortunately, the yield from *o2* seed was reduced, and their soft endosperm made them susceptible to microbial infection and mechanical damage. Current approaches to remedy these problems include the development of Quality Protein Maize (QPM) lines by conventional breeding (Vasal et al. 1980). These modified *o2* lines retain the higher percentage of lysine, but produce higher yields and harder kernels. Unfortunately, the QPM trait is polygenic, and therefore cannot be effectively introduced into standard inbred lines.

Aside from the agronomic utility of the *o2* mutant, much has been learned about zein gene regulation by characterizing the product of the *o2* locus. As mentioned above, mutations in the *O2* gene cause a reduction in primarily the 22-kDa zein polypeptides (reviewed in Motto et al. 1989). The *o2* locus has been mapped to chromosome 7, whereas many of the 22-kDa zein polypeptides affected by *o2* are located on other chromosomes, suggesting that the *O2* gene product is a *trans*-acting regulator (Soave et al. 1978). The accumulation of transcripts corresponding to the 22-kDa zeins is reduced in *o2* mutants (Pedersen et al. 1980; Burr and Burr 1982; Langridge et al. 1982a; Marks et al. 1985b), and this reduction in zein message has been correlated with a lower rate of zein gene transcription in *o2* mutants (Kodrzycki et al. 1989). Taken together, these data suggest that the product of the *o2* locus is a transcriptional activator of the zein genes.

Transposon tagging experiments with the maize transposable elements *Suppressor-mutator* (*Spm*) or *Activator* (*Ac*) identified *o2*-mutable plants containing these respective elements inserted at the *o2* locus (Schmidt et al. 1987; Motto et al. 1988). The availability of both *Spm* and *Ac* probes facilitated the isolation of maize genomic DNA sequences flanking the inserted elements, which were subsequently used to obtain genomic clones of *O2*. Analysis of *O2* transcript accumulation indicated that *O2* is expressed exclusively in the endosperm of maize seed, consistent with its proposed role in tissue-specific zein gene regulation (Schmidt et al. 1987; Motto et al. 1988). *O2* message is absent in endosperms homozygous for a recessive *o2* allele, but is present in normal amounts in both the *o7* and *fl2* mutants. This result indicates that the *o7* and *fl2* mutations do not affect *O2* transcript accumulation.

Sequence analysis of the *O2* cDNA (Hartings et al. 1989; Schmidt et al. 1990) indicated that the O2 protein contains a basic domain/leucine zipper (bZIP) motif present in many transcription factors (Landschulz et al. 1988). The bZIP motif (Fig. 2) consists of a stretch of primarily basic amino acids (basic domain) directly abutted to a segment containing a periodic repeat of leucines every seven amino acids (leucine zipper). Copious experimental evidence indicates that the basic domain of this motif directly binds to DNA, whereas the leucine zipper provides a dimerization interface (Kouzarides

```
                BASIC  DOMAIN              LEUCINE  HEPTAD  REPEATS
                                           1        2       3       4
O2      ...RVRKRKESNRESARRSRYRKAAHLKE LEDQVAQ LKAENSC LLRRIAAL
o2-676  ...----------------K--------- ------- ------- --------
OHP1    ...-LQR--Q----------S------N- --A---- -RV---S ----L-DV


                5       6       7
O2      NQKYND ANVDNRV LRADMET LRAKVKMGEDSLKRV...
o2-676  ------ ------- ------- ---------------...
OHP1    ---F-E -A----- -K--V-- -------A---V---...
```

Fig. 2. Comparison of the bZIP domains of *O2*, *o2-676*, *OHP1*. Amino acid sequences are shown in single-letter format. Residues identical to those in O2 are designated by *dashes* in o2-676 and OHP1. *Numbers above the sequence* refer to hydrophobic residues (L, A, or V) which occur at the first position of every heptad repeat. *Underlined sequence* is the basic domain

and Ziff 1988; Bos et al. 1989; Gentz et al. 1989; Landschulz et al. 1989; Neuberg et al. 1989; Ransone et al. 1989; Schuermann et al. 1989; Turner and Tjian 1989). The leucine repeat region has been found to form an amphipathic α-helix which associates with a second leucine repeat region in a parallel orientation (Gentz et al. 1989; O'Shea et al. 1989). This association facilitates the dimerization of two bZIP-containing proteins, and results in a forked juxtapositioning of their adjacent basic domains, similar to the arms of the letter Y (Vinson et al. 1989). Dimerization mediated by the leucine repeat region is absolutely required for DNA binding (Kouzarides and Ziff 1988; Gentz et al. 1989; Landschulz et al. 1989; Ransone et al. 1989; Turner and Tjian 1989). A protein/DNA crystal structure for this class of DNA-binding protein has recently been obtained (Ellenberger et al. 1992).

8 Analysis of Opaque-2 Function

The occurrence of a bZIP domain in the O2 protein predicts that it should bind to DNA in a sequence-specific manner. Consistent with this prediction is the observation that the O2 protein is localized to endosperm nuclei (Varagona et al. 1991). One likely target for DNA binding by O2 would be the promoters of genes encoding the 22-kDa zein genes. Indeed, we have demonstrated that the O2 protein binds to the promoter region of genes encoding the 22-kDa zeins, and that the bZIP domain of O2 mediates this binding (Schmidt et al. 1990). We have also identified a specific arginine residue in the bZIP domain that is critical to the DNA-binding activity of the O2 protein (Aukerman et al. 1991). This arginine is completely conserved among all functional bZIP proteins found thus far. In a particular o2 mutant protein, o2-676, this arginine is changed to lysine (Fig. 2), resulting in a total elimination of DNA-binding activity (Aukerman et al. 1991). The inability of the o2-676 protein to bind to 22-kDa zein promoters is correlated with reduced levels of 22-kDa zein expression in the *o2-676* mutant, providing strong evidence for a model in which the wild-type O2 protein directly regulates the 22-kDa zein genes by binding to their promoters and activating transcription.

The target site recognized by the O2 protein, 5′-TCCACGTAGA-3′, is present in 22-kDa zein promoters and absent in promoters of other zein genes that are independent of O2 control (Schmidt et al. 1992). This site is similar to the recognition sequence of a number of plant bZIP proteins (Schmidt et al. 1992). Interestingly, some cloned 22-kDa zein genes lack the O2 box. As discussed previously (Schmidt et al. 1992), these same clones contain one or more in-frame stop codons in their coding sequence, suggesting that they represent pseudogene members of this family that have sustained mutations in the promoter as well as in the coding sequence. Another group has reported binding of O2 to the promoter of the gene encoding the maize-soluble cytoplasmic protein b-32 (Lohmer et al. 1991). In their study, O2 was found to bind five similar but not identical sequences in the b-32 promoter with the following consensus: 5′-GATGAPyPuTGPu-3′. This sequence bears little similarity to the O2 target site identified in the 22-kDa zein promoter. One of these sites, B5 (GTTGACGTGA), did contain the ACGT core motif. Both studies involved the use of large quantities of O2 protein overexpressed in bacteria to perform nuclease protection experiments. It is possible that at such high concentrations of O2, weaker inter-

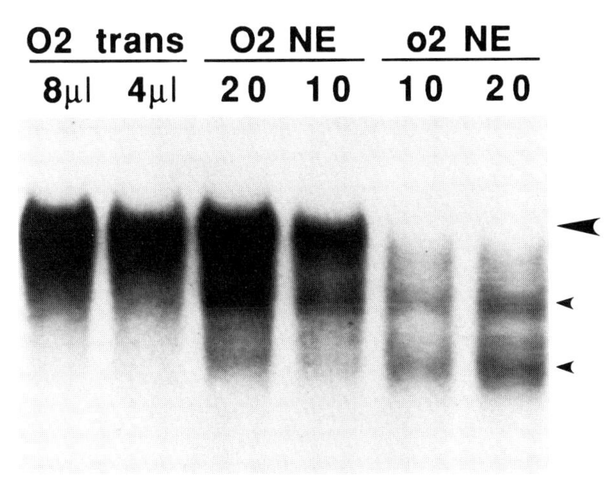

Fig. 3. Mobility shift assay showing the binding of other proteins besides O2 to the O2 box sequence. The assay utilizes in vitro translated O2 and endosperm nuclear extracts. 8 or 4 ul of translation (*O2-trans*) was assayed, alongside 20 or 10 μg of wild-type (*O2*) or the o2 mutant (*o2*) nuclear extracts (*NE*). The labeled DNA was a 32-bp oligonucleotide containing the O2 binding site sequence, TCCACGTAGA. *Large arrowhead* indicates uppermost complex in O2 NE which comigrates with complex in O2 translation and is absent in o2 NE. The two *small arrowheads* mark the two lower complexes present in both the O2 and o2 NEs. The unbound label is not shown

actions may occur with sequences that are not physiologically relevant with regard to promoter regulation in vivo. It is crucial, therefore, to demonstrate that O2 can bind to these sequences under conditions which more closely approximate those found in endosperm nuclei. In this regard, we have observed that the O2 protein present in wild-type endosperm nuclear extracts will specifically bind a 37-bp zein promoter fragment containing the TCCACGTAGA sequence (Schmidt et al. 1992; Fig. 3). It has not yet been demonstrated that the O2 protein in endosperm nuclear extracts can interact with the sites in the b-32 promoter that were identified utilizing O2 expressed in bacteria.

A possible explanation for these apparently conflicting observations has emerged from the recent studies of Izawa et al. (1993). In this work the affinity of a number of different plant bZIP proteins for degenerate sites containing an ACGT core motif was tested. Their data indicate that the TCCACGTAGA site in 22-kDa zein genes is predicted to be a high affinity site, while TTGACGTGA in the b-32 promoter is a comparatively weak site for O2 binding. It is possible that the interaction of O2 with b-32, although weak, becomes physiologically relevant through protein-protein contacts afforded by the close proximity of a number of comparatively weak binding sites, whereas a single high affinity site in 22-kDa zein promoters is sufficient for a strong protein-DNA interaction. Alternatively, the sites in the b-32 promoter may be poor binding sites for the O2 homodimer, but relatively good sites for a heterodimer between O2 and some other, yet unidentified, bZIP protein.

Utilizing an endosperm cell culture system, Ueda et al. (1992) demonstrated that O2 activates the transcription of a chimeric gene containing a 22-kDa zein gene promoter upstream of a reporter gene. The TCCACGTAGA site is most likely the sequence that mediates transcriptional regulation by the O2 protein. This is evidenced by the fact that O2 can transactivate a promoter containing a multimer of this site (Schmidt et al. 1992; Ueda et al. 1992), but cannot transactivate a promoter containing a multimer that is mutated in this site (TCCACATAGA) (Ueda et al. 1992). This lack of transactivation is correlated with the inability of O2 to bind to the mutant target site (Schmidt et al. 1992). Furthermore, a promoter from a γ-zein, which is normally not regulated by O2, can be converted to an O2-regulated promoter by simply changing a TTTACGTAGA sequence in the promoter to TCCACGTAGA (Ueda et al. 1992). The O2 protein will only bind strongly to the latter sequence and not the former, further demonstrating that regulation by O2 is strictly correlated with its ability to recognize the TCCACGTAGA motif. Curiously, the closely related α-coixin storage protein gene (Ottoboni et al. 1993) does not have the exact O2 box sequence in its promoter (Fig. 1B), but does respond in transient assays to O2 as well as to the *Coix* homologue of O2 (P. Arruda, pers. comm.). Since the bZIP domain in the *Coix* homologue is nearly identical to O2 from maize (P. Arruda, pers. comm.), it seems likely that it will bind to a very similar *cis* element. A related sequence, TCCAATAGA, is present in the α-coixin promoter beginning 11 bp down-stream of the position of the O2 box in 22-kDa zein genes. It will be interesting to see if this site has evolved as the binding site in *Coix* seed storage protein genes for the *Coix* O2 homologue.

The O2 protein has also been shown to activate the b-32 promoter in tobacco mesophyll cells (Lohmer et al. 1991). Transactivation of b-32 by O2 has recently been utilized as an indirect assay for O2 protein levels in order to study the regulatory role of the leader sequence present in the O2 message (Lohmer et al. 1993). Contained within the O2 leader sequence are three short open reading frames (ORFs) upstream of the ORF encoding the O2 protein. This unusual feature has been observed in other eukaryotic messages, most notably the yeast bZIP protein GCN4. The upstream ORFs in GCN4 play a role in regulating the translation of GCN4 in response to amino acid starvation (reviewed in Hinnebusch 1988). Likewise, the upstream ORFs in O2 appear to have a regulatory role. If the entire leader sequence containing the upstream ORFs is deleted, O2 expression is enhanced four- to five-fold relative to full-length O2 when assayed in yeast (Schmidt et al. 1992) and tobacco mesophyll cells (Lohmer et al. 1993). If the AUG initiation codons of all three upstream ORFs are mutated but the overall leader sequence remains intact, the same increase in O2 expression is observed (Lohmer et al. 1993). These results are consistent with the idea that the upstream ORFs in the O2 leader sequence act to inhibit translation of the downstream ORF corresponding to the O2 polypeptide. How this inhibition occurs is not known; however, Lohmer et al. (1993) have shown that the inhibition is a *cis*-dominant effect and therefore probably does not involve a diffusible peptide(s) encoded by one or more of the upstream ORFs. The physiological purpose of translational regulation of O2 is also unknown. Perhaps O2 expression is, like GCN4 expression, regulated by amino acid availability; this possibility has not yet been rigorously tested.

9 Proteins That Interact with Opaque-2

Several lines of evidence point toward the possibility that O2 may interact with one or more other proteins also present in maize endosperm nuclei. The first clue came from studies on the *o2-676* mutant (Aukerman et al. 1991). The o2-676 mutant protein does not bind to the O2 target site in vitro, and therefore *o2-676* might be expected to behave as a null allele in vivo. However, the levels of 22-kDa zein expression observed in vivo for *o2-676* were 50% of the levels observed in wild type and two to four times greater than the levels measured in an *o2* null line. This indicated that the *o2-676* allele is partially functional in vivo. One model to resolve the partial activity of *o2-676* with the inability of the o2-676 protein to bind DNA is to imagine that the O2 protein normally forms a heterodimer with another bZIP protein in endosperm nuclei. In the *o2-676* mutant, this heterodimer formation might partially compensate for the defect in the o2-676 protein, perhaps by facilitating an interaction of the heterodimer with the O2 target site.

Further evidence for the existence of proteins that interact with O2 was supplied by mobility shift experiments with endosperm nuclear extracts (Schmidt et al. 1992). These studies identified other proteins in endosperm nuclei besides O2 that can bind to the O2 target site. At least three binding complexes were revealed, and two of the complexes persisted even in the complete absence of the O2 protein. An example of these results is shown in Fig. 3. Experiments utilizing O2 antibodies and an *o2* null mutant suggested that O2 is present in at least two different complexes (Schmidt et al. 1992), consistent with the idea that at least some of the O2 polypeptides in endosperm nuclei bind the O2 target site as a heterodimer with another bZIP protein.

One approach to identify bZIP proteins which have the potential to interact with O2 is to identify other bZIP proteins expressed in the same tissue as O2, i.e., in the endosperm. In this manner, we have identified a protein, OHP1, which contains a bZIP domain that is strikingly similar to the bZIP domain of O2 (Fig. 2). Unlike O2, which is expressed specifically in the endosperm, OHP1 appears to be expressed in many different tissues (Pysh et al. 1993). We have determined that the OHP1 protein binds to the zein promoter target site, either as a homodimer or as a heterodimer with O2 (Fig. 4). Currently, we are attempting to determine the specific regulatory function of OHP1 with regard to zein gene regulation, as well as study the implications of its interaction with O2. The fact that *o2* mutations produce a recognizable phenotype indicates that OHP1 does not merely substitute for O2; that is, OHP1 is somehow functionally distinct from O2. One possibility is that OHP1 is not an activator of zein genes but a repressor that keeps zein genes transcriptionally quiet until O2 is produced. The role of O2 would then be twofold, first to titrate (through dimerization with OHP1) the repressor from the promoter and then to bind as a homodimer to provide high levels of activation. Two lines of preliminary analyses support this model. OHP1 is not a strong activator relative to O2, and the abundance of the OHP1 protein in endosperm is much lower than O2 (L. Pysh and R. Schmidt, unpubl.).

Because the o2-676 mutant protein appears to be partially active in vivo even though it cannot bind DNA on its own in vitro (Aukerman et al. 1991), we have considered the possibility that a heterodimer between o2-676 and OHP1 might provide o2-676 with partial DNA-binding activity. Based on our results, it appears that the OHP1/o2-676 heterodimer does not bind to the target site; instead, the o2-676 protein seems to act as a dominant negative inhibitor by forming heterodimers with OHP1 and effectively decreasing the amount of OHP1/OHP1 homodimers available to bind the target site (Fig. 4). The o2-676 protein also inhibits the binding of the wild-type O2 protein (Aukerman and Schmidt, unpubl.). The observation that

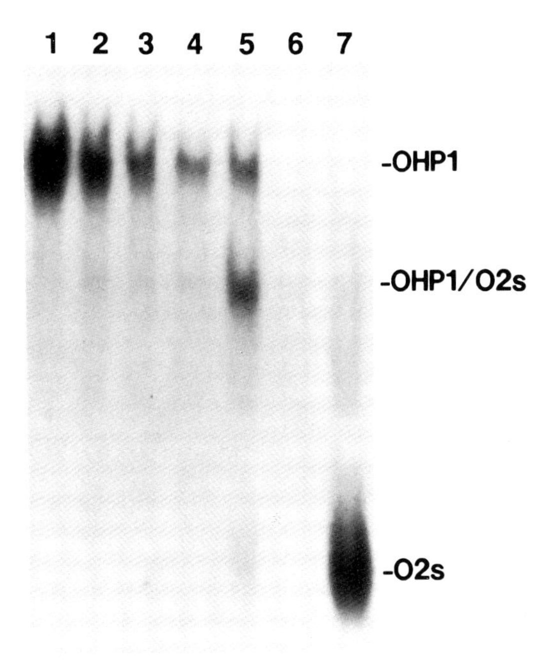

Fig. 4. Heterodimerization between *OHP1* and *O2*, and inhibition of OHP1 DNA binding by o2-676. Increasing amounts of a truncated form of o2-676 were mixed with a constant amount of OHP1, and the mixture assayed by mobility shift for binding to the O2 target site. *Lane 1* OHP1 alone. *Lanes 2–4* OHP1 plus o2-676 at 1:1, 1:2, and 1:4 molar ratios. *Lane 5* OHP1 plus a truncated form of O2. *Lane 6* o2-676 alone. *Lane 7* The truncated form of O2 alone. DNA binding probe was the same as that used in Fig. 3. The unbound label is not shown

the o2-676 protein can inhibit the DNA-binding activity of other bZIP proteins may have some relevance to its partial activity in vivo. We can now envisage two scenarios to explain the intermediate zein levels observed for *o2-676* (Fig. 5). In the first, o2-676 (and O2) inhibits a bZIP protein (OHP1?) which normally acts as a negative regulator of the 22-kDa zein genes, resulting in a derepression of the zein genes that leads to intermediate levels of expression. Since o2-676 cannot bind as a homodimer, wild-type levels of zein expression never occur. In the second, o2-676 binds weakly as a heterodimer with another bZIP protein (other than OHP1) to achieve partial activation of the zein genes. Although we have not obtained any in vitro evidence to support the second model, we cannot rule it out entirely. Other possibilities can be imagined that are also consistent with the available data. For instance, the o2-676 homodimer may provide partial activation as a consequence of stabilization through contacts with other proteins that bind the zein promoter in the vicinity of O2.

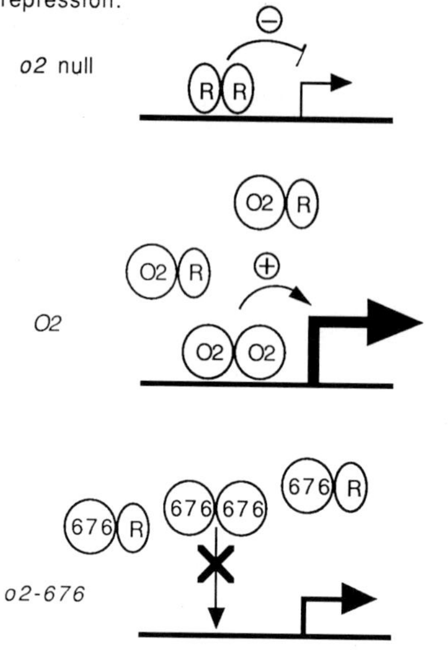
Model I: A bZIP repressor "R" keeps zein genes repressed. O2 alleviates suppression by titrating the repressor (providing partial activation) and binding as a homodimer, to provide high levels of activation. O2-676 provides partial activation by relieving repression.
o2 null
R
R
O2
O2
R
O2
R
O2
O2
o2-676
676
R
676
676
676
R

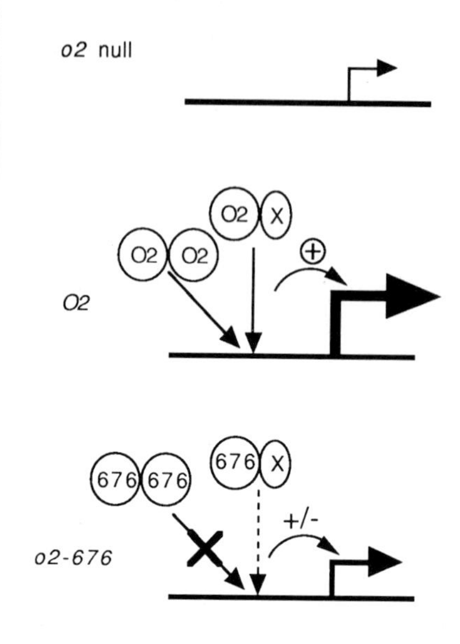
Model II: O2 activates zein gene expression by binding as a homodimer or as a heterodimer with another bZIP protein. O2-676 provides partial activation upon binding (weakly) as a heterodimer.
o2 null
O2
O2
O2
O2
X
o2-676
676
676
676
X
+/-

Fig. 5. Two alternative models to explain the *o2-676* phenotype. Each side-by-side pair of figures indicates possible interactions at the 22-kDa zein promoter in the *o2* null (*top*), wild-type *O2* (*middle*), and *o2-676* (*bottom*) backgrounds. The *bent arrows* depict zein gene transcription, and the relative sizes correlate with the relative levels of transcription observed in each background

10 Future Directions

The models suggested above are intended merely to provide a starting point for further investigations into the *o2-676* paradox and the regulation of the zein genes by O2 and OHP1. To continue studies with the *o2-676* mutant, DNA-binding studies with *o2-676* nuclear extracts should be performed utilizing zein promoter fragments that contain additional sequences flanking the O2 target site. It is possible that these larger zein promoter fragments contain binding sites for other protein complexes which can stabilize the interaction of o2-676 with the O2 target site. In particular, DNA binding by o2-676 (and wild-type O2) may be stabilized by proteins bound at the prolamin box, located 20 bp upstream of the O2 target site in the zein promoter (Schmidt et al. 1992). Interactions between O2 and prolamin box proteins may lead to transcriptional synergism, which could be determined with an endosperm culture transcriptional activation assay (Ueda et al. 1992).

To ascertain the role (if any) of OHP1 in zein gene regulation, we must determine whether OHP1 is a repressor or an activator (or neither) of the zein genes. We are currently pursuing this in both heterologous (yeast) and homologous (endosperm culture) systems, by determining the effect of coexpressing OHP1 along with reporter genes fused to promoters containing one or several O2 target sites. Once the nature of regulation by OHP1 is determined, the consequences of its interactions with O2 and o2-676 can be studied in these systems. Another layer of complexity regarding OHP1 is that there is a closely related gene, OHP2, in maize (Pysh et al. 1993). If both OHP isoforms can bind the O2 target site as well as interact with O2, then zein gene regulation could be much more complex than previously thought. It will therefore be important to sort out the specific roles of each of the OHP proteins by determining their individual expression patterns and transcriptional regulatory properties.

It is clear that other proteins in endosperm nuclei besides O2 can bind the O2 target site (Schmidt et al. 1992). We have identified one, OHP1 (Pysh et al. 1993), but recent evidence suggests that this protein only accounts for a small percentage of the binding activity seen in the absence of O2 (M. Aukerman, unpubl. observ.). Therefore, it will be critical to determine the identity of these other zein promoter binding factors (ZBFs). One method that could be used to directly isolate the ZBF proteins would be to screen an endosperm λgt11 expression library for proteins which can bind to the radiolabeled O2 target site.

In order to obtain a comprehensive understanding of zein gene regulation, it may be necessary to return to the approach which has proven so successful for *O2*, i.e., the isolation of regulatory genes which, when mutated, give rise to an observable phenotype. Mutations in the *fl2*, *o7*, *DeB30*, and *Mc* genes confer a phenotype to the seed which is similar to that of *o2*, and a reduction in zein protein and mRNA has been correlated with this mutant phenotype for some of these genes. These genes may play an important role in zein regulation, and it is possible that they could be isolated, like *O2*, by transposon tagging. Because of the extensive physiology and genetics that have been performed on these mutants (reviewed in Motto et al. 1989), it may turn out to be relatively straightforward to incorporate these genes into a multifactorial model for zein gene regulation.

A significant portion of the studies described here suggests that multiple proteins are involved in the regulation of the zein gene family. In order to understand the complexities of this process, it will be critical to isolate the relevant molecules, test their specific regulatory properties, and determine the mechanisms by which these molecules regulate the zeins in a combinatorial manner. By acquiring a fuller understanding of zein gene regulation, we may someday be able to utilize this knowledge to express genetically engineered zein genes in maize, an achievement with implications for worldwide nutrition and agriculture.

Acknowledgments. We thank Karen Archer for assistance with the graphics. We also thank Dr. Nam-Hai Chua and Dr. Paulo Arruda for communicating their unpublished data. This work was supported by grants to RJS from the McKnight Foundation and the National Institutes of Health (GM41286).

References

Argos P, Pedersen K, Marks MD, Larkins BA (1982) A structural model for maize zein proteins. J Biol Chem 257:9984–9990

Aukerman MJ, Schmidt RJ, Burr B, Burr FA (1991) An arginine to lysine substitution in the bZIP domain of an *opaque-2* mutant in maize abolishes specific DNA binding. Genes Dev 5:310–320

Bass HW, Webster C, O'Brian GR, Roberts JKM, Boston RS (1992) A maize ribosome-inactivating protein is controlled by the transcriptional activator *opaque-2*. Plant Cell 4:225–234

Bianchi MW, Viotti A (1988) DNA methylation and tissue-specific transcription of the storage protein genes of maize. Plant Mol Biol 11:203–214

Bos TJ, Rauscher FJ III, Curran T, Vogt PK (1989) The carboxy terminus of the viral Jun oncoprotein is required for complex formation with the cellular Fos protein. Oncogene 4:123–126

Boston RS, Fontes EBP, Shank BB, Wrobel RL (1991) Increased expression of the maize immunoglobulin binding protein homolog b-70 in three zein regulatory mutants. Plant Cell 3:497–505

Brown JWS, Wandelt C, Feix G, Neuhaus G, Schweiger HG (1986) The upstream regions of zein genes: sequence analysis and expression in the unicellular alga *Acetabularia*. Eur J Cell Biol 42:161–170

Burr B, Burr F (1976) Zein synthesis in maize endosperm by polyribosomes attached to protein bodies. Proc Natl Acad Sci USA 73:515–519

Burr FA, Burr B (1981) In vitro uptake and processing of prezein and other maize preproteins by maize membranes. J Cell Biol 90:427–434

Burr FA, Burr B (1982) Three mutations in *Zea mays* affecting zein accumulation: a comparison of zein polypeptides, in vitro synthesis and processing, mRNA levels, and genomic organization. J Cell Biol 94:201–206

Burr B, Burr FA, St. John TP, Thomas M, Davis RW (1982) Zein storage protein gene family of maize. J Mol Biol 154:33–49

Colot V, Robert LS, Kavanagh TA, Bevan MW, Thompson RD (1987) Localization of sequences in wheat endosperm protein genes which confer tissue-specific expression in tobacco. EMBO J 6:3559–3564

DeRose RT, Ma D-P, Kwon I-S, Hasnain SE, Klassy RC, Hall TC (1989) Characterization of the kafirin gene family from sorghum reveals extensive homology with zein from maize. Plant Mol Biol 12:245–256

Di Fonzo N, Fornasari E, Salamini F, Reggiani R, Soave C (1980) Interaction of maize mutants *floury-2* and *opaque-7* with *opaque-2* in the synthesis of endosperm proteins. J Hered 71:397–402

Di Fonzo N, Hartings H, Brembilla M, Motto M, Soave C, Navarro E, Palau J, Rhode W, Salamini F (1988) The b-32 protein from maize endosperm, an albumin regulated by the *o2* locus: nucleic acid (cDNA) and amino acid sequences. Mol Gen Genet 212:481–487

Dolfini SF, Landoni M, Tonelli C, Bernard L, Viotti A (1992) Spatial regulation in the expression of structural and regulatory storage-protein genes in *Zea mays* endosperm. Dev Genet 13:264–276

Ellenberger TE, Brandl CJ, Struhl K, Harrison SC (1992) The GCN4 basic region leucine zipper binds DNA as a dimer of uninterrupted α helices: crystal structure of the protein-DNA complex. Cell 71:1223–1237

Esen AA (1986) Separation of alcohol-soluble proteins (zeins) from maize into three fractions by differential solubility. Plant Physiol 80:623–627

Fontes EBP, Shank BB, Wrobel RL, Moose SP, O'Brian GR, Wurtzel ET, Boston RS (1991) Characterization of an immunoglobulin binding protein homolog in the maize *floury-2* endosperm mutant. Plant Cell 3:483–496

Forde BG, Hayworth A, Pywell J, Kries M (1985) Nucleotide sequence of a B1 hordein gene and the identification of possible upstream regulatory elements in endosperm storage protein genes from barley, wheat, and maize. Nucleic Acids Res 13:7327–7339

Fornasari E, Di Fonzo N, Salamini F, Reggiani R, Soave C (1982) *Floury-2* and *opaque-7* interaction in the synthesis of zein polypeptides. Maydica 27:185–189

Galili G, Kawata E, Larkins B (1987) Characterization of polypeptides corresponding to clones of maize zein mRNAs. Plant Physiol 84:291–297

Gentz R, Rauscher FJ III, Abate C, Curran T (1989) Parallel association of Fos and Jun leucine zippers juxtaposes DNA binding domains. Science 243:1695–1699

Geraghty D, Peifer MA, Rubenstein I, Messing J (1981) The primary structure of a plant storage protein: zein. Nucleic Acids Res 9:5163–5174

Goldsbrough PB, Gelvin SB, Larkins BA (1986) Expression of maize zein genes in transformed sunflower cells. Mol Gen Genet 202:374–381

Hagen G, Rubenstein I (1980) Two-dimensional gel analysis of the zein proteins in maize. Plant Sci Lett 19:217–223

Hagen G, Rubenstein I (1981) Complex organization of zein genes in maize. Gene 13:239–249

Hartings H, Maddaloni M, Lazzaroni N, Di Fonzo N, Motto M, Salamini F, Thompson R (1989) The *O2* gene which regulates zein deposition in maize endosperm encodes a protein with structural homologies to transcriptional activators. EMBO J 8:2795–2801

Heidecker G, Messing J (1986) Structural analysis of plant genes. Annu Rev Plant Physiol 37:439–466

Heidecker G, Chaudhuri S, Messing J (1991) Highly clustered zein gene sequences reveal evolutionary history of the multigene family. Genomics 10:719–732

Higgins TJV (1984) Synthesis and regulation of major proteins in seeds. Annu Rev Plant Physiol 35:191–221

Hinnebusch AG (1988) Novel mechanisms of translational control in *Saccharomyces cerevisiae*. Trends Genet 4:169–174

Izawa T, Foster R, Chua N-H (1993) Plant bZIP protein DNA binding specificity. J Mol Biol 230:1131–1144

Jones RA (1978) Effects of *floury-2* locus on zein accumulation and RNA metabolism during maize endosperm development. Biochem Genet 16:27–380

Jones RA, Larkins BA, Tsai CY (1977a) Storage protein synthesis in maize II. Reduced synthesis of a major zein component by the *opaque-2* mutant of maize. Plant Physiol 59:525–529

Jones RA, Larkins BA, Tsai CY (1977b) Storage protein synthesis in maize. III. Developmental changes in membrane-bound polyribosome composition and in vitro protein synthesis of normal and *opaque-2* maize. Plant Physiol 59:733–737

Klein TM, Wolf ED, Wu R, Sanford JC (1987) High-velocity microprojectiles for delivering nucleic acids into living cells. Nature 327:70–73

Kouzarides T, Ziff E (1988) The role of the leucine zipper in the fos-jun interaction. Nature 336:646–651

Kodrzycki R, Boston RS, Larkins BA (1989) The *opaque-2* mutation of maize differentially reduces zein gene transcription. Plant Cell 1:105–114

Kridl JC, Vieira J, Rubenstein I, Messing J (1984) Nucleotide sequence analysis of zein genomic clone with a short open reading frame. Gene 28:113–118

Kriz AL, Boston RS, Larkins BA (1987) Structural and transcriptional analysis of DNA sequences flanking genes that encode 19 kilodalton zeins. Mol Gen Genet 207:90–98

Landschulz WH, Johnson PF, McKnight SL (1988) The leucine zipper: a hypothetical structure common to a new class of DNA binding proteins. Science 240: 1759–1764

Landschulz WH, Johnson PF, McKnight SL (1989) The DNA binding domain of the rat liver nuclear protein C/EBP is bipartite. Science 243:1681–1688

Langridge P, Feix G (1983) A zein gene of maize is transcribed from two widely separated promoter regions. Cell 34:1015–1022

Langridge P, Pintor-Toro JA, Feix G (1982a) Transcriptional effects of the *opaque-2* mutation of *Zea mays* L. Planta 156:166–170

Langridge P, Pintor-Toro JA, Feix G (1982b) Zein precursor mRNAs from maize endosperm. Mol Gen Genet 187:432–438

Langridge P, Eibel H, Brown JWS, Feix G (1984) Transcription from maize storage protein gene promoters in yeast. EMBO J 3:2467–2471

Larkins BA, Hurkman WI (1978) Synthesis and deposition of zein in protein bodies of maize endosperm. Plant Physiol 62:256–263

Lee KH, Jones RA, Dalby A, Tsai CY (1976) Genetic regulation of storage protein content in maize endosperm. Biochem Genet 14:641–650

Leite A, Ottoboni LMM, Targon MLPN, Silva MJ, Turcinelli SR, Arruda P (1990) Phylogenetic relationship of zeins and coixins as determined by immunological cross-reactivity and Southern blot analysis. Plant Mol Biol 14:743–751

Lending CR, Larkins BA (1989) Changes in the zein composition of protein bodies during maize endosperm development. Plant Cell 1:1011–1023

Lending CR, Larkins BA (1992) Effect of the *floury-2* locus on protein body formation during maize endosperm development. Protoplasma 171:123–133

Liu CN, Rubenstein I (1992a) Genomic organization of an alpha-zein gene cluster in maize. Mol Gen Genet 231:304–312

Liu C-N, Rubenstein I (1992b) Molecular characterization of two types of 22 kilodalton α-zein genes in a gene cluster in maize. Mol Gen Genet 234:244–253

Lohmer S, Maddaloni M, Motto M, Di Fonzo N, Hartings H, Salamini F, Thompson RD (1991) The maize regulatory locus *opaque-2* encodes a DNA-binding protein which activates the transcription of the *b-32* gene. EMBO J 10:617–624

Lohmer S, Motto M, Salamini F, Thompson RD (1993) Translation of the mRNA of the maize transcriptional activator *opaque-2* is inhibited by upstream open reading frames present in the leader sequence. Plant Cell (in press)

Maier U, Brown JWS, Toloczyki C, Feix G (1987) Binding of a nuclear factor to a consensus sequence in the 5′ flanking region of zein genes from maize. EMBO J 6:17–22

Marks MD, Larkins BA (1982) Analysis of sequence microheterogeneity among zein messenger RNAs. J Biol Chem 257:9976–9983

Marks MD, Lindell JS, Larkins BA (1985a) Nucleotide sequence analysis of zein mRNAs from maize endosperm. J Biol Chem 260:16451–16459

Marks MD, Lindell JS, Larkins BA (1985b) Quantitative analysis of the Accumulation of zein mRNA during maize endosperm development. J Biol Chem 260:16445–16450

Marocco A, Santucci A, Cerioli S, Motto M, Di Fonzo N, Thompson R, Salamini F (1991) Three high-lysine mutations control the level of ATP-binding HSP70-like proteins in the maize endosperm. Plant Cell 3:507–515

Matzke AJM, Stöger EM, Schernthaner JP, Matzke MA (1990) Deletion analysis of a zein gene promoter in transgenic tobacco plants. Plant Mol Biol 14:323–332

Mertz ET, Bates LS, Nelson OE (1964) Mutant gene that changes protein composition and increases lysine content of maize endosperm. Science 145:279–280

Messing J (1987) The genes encoding seed storage proteins in higher plants. In: Rigby PWJ (ed) Genetic engineering, vol 6. Academic Press, Orlando, pp 1–46

Motto M, Maddaloni M, Ponziani G, Brembilla M, Marotta R, Di Fonzo N, Soave C, Thompson R, Salamini F (1988) Molecular cloning of the *o2-m5* allele of *Zea mays* using transposon marking. Mol Gen Genet 212:488–494

Motto M, Di Fonzo N, Hartings H, Maddaloni M, Salamini F, Soave C, Thompson RD (1989) Regulatory genes affecting maize storage protein synthesis. Oxford Surv Plant Mol Cell Biol 6:87–114

Nelson OE (1979) Inheritance of amino acid content in cereals. In: IAEA (eds) Seed protein improvement in cereals and grain legumes, vol 1. International Atomic Energy Agency, Vienna, pp 79–88

Nelson OE (1980) Genetic control of polysaccharide and storage protein synthesis in the endosperms of barley, maize, and sorghum. In: Pomeranz Y (eds) Advances in cereal science and technology, vol 3. American Association of Cereal Chemists, St Paul, pp 41–71
Nelson OE, Mertz ET, Bates LS (1965) Second mutant gene affecting the amino acid pattern of maize endosperm proteins. Science 150:1469–1470
Osborne TB (1924) The vegetable proteins. Longmans, Green, London
O'Shea EK, Rutkowski R, Kim PS (1989) Evidence that the leucine zipper is a coiled coil. Science 243:538–542
Ottoboni LMM, Leite A, Yunes JA, Targon ML, de Souza Filho GA, Arruda P (1993) Sequence analysis of 22 kDa-like α-coixin genes and their comparison with homologous zein and kafirin genes reveals highly conserved protein structure and regulatory elements. Plant Mol Biol 21:765–778
Pedersen K, Bloom KS, Anderson JN, Glover DV, Larkins BA (1980) Analysis of the complexity and frequency of zein genes in the maize genome. Biochemistry 19:1644–1650
Pedersen K, Devereux J, Wilson DR, Sheldon E, Larkins BA (1982) Cloning and sequence analysis reveal structural variation among related zein genes in maize. Cell 29:1015–1026
Pedersen K, Argos P, Naravana SV, Larkins BA (1986) Sequence analysis and characterization of a maize gene encoding a high-sulfur zein protein of Mr 15 000. J Biol Chem 261:6279–6284
Pysh L, Aukerman MJ, Schmidt RJ (1993) OHP1: a maize bZIP protein that interacts with opaque-2. Plant Cell 5:227–236
Quattrocchio F, Tolk MA, Coraggio I, Mol JN, Viotti A, Koes RE (1990) The maize zein gene zE19 contains two distinct promoters which are independently activated in endosperm and anthers of transgenic *Petunia* plants. Plant Mol Biol 15:81–93
Quayle T, Feix G (1992) Functional analysis of the −300 region of maize zein genes. Mol Gen Genet 231:369–374
Quayle TJA, Hetz W, Feix G (1991) Characterization of a maize endosperm culture expressing zein genes and its use in transient transformation assays. Plant Cell Rep 9:544–548
Ransone LJ, Visvader J, Sassone CP, Verma IM (1989) Fos-Jun interaction: mutational analysis of the leucine zipper domain of both proteins. Genes Dev 3: 770–781
Roussell DL, Boston RS, Goldsbrough PB, Larkins BA (1988) Deletion of DNA sequences flanking an Mr 19 000 zein gene reduces its transcriptional activity in heterologous plant tissues. Mol Gen Genet 211:202–209
Rubenstein I, Geraghty DF (1986) The genetic organization of zein. In: Pomeranz Y (ed) Advances in cereal science and technology, vol 8. American Association of Cereal Chemists, St Paul, pp 297–315
Salamini F, Di Fonzo N, Gentinetta E, Soave C (1979) A dominant mutation interfering with protein accumulation in maize seeds. In: IAEA (eds) Seed protein improvement in cereals and grain legumes, vol 1. International Atomic Energy Agency, Vienna, pp 97–108
Salamini F, Di Fonzo N, Fornasari E, Gentinetta E, Reggiani R, Soave C (1983) Mucronate, *Mc*, a dominant gene of maize which interacts with *opaque-2* to suppress zein synthesis. Theor Appl Genet 65:123–128
Schernthaner JP, Matzke MA, Matzke AJM (1988) Endosperm-specific activity of a zein gene promoter in transgenic tobacco plants. EMBO J 7:1249–1255
Schmidt RJ (1992) Opaque-2 and zein gene expression. In: Verma DP (ed) Control of plant gene expression. CRC, London, pp 337–355
Schmidt RJ, Burr FA, Burr B (1987) Transposon tagging and molecular analysis of the maize regulatory locus *opaque-2*. Science 238:960–963
Schmidt RJ, Burr FA, Aukerman MJ, Burr B (1990) Maize regulatory gene *opaque-2* encodes a protein with a "leucine-zipper" motif that binds to zein DNA. Proc Natl Acad Sci USA 87:46–50

Schmidt RJ, Ketudat M, Aukerman MJ, Hoschek G (1992) Opaque-2 is a transcriptional activator that recognizes a specific target site in 22-kD zein genes. Plant Cell 4:689–700

Schuermann M, Neuberg M, Hunter JB, Jenuwein T, Ryseck RP, Bravo R, Muller R (1989) The leucine repeat motif in Fos protein mediates complex formation with Jun/AP-1 and is required for transformation. Cell 56:507–516

Schwall M, Feix G (1988) Zein promoter activity in transiently transformed protoplasts from maize. Plant Sci 56:161–166

Shewry PR, Tatham AS (1990) The prolamin storage proteins of cereal seeds: structure and evolution. Biochem J 267:1–12

Soave C, Salamini F (1984) Organization and regulation of zein genes in maize endosperm. Philos Trans R Soc Lond 304:341–347

Soave C, Suman N, Viotti A, Salamini F (1978) Linkage relationships between regulatory and structural gene loci involved in zein synthesis in maize. Theor Appl Genet 52:263–267

Soave C, Reggiani R, Di Fonzo N, Salamini F (1981a) Clustering of genes for 20 kD zein subunits in the short arm of maize chromosome 7. Genetics 97:363–377

Soave C, Tardani L, Di Fonzo N, Salamini F (1981b) Zein level in maize endosperm depends on a protein under control of the *opaque-2* and *opaque-6* loci. Cell 27:403–410

Spena A, Viotti A, Pirrotta V (1982) A homologous repetitive block structure underlies the heterogeneity of heavy and light chain zein genes. EMBO J 1:1589–1594

Spena A, Viotti A, Pirrotta V (1983) Two adjacent genomic zein sequences: structure, organization and tissue-specific restriction pattern. J Mol Biol 169:799–811

Thompson GA, Larkins BA (1989) Structural elements regulating zein gene expression. Bio Essays 10:108–111

Thompson GA, Boston RS, Lyznik LA, Hodges TK, Larkins BA (1990) Analysis of promoter activity from an alpha-zein gene 5′ flanking sequence in transient expression assays. Plant Mol Biol 15:755–764

Thompson GA, Siemieniak DR, Sieu LC, Slightom JL, Larkins BA (1992) Sequence analysis of linked maize 22 kDa alpha-zein genes. Plant Mol Biol 18:827–833

Turner R, Tjian R (1989) Leucine repeats and an adjacent DNA binding domain mediate the formation of functional cFos-cJun heterodimers. Science 243: 1689–1694

Ueda T, Messing J (1991) A homologous expression system for cloned zein genes. Theor Appl Genet 82:93–100

Ueda T, Waverczak W, Ward K, Sher N, Ketudat M, Schmidt RJ, Messing J (1992) Mutations of the 22- and 27-kD zein promoters affect transactivation by the opaque-2 protein. Plant Cell 4:701–709

Ueng P, Galili G, Sapanara V, Goldsborough PB, Dube P, Beachy RN, Larkins BA (1988) Expression of a maize storage protein gene in petunia plants is not restricted to seeds. Plant Physiol 86:1281–1285

Varagona MJ, Schmidt RJ, Raikhel NV (1991) Monocot regulatory protein opaque-2 is localized in the nucleus of maize endosperm and transformed tobacco plants. Plant Cell 3:105–113

Vasal SK, Villegas E, Bjarnason M, Gelaw B, Goertz P (1980) Genetic modifiers and breeding strategies in developing hard endosperm *opaque-2* materials. In: Pollmer WG, Phillips RH (eds) Improvement of quality traits of maize for grain and silage use. Nijhoff, London, pp 37–73

Vinson CR, Sigler PB, McKnight SL (1989) Scissors-grip model for DNA recognition by a family of leucine zipper proteins. Science 246:911–916

Viotti A, Sala E, Marotta R, Alberi P, Balducci C, Soave C (1979) Genes and mRNAs coding for zein polypeptides in *Zea mays*. Eur J Biochem 102:211–222

Viotti A, Abildsten D, Pogna N, Sala E, Pirrotta V (1982) Multiplicity and diversity of cloned zein cDNA sequences and their chromosomal localization. EMBO J 1:53–58

Vitale A, Soave C, Galante E (1980) Peptide mapping of IEF zein components from maize. Plant Sci Lett 18:57–64
Wallace JC, Galili G, Kawata EE, Cuellar RE, Shotwell MA, Larkins BA (1988) Aggregation of lysine-containing zeins into protein bodies in *Xenopus* oocytes. Science 240:662–664
Wandelt C, Feix G (1989) Sequence of a 21 kd zein gene from maize containing an in-frame stop codon. Nucleic Acids Res 17:2354
Wilson CM (1991) Multiple zeins from maize endosperms characterized by reversed-phase high performance liquid chromatography. Plant Physiol 95:777–786
Wilson DR, Larkins BA (1984) Zein gene organization in maize and related grasses. J Mol Evol 20:330–340
Zhang F, Boston RS (1992) Increases in binding protein (BiP) accompany changes in protein body morphology in three high-lysine mutants of maize. Protoplasma 171:142–152

11 Control of Floral Organ Identity by Homeotic MADS-Box Transcription Factors

Brendan Davies and Zsuzsanna Schwarz-Sommer

1 Introduction

In recent years the combined use of genetics and molecular biology has resulted in dramatic progress in the field of developmental biology, particularly in the case of *Drosophila*. These complementary approaches are also being applied to the problem of flower development and initial results already demonstrate that, despite the complexity of the system, there may be an underlying unity with the animal kingdom. The purpose of this chapter is to review the current state of our knowledge of how a large family of evolutionarily conserved, related transcription factors, the MADS-box proteins, is involved in telling the plant where, when and how to form a flower.

1.1 The Use of Homeotic Mutants

A successful approach towards the understanding of normal organ development in higher plants has been the study of homeotic mutants in which more or less apparently normal organs develop at the wrong place. The conversion of one organ type to another in homeotic mutants has been recognized for more than 200 years, as has the idea that from their study the normal processes of development may be better understood (for review, see Meyerowitz et al. 1989). Heritable floral homeotic mutants are known in many plant species, but at the molecular level most of the recent progress has been confined to two species; *Antirrhinum majus* (snapdragon) and *Arabidopsis thaliana* (thale cress). In these two species three main types of homeotic mutants have been found whose phenotypes indicate the involvement of the corresponding genes in the control of floral organogenesis (see Fig. 1). Flowers of mutants belonging to either of the three types display common features, such as the kinds of homeotic transformations of organs in two adjacent whorls. These similarities have lent support to the idea that the homeotic functions may be encoded by homologous genes in the two species and that the basic mechanisms which control floral organogenesis

Max-Planck-Institut für Züchtungsforschung, Carl-von-Linné-Weg 10, D-50829 Köln, FRG

Results and Problems in Cell Differentiation 20
L. Nover (Ed.)
Plant Promoters and Transcription Factors

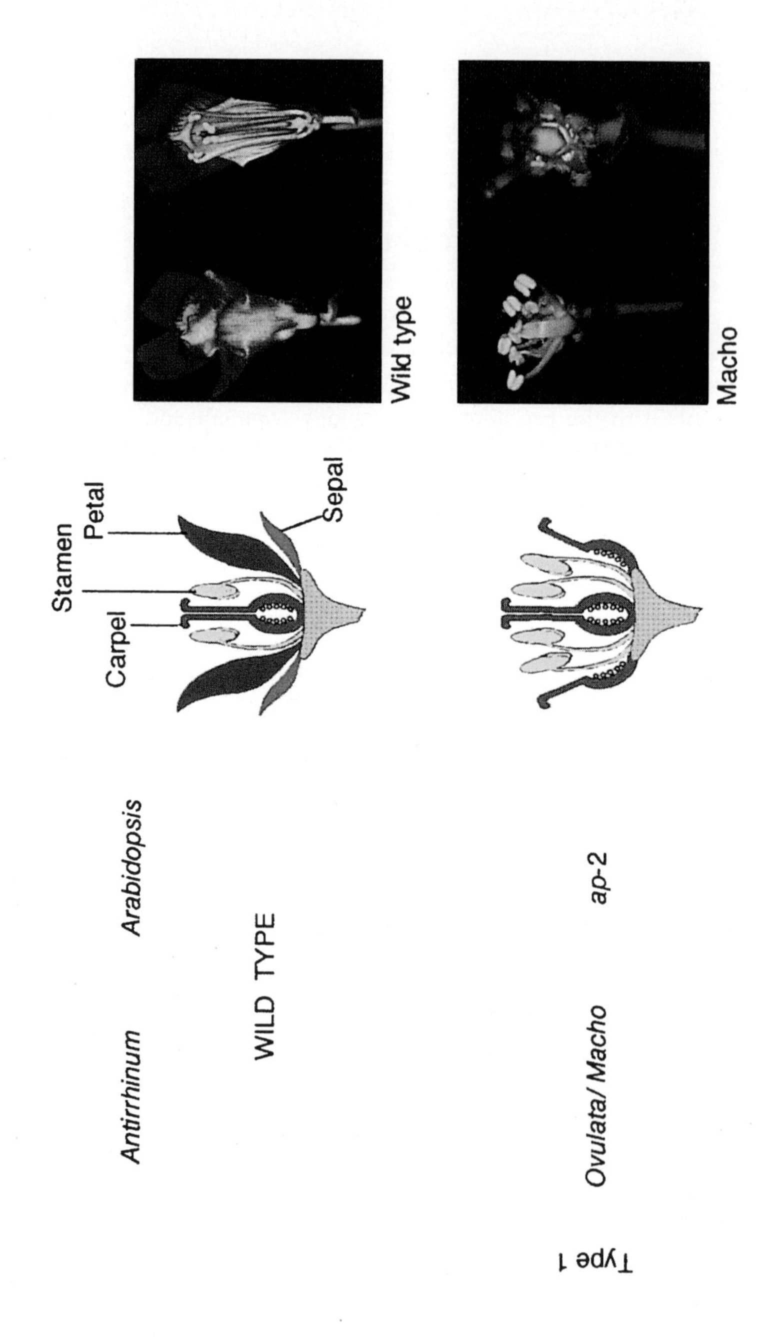
Antirrhinum
Arabidopsis
WILD TYPE
Carpel
Stamen
Petal
Sepal
Wild type
Type 1
Ovulata/ Macho
ap-2
Macho

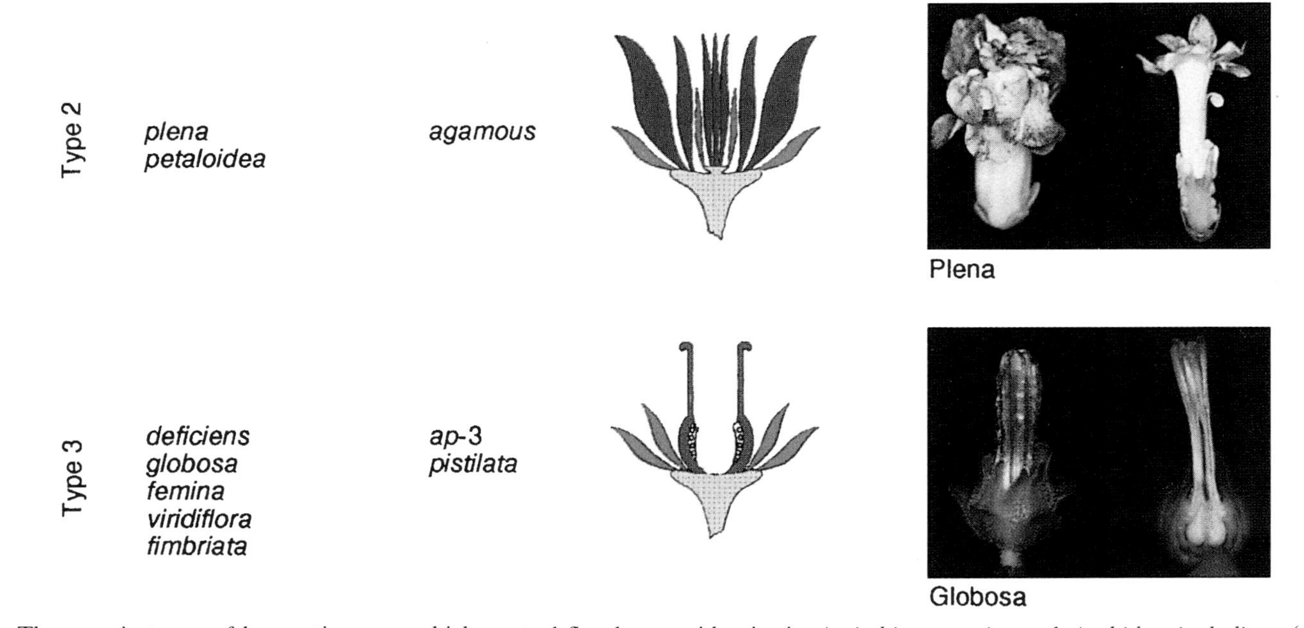

Fig. 1. Three main types of homeotic genes which control floral organ identity in *Antirrhinum majus* and *Arabidopsis thaliana* (modified and updated after Schwarz-Sommer et al. 1990). Genotypes are indicated at the *left* and the corresponding phenotypes of the flowers are depicted by a schematic drawing together with a photograph of *Antirrhinum* flowers representative of each type. The photographs on the *right* show the internal part of the flower after removing some parts of the corolla or after preparing a cross section to reveal the reproductive organs. Notice that the figure does not account for all aberrant features displayed by a given mutant, nor does it contain information on the differences revealed by mutants of the two species, such as altered organ number in some of the *Arabidopsis* mutants

may also be similar. As will be discussed later, molecular analyses of the control processes in different plant species suggest that there may be subtle differences in the exact mechanisms employed (Angenent et al. 1992; Jack et al. 1992; Tröbner et al. 1992).

1.2 The Discovery of Plant MADS-Box Genes

The isolation of the first plant homeotic gene, the *DEFICIENS* gene (*DEF*) of *Antirrhinum majus* (see Fig. 1), revealed an interesting homology to known human and yeast transcription factors (Sommer et al. 1990). Perhaps surprisingly, the subsequent isolation of eight more genes whose mutants give rise to homeotic phenotypes revealed that, in all cases except two (Coen et al. 1990; Weigel et al. 1992), they are members of the same family of transcription factors (Yanofsky et al. 1990; Huijser et al. 1992; Jack et al. 1992; Mandel et al. 1992b; Tröbner et al. 1992; Bradley et al. 1993). On the basis of the homology between the predicted amino acid sequences of parts of the two first isolated plant homeotic genes (*AGAMOUS* from *Arabidopsis* and *DEF* from *Antirrhinum*) and the DNA-binding domains of the two well-characterized transcription factors MCM1 (yeast) and SRF (human), an acronym, the MADS-box (from *M*CM1, *A*G, *D*EF A and *S*RF), was suggested to describe this conserved domain. Although only two MADS-box genes are known in the yeast *Saccharomyces cerevisiae*, larger families of MADS-box containing genes are present in humans (Pollock and Treisman 1991; Yu et al. 1992) and plants. At present, 22 MADS-box genes have been described from 5 different plant species (9 *Arabidopsis*, 5 tomato, 4 *Antirrhinum*, 2 petunia, 1 *Brassica napus* and 1 tobacco), most of which were isolated by utilizing their homology to either *DEF* or *AG* (see Table 1). In about one-third of these cases a mutant phenotype has been associated with the lack of gene function, and it is from these genes that our ideas on the role of MADS-box genes in floral morphogenesis are largely

Table 1. Survey of plant MADS-box proteins

Name	Species	Phenotype	Isolation	Reference
DEF A	*Antirrhinum majus*	Second whorl sepaloid petals. Third whorl carpeloid stamens.	Differential screening. cDNA subtraction. Transposon tagging.	Sommer et al. (1990)
AG	*Arabidopsis thaliana*	Third whorl petaloid stamens. Fourth whorl carpels become new mutant flower. Repetition of mutant flower.	T-DNA tagging.	Yanofsky et al. (1990)
AGL1	*Arabidopsis thaliana*	?	Degenerate MADS-box oligonucleotide library screening.	Ma et al. (1991)
AGL2	*Arabidopsis thaliana*	?	Degenerate MADS-box oligonucleotide library screening.	Ma et al. (1991)

Table 1. *Continued*

Name	Species	Phenotype	Isolation	Reference
AGL3	*Arabidopsis thaliana*	?	Degenerate MADS-box oligonucleotide library screening.	Ma et al. (1991)
AGL4	*Arabidopsis thaliana*	?	Degenerate MADS-box oligonucleotide library screening.	Ma et al. (1991)
AGL5	*Arabidopsis thaliana*	?	Homology to AGL 1.	Ma et al. (1991)
AGL6	*Arabidopsis thaliana*	?	Homology to *AGAMOUS*.	Ma et al. (1991)
TM3	Tomato	?	Homology to *DEFICIENS*.	Pnueli et al. (1991)
TM4	Tomato	?	Homology to *DEFICIENS*.	Pnueli et al. (1991)
TM5	Tomato	?	Homology to *DEFICIENS*.	Pnueli et al. (1991)
TM6	Tomato	?	Homology to *DEFICIENS*.	Pnueli et al. (1991)
TM8	Tomato	?	Homology to *DEFICIENS*.	Pnueli et al. (1991)
AP1	*Arabidopsis thaliana*	First whorl sepals become leaf-like. Secondary and tertiary flowers formed in axils of first whorl leaves.	Homology to *AGAMOUS*.	Mandel et al. (1992b)
AP3	*Arabidopsis thaliana*	Second whorl sepaloid petals. Third whorl carpeloid stamens.	Homology to *DEFICIENS*.	Jack et al. (1992)
SQUA	*Antirrhinum majus*	Flower often replaced by secondary shoot.	Transposon tagging and homology to *DEFICIENS*.	Huijser et al. (1992)
fbp1	*Petunia hybrida*	?	Degenerate MADS-box oligonucleotide library screening.	Angenent et al. (1992)
fbp2	*Petunia hybrida*	?	Degenerate MADS-box oligonucleotide library screening.	Angenent et al. (1992)
BAG	*Brassica napus*	?	Homology to *AGAMOUS*.	Mandel et al. (1992a)
GLO	*Antirrhinum majus*	Second whorl sepaloid petals. Third whorl carpeloid stamens.	Transposon tagging and homology to *DEFICIENS*.	Tröbner et al. (1992)
PLE	*Antirrhinum majus*	Third whorl petaloid stamens. Fourth whorl sepal/petaloid carpels. Repetition of inner whorls.	Transposon tagging and homology to *AGAMOUS*.	Bradley et al. (1993)
NTGLO	Tobacco	?	Homology to *GLOBOSA*	Hansen et al. (1993)

based. Efforts are currently being made to assign mutant phenotypes to the other members of the growing list of MADS-box genes. The plant MADS-box genes share common features both with each other and with similar genes in vertebrates and yeast (see Fig. 2A,B). There are striking parallels between the yeast and human MADS-box genes, both at the level of their

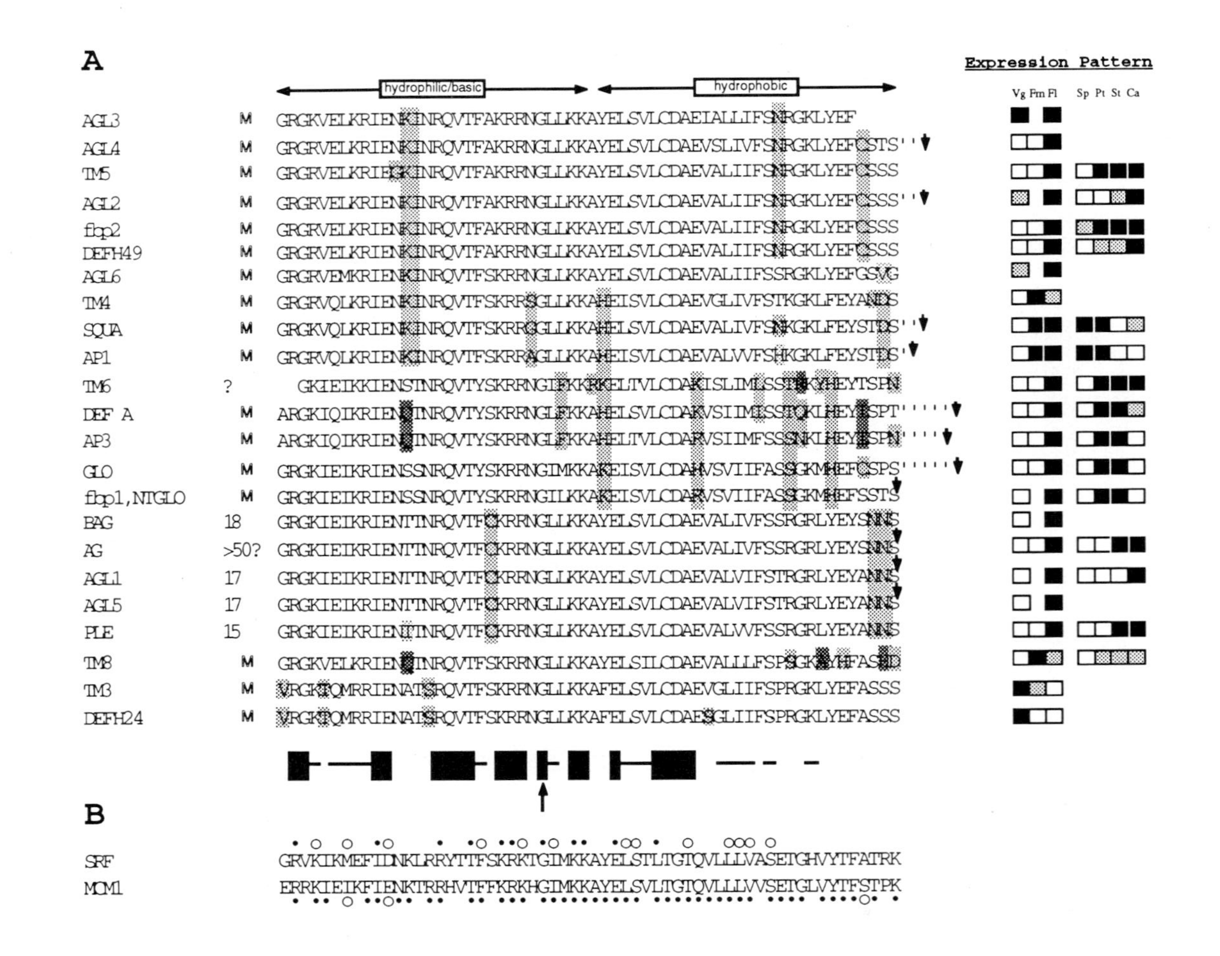
A
hydrophilic/basic
hydrophobic
Expression Pattern
Vg Fm Fl
Sp Pt St Ca
AGL3 M GRGKVELKRIENKINRQVTFAKRRNGLLKKAYELSVLCDAEIALLIFSNRGKLYEF
AGL4 M GRGRVELKRIENKINRQVTFAKRRNGLLKKAYELSVLCDAEVSLIVFSNRGKLYEFCSTS
TM5 M GRGRVELKRIEGKINRQVTFAKRRNGLLKKAYELSVLCDAEVALIIFSNRGKLYEFCSSS
AGL2 M GRGRVELKRIENKINRQVTFAKRRNGLLKKAYELSVLCDAEVALIIFSNRGKLYEFCSSS
fbp2 M GRGRVELKRIENKINRQVTFAKRRNGLLKKAYELSVLCDAEVALIIFSNRGKLYEFCSSS
DEFH49 M GRGRVELKRIENKINRQVTFAKRRNGLLKKAYELSVLCDAEVALIIFSNRGKLYEFCSSS
AGL6 M GRGRVEMKRIENKINRQVTFSKRRNGLLKKAYELSVLCDAEVALIIFSSRGKLYEFGSVG
TM4 M GRGRVQLKRIENKINRQVTFSKRRSGLLKKAHEISVLCDAEVGLIVFSTKGKLFEYANDS
SQUA M GRGKVQLKRIENKINRQVTFSKRRGGLLKKAHELSVLCDAEVALIVFSNKGKLFEYSTDS
AP1 M GRGRVQLKRIENKINRQVTFSKRRAGLLKKAHEISVLCDAEVALVVFSHKGKLFEYSTDS
TM6 ? GKIEIKKIENSTNRQVTYSKRRNGIFKKRKELTVLCDAKISLIMLSSTRKYHEYTSPN
DEF A M ARGKIQIKRIENQTNRQVTYSKRRNGLFKKAHELSVLCDAKVSIIMISSTQKLHEYISPT
AP3 M ARGKIQIKRIENQTNRQVTYSKRRNGLFKKAHELTVLCDARVSIIMFSSSNKLHEYISPN
GLO M GRGKIEIKRIENSSNRQVTYSKRRNGIMKKAKEISVLCDAHVSVIIFASSGKMHEFCSPS
fbp1, NTGLO M GRGKIEIKRIENSSNRQVTYSKRRNGILKKAKEISVLCDARVSVIIFASSGKMHEFSSTS
BAG 18 GRGKIEIKRIENTTNRQVTFCKRRNGLLKKAYELSVLCDAEVALIVFSSRGRLYEYSNNS
AG >50? GRGKIEIKRIENTTNRQVTFCKRRNGLLKKAYELSVLCDAEVALIVFSSRGRLYEYSNNS
AGL1 17 GRGKIEIKRIENTTNRQVTFCKRRNGLLKKAYELSVLCDAEVALVIFSTRGRLYEYANNS
AGL5 17 GRGKIEIKRIENTTNRQVTFCKRRNGLLKKAYELSVLCDAEVALVIFSTRGRLYEYANNS
PLE 15 GRGKIEIKRIENTTNRQVTFCKRRNGLLKKAYELSVLCDAEVALVVFSSRGRLYEYANNS
TM8 M GRGKVELKRIENQTNRQVTFSKRRNGLLKKAYELSILCDAEVALLLFSPSGKAYHFASHD
TM3 M VRGKTQMRRIENATSRQVTFSKRRNGLLKKAFELSVLCDAEVGLIIFSPRGKLYEFASSS
DEFH24 M VRGKTQMRRIENATSRQVTFSKRRNGLLKKAFELSVLCDAESGLIIFSPRGKLYEFASSS
B
SRF GRVKIKMEFIDNKLRRYTTFSKRKTGIMKKAYELSTLTGTQVLLLVASETGHVYTFATRK
MCM1 ERRKIEIKFIENKTRRHVTFFKRKHGIMKKAYELSVLTGTQVLLLVVSETGLVYTFSTPK

direct interactions and their positions and functions in the regulatory hierarchy, which suggest that a similar organization may also exist in plants. Since much more is currently known about the biochemistry of mammalian and yeast MADS-box genes, it is worth examining the available information about these genes and attempting to relate this to the control of flowering in plants. Such comparisons will be made in the following sections.

2 Structural Features of the MADS-Box Genes

2.1 The MADS-Box and Interactions with DNA

The MADS box is an evolutionarily conserved DNA-binding domain which appears to be unrelated to other known DNA-binding motifs. The MADS box comprises 60 amino acids that are highly conserved between the plant genes and the DNA-binding domains of the above-mentioned yeast and vertebrate transcription factors (see Fig. 2). Nineteen amino acids are identical between all the plant genes, with a further 16 residues showing relatively conservative changes. Most of this homology is located in the N-terminal two-thirds of the MADS box, with the C-terminus showing more variability. The MADS boxes can be divided into two approximately equal halves: an N-terminal basic region and a C-terminal hydrophobic domain

←

Fig. 2. The MADS box. The identity and conservation of amino acids in the plant MADS boxes (**A**) is indicated by *black bars* beneath the alignment. A *thick black bar* denotes identity and a *thin black bar* signifies a conservative amino acid substitution. Conservative amino acid substitutions were considered to be within the following groups: (G, A, S, T, P)(Q, E, N, D)(K, R, H)(L, I, V, M)(F, Y, W). Nonconservative changes have been grouped by different degrees of *shading*. A *black arrow* marks the absolutely conserved G-residue which is converted to a D-residue in the *def*-nic allele. The basic-hydrophilic and hydrophobic domains, referred to in the text, are indicated *above* the sequences. *M* indicates the position of the initiating methionine. Where the first amino acid of the MADS box does not immediately follow the methionine, the position within the sequence is indicated by a *number*, if known. Where it has been reported, the site of the splice donor which follows the MADS box is indicated by a downward pointing *arrowhead* with *dots* indicating bases from the last codon. The expression pattern of each gene is shown on the *right*. *Boxes* representing vegetative tissues (*Vg*), floral meristem (*Fm*), flower (*Fl*), sepal (*Sp*), petal (*Pt*), stamen (*St*), and carpel (*Ca*) are drawn with the amount of shading corresponding to relative levels of transcript in the different organs for each gene. NTGLO is the tobacco homologue of GLO (Hansen et al. 1993). BAG is the *Brassica napus* homologue of AG (Mandel et al. 1992a). DefH49 and DefH24 are unpublished *Antirrhinum* MADS-box proteins (H. Sommer pers. comm.). In **B** the MADS boxes of SRF and MCM1 are presented for comparison. *Above* the sequences is an indication of the homology between SRF, MCM1 and the plant proteins. A *black dot* denotes a residue which is identical in all the MADS boxes. An *open circle* marks a position at which only conservative changes are observed. The same system is used beneath the genes to show the homology between SRF and MCM1 alone (see Fig. 2). Such an organization also exists in SRF, MCM1 and ARG80, although in these cases the homology can be extended further to the C-terminus beyond the MADS box (Pollock and Treisman 1991).

The importance for DNA-protein and protein-protein interactions of the various components of the MADS box in SRF and MCM1 has been analyzed in detail. SRF, together with the RSRFs, which are also human MADS-box proteins, and yeast MCM1, all show slightly different binding site preferences (Pollock and Treisman 1991; see below). The basic N-terminal portion of SRF makes contacts with DNA and corresponds to the site-specific DNA recognition domain (Pollock and Treisman 1991). It is likely that whilst the N-terminal basic domain sets the DNA specificity, the rest of the MADS box and C-terminal extension determines the overall affinity (Mueller and Nordheim 1991). At the C-terminus of the MADS box the homology breaks down and the different genes fall into separate families. The human RSRF genes and MEF2 genes form one family (Pollock and Treisman 1991; Yu et al. 1992) and SRF, MCM1 and ARG80 form another. These homologies extend for approximately 30 amino acids beyond the C-terminus of the MADS box, and since the C-terminal borders of these extended regions of MADS-box homology approximately coincide, it has been suggested that they define a functional domain (Pollock and Treisman 1991). This C-terminal extended MADS box is both necessary and sufficient for high affinity DNA-binding of SRF and MCM1 and also contains residues involved in dimerization and ternary complex formation (Norman et al. 1988; Ammerer 1990; Mueller and Nordheim 1991; Pollock and Treisman 1991). It has been demonstrated that the substitution of just four amino acids in this region of MCM1, ARG80 or SRF is sufficient to alter the specificity of ternary complex formation (Mueller and Nordheim 1991). The C-terminal domain of the SRF MADS box, including the C-terminal extension, is sufficient to promote homodimer formation, although such dimers are incapable of binding DNA (Norman et al. 1988).

Since the MADS boxes of SRF and MCM1 are known to be involved in DNA binding and dimerization, it is likely that the plant MADS boxes have a similar function. Currently, little is known about the mechanisms involved in the action of the plant MADS boxes; it is possible that the invariant amino acids are required for a function such as DNA binding, whilst the variable residues define the specificity of such a function. In this respect it is noteworthy that in the mutant *DEF* allele *def*-nicotianoides (*def*-nic) the glycine residue (G), which is absolutely conserved at the same position in all plant and animal MADS-box proteins (arrow in Fig. 2), is converted by a point mutation to an aspartic acid residue (D).

The plant MADS-box proteins do not contain an obvious block of homology extending 30 amino acids beyond the MADS box, as is present in the human genes. Certain MADS-box proteins, presumably interspecies homologues of one another, such as PLENA (*Antirrhinum*) and AGAMOUS (*Arabidopsis*), share high levels of homology throughout their entire amino acid sequences. In other cases (e.g., AGL1 and AGL5) pairs of genes are found within a species which also show extended homology throughout their lengths. Other pairs of plant proteins show no obvious homology to each other except within the MADS box and, to a lesser extent, the K-box (see below). It is possible, therefore, that distinct differences exist in the mechanisms of the interactions between the plant MADS-box proteins and the yeast and mammalian ones. It may be that the protein-protein interaction specificity of SRF and MCM1 is determined by the extended MADS box, whereas that of the plant genes also requires the K-box, a domain which is lacking from the vertebrate and yeast MADS-box proteins (see below). Evidence to support this comes from the finding that SRF, MCM1 and ARG80 can all form heterodimers with each other but not with the plant proteins DEF A or AG (Mueller and Nordheim 1991). Another suggestion that the plant MADS boxes may define shorter structural domains than those of other organisms comes from the intron-exon organization of these genes (see below).

In every case in which it has been tested these proteins have been shown to bind DNA as dimers. The optimal DNA-binding sites for SRF, MCM1 and one of the RSRF proteins have been determined and are slightly different (Pollock and Treisman 1991; Dalton and Treisman 1992). The optimal binding site for SRF is $CC(A/T)_6GG$, whilst for RSRFC4 it is $CTA(A/T)_4TAG$ (Pollock and Treisman 1991). Sequences similar to the SRF consensus are also called CArG boxes (CC-A rich-GG). The binding site specificities of the plant MADS-box proteins have not yet been determined. In addition, it may be that the association of MADS-box proteins with specific accessory factors influences their binding site specificity as was shown for ARG80 (Dubois and Messenguy 1991).

2.2 The K-Box and Interactions with Proteins

The K-box is an approximately 70 amino acid region with some similarity to the coiled-coil domain of keratin (Ma et al. 1991), located at an equivalent position in all the plant MADS-box proteins but not in SRF or MCM1. The conservation of the K-box between the plant MADS-box genes is not particularly striking at the primary sequence level (Fig. 3A); rather, it is the predicted potential of this region to form two or three amphipathic helices which is remarkable. It has been suggested that the proteins form α-helices which bring regularly spaced hydrophobic residues together onto one face of the helix, as demonstrated in the helical wheel examples shown in Fig. 3B. Such amphipathic helices could then potentially interact with those of another K-box-containing protein to promote dimerization. A similar type of interaction is envisaged for leucine zipper proteins (Landschulz et al. 1988). Several mutant alleles of MADS-box genes are known which demonstrate the functional importance of this region. A mutant allele of *DEF*, *defA*-101, contains a deletion of three bases which removes a lysine residue at the N-terminus of the K-box. The *AP3*-1 allele of *Arabidopsis* produces a mutant

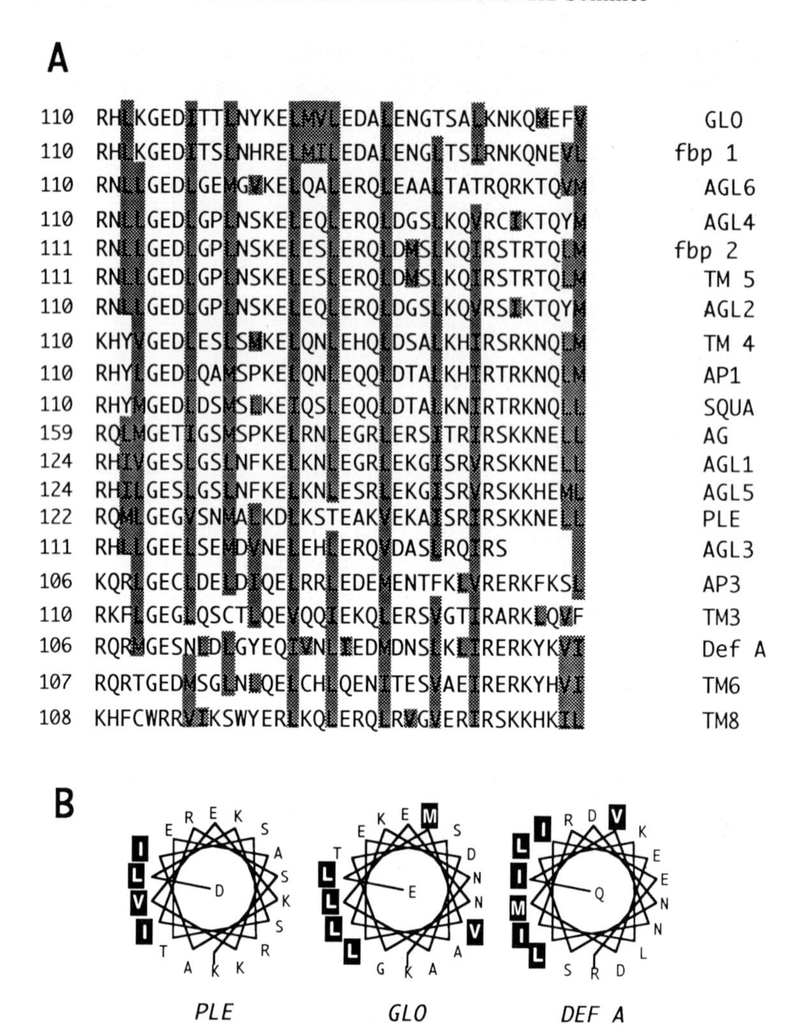

Fig. 3. The K-box. Parts of the K-boxes of the plant MADS-box proteins are aligned in **A**. The *numbers on the left* refer to the amino acid sequence. The hydrophobic residues (*L*, *I*, *V*, *M*) are *shaded*. The regular spacing of these residues is apparent. Example helical wheels are presented for part of the aligned sequences of the K-boxes of PLE, GLO and DEF A (**B**). The diagram represents five turns of an ideal α-helix containing approximately 3.5 residues per turn. Hydrophobic residues are shown in *black boxes*. It can be seen that the regular spacing of these residues, noted above, results in the potential to form a helix bearing a cluster of hydrophobic residues on one face and an opposite face of charged residues. The implications of such a structure are discussed in the text

form of AP3 in which a point mutation converts a conserved lysine residue at the C-terminus of the K-box to a methionine (Jack et al. 1992). Interestingly, both of these small mutations give rise to temperature-sensitive phenotypes which is consistent with this region being involved in protein-

protein interactions. Additional evidence is provided by the *AP3*-5 allele which produces a protein which is truncated immediately before the K-box. Despite having an intact MADS box and 20 amino acids C-terminal to the MADS box, the phenotype of this mutant is similar to the *AP3*-4 and *AP3*-3 alleles which are more severely truncated (Jack et al. 1992). Whilst most of the plant MADS-box proteins may form two such helices (Pnueli et al. 1991; Jack et al. 1992), GLO (from the *Antirrhinum* gene *GLOBOSA*), FBP1 (the product of the petunia gene fbp1) and NTGLO (the product of the tobacco *GLOBOSA* homologue) have the potential to form three (Angenent et al. 1992; Tröbner et al. 1992; Hansen et al. 1993).

2.3 Gene Organization

In all but five of the plant MADS-box genes the MADS box is located at the extreme N-terminus of the predicted proteins (see Fig. 2). In the cases of TM6 (Pnueli et al. 1991) and AG (Yanofsky et al. 1990) no ATG initiation codon was discovered and hence the exact position of the MADS box within the protein is unknown. The presumed *AGAMOUS* homologue from *Brassica napus* has been isolated and, in common with AGL1 and AGL5, the predicted protein has 17 amino acids at its N-terminus before the MADS box (Mandel et al. 1992a). The *Antirrhinum* gene PLENA has 15 amino acids before the MADS box in its predicted protein sequence (Bradley et al. 1993). Thus AG, PLE, AGL1 and AGL5 differ from all the other known plant MADS-box proteins in that their MADS boxes are located in a more internal position. AG, PLE and AGL1 are all strongly expressed in the carpels and it will be interesting to discover whether AGL5 shows a similar expression pattern. The yeast MADS-box proteins also contain several amino acids at their N-terminal ends before the MADS box, as does human SRF. However, the other members of the human MADS-box family have the MADS-box domain at their extreme N-terminus (Pollock and Treisman 1991; Yu et al. 1992).

The positions of the splice donor sites, where this has been determined, are indicated in Fig. 2. It is noticeable that in every case there is a splice donor site at or very close to the C-terminal end of the MADS box, thus the entire MADS-box domain is encoded within one exon. The splice donor sites that follow the human MADS-box genes MEF2/aMEF2 do not occur at the exact C-terminus of the MADS box, but rather are located immediately C-terminal to the extended MADS-box homology found in these genes. This organizational difference between the plant and mammalian genes could reflect a difference in the functional unit of the MADS box. Whilst both plant and mammalian MADS-box proteins appear to use a similar DNA binding domain, it is possible that this acts in conjunction with the K-box in the plant proteins and the extended MADS box in the yeast and mammalian proteins.

2.4 Homology Between Plant MADS-Box Proteins

The MADS-box proteins can be grouped by homology based on their amino acid sequences. Within the MADS box some groups are apparent in Fig. 2A. One such group would comprise BAG, AG, AGL1, AGL5 and PLE, which share three nonconservative changes compared to the other proteins; for example, they all have a C-residue at position 21 in the alignment where the others have S or T. It is noticeable that such homologous groups also share similar expression patterns. Thus *AG*, *AGL1* and *PLE* are all only transcribed in the reproductive organs. Another example of such an homologous group having a common expression pattern is provided by TM3 and DEFH24. This pair of proteins shares three common nonconservative changes at positions 1, 5 and 15 in the alignment in Fig. 2A, which sets them apart from the other proteins. *TM3* and *DEFH24* are unusual since their transcripts are found primarily in vegetative tissues (Pnueli et al. 1991; H. Sommer pers. comm.). Homologous proteins such as GLO, NTGLO and fbp1 also show amino acid homology outside the MADS and K-boxes. It has been noted that the 19 C-terminal amino acids of GLO and NTGLO are identical (Hansen et al. 1993). The final 18 of these amino acids are also conserved in fbp1 (Angenent et al. 1992). Such blocks of C-terminal homology are observed in some other pairs of proteins with similar expression patterns such as SQUA (Huijser et al. 1992) and AP1 (Mandel et al. 1992b). Some evidence for the functional importance of the C-terminal end of DEF is provided by the mutant allele *defA*-23 in which the final 12 amino acids of DEF have been replaced by 7 from a transposon (Schwarz-Sommer et al. 1992). However, other explanations for this mutant function have not yet been excluded and further analysis is required to determine whether the C-terminus of the MADS-box proteins is required for their function.

2.5 Posttranslational Modifications

It has been noted that a potential site for calmodulin-dependent phosphorylation [RXX(S/T)] is conserved within the MADS box of all the plant MADS-box proteins (RQVT) and is also found in SRF (RRYT) and in MCM1 and ARG80 (RHVT). It remains to be seen whether this site is phosphorylated in vivo and whether such a modification has any effect on protein-protein or protein-DNA interactions. Since calmodulin activity and the associated Ca^{2+} concentration have been linked to environmental and hormonal influences on plant development, it has been suggested that such a phosphorylation site might mediate the integration of genetic and environmental control of developmental pathways (Sommer et al. 1990). There have been several investigations of the effects of phosphorylation on the function of SRF and apparently conflicting results have been reported (see Marais et al. 1992 and refs. therein). However, it appears that phosphorylation does not affect the affinity of SRF DNA-binding, but increases the rates of association and dissociation (Marais et al. 1992). In this case the major site of phosphorylation is at a casein kinase II site outside the MADS box in a region which is not conserved in the plant genes. In common with many RNA polymerase II transcription factors, SRF is glycosylated, although the significance of this modification is unknown. Potential glycosylation sites also exist in the plant MADS-box proteins, and it will be interesting to discover whether these too are glycoproteins. It remains possible that exogenous and/or intrinsic signals may modulate the activity of plant MADS-box proteins by posttranslational modifications.

3 Functions of the MADS-Box Proteins

There are two genes known in the yeast *Saccharomyces cerevisiae* which show homology to the MADS box: MCM1 and ARG80. ARG80 is involved as a repressor in the control of arginine metabolism and is ~70% identical to MCM1 through its DNA-binding domain. ARG80 mutants can be partially complemented by the overexpression of MCM1 (Bercy et al. 1987). MCM1 is an ubiquitous transcription factor which is involved as an activator or repressor in many aspects of cell growth, division and specialization. In particular, the role of MCM1 in the determination of mating type has been well characterized, thus integrating a genetically programmed developmental pathway with a pathway which responds to environmental signals (Herskowitz 1989; Dolan and Fields 1991). There are three types of yeast cells: the two haploid mating types (**a** and α) and the diploid cell (**a**/α) which sporulates following meiosis. Despite being expressed in all three cell types, MCM1 is required for cell-type determination. In **a**-cells MCM1 binds upstream of, and activates, **a**-specific genes. In α-cells MCM1 binds together with the cell-type-specific factors α1 or α2 to cause activation of the α-specific genes and inactivation of the **a**-specific genes. MCM1 is also involved in the response to extracellular pheromone signaling which allows the two haploid cell types to become competent for mating. The polypeptide pheromones, **a**-factor and α-factor, have cognate receptors on the opposite cell type. These receptors share a common intracellular signal transduction pathway, involving a tripartite G-protein followed by a cascade of protein kinases, the eventual targets of which include STE12. STE12 interacts with MCM1 and indeed it has been demonstrated that the binding of STE12 upstream of an **a**-specific gene requires MCM1 (Errede and Ammerer 1989). So, by virtue of its ability to interact with different accessory proteins and to allow those interactions to modulate its activity in a cell-type-specific and/or DNA context-specific manner, MCM1 is at the center of different developmental pathways in yeast. It should also be pointed out that at least one of the MCM1 ternary-complex-forming accessory proteins, a2, is related to another class of homeotic genes in *Drosophila*, the homeobox genes.

There is also a family of MADS-box genes in vertebrates, the prototypical member of which is SRF. SRF (serum response factor) was identified as a transcription factor binding to the serum response element (SRE), a short sequence motif present in the promoters of many mammalian, immediate early genes such as c-*fos*, which is required to confer serum/growth factor inducibility on such genes (Treisman 1990). Binding sites for SRF are also found in the promoters of several muscle-specific genes, suggesting that the specificity of activation of the different types of genes is determined by cell-type-specific and promoter-specific factors which interact with SRF (Pollock and Treisman 1991). Some proteins that form ternary complexes with SRF have been identified (e.g., $p62^{TCF}$, SAP-1), and it is possible that there may be several different proteins capable of modulating SRF binding and activity (Dalton and Treisman 1992). This arrangement is reminiscent of the various interactions of different ternary complex forming proteins with MCM1 described above. Other human MADS-box genes have been isolated by low stringency library screening (RSRFs; Pollock and Treisman 1991) or by screening an expression library for proteins capable of binding an essential element of muscle promoters (MEF2s; Yu et al. 1992). There appear to be at least four different human MADS-box genes, and tissue-specific alternative splicing expands the isoform diversity of some of these (Yu et al. 1992). There is evidence that at least some of these genes are involved in developmental pathways (Yu et al. 1992). There is also evidence that these MADS-box proteins, like MCM1 and SRF, can act, both with and without the assistance of other accessory proteins, to promote transcription of genes which contain the appropriate DNA-binding sites in their promoters.

So, from a study of yeast and human MADS-box proteins, a common picture emerges. They are DNA-binding activators (or repressors) of transcription which bind to slightly different recognition sites as homodimers or heterodimers. They can form ternary complexes with a variety of different unrelated proteins and recruit these proteins to their binding sites. It is also likely that the accessory factors that are recruited to the DNA can be the final targets of signal transduction pathways (Dalton and Treisman 1992).

Less is known about the plant MADS-box proteins since few of them have been characterized at the DNA-binding level. In vitro DNA binding has been demonstrated for truncated AG (Mueller and Nordheim 1991) and for intact DEF A and GLO binding as a heterodimer (Schwarz-Sommer et al. 1992). In the latter case DEF A and GLO heterodimers bind to a number of slightly different sites, including motifs found upstream of the *DEFICIENS* (Schwarz-Sommer et al. 1992) and *GLOBOSA* coding regions (Tröbner et al. 1992), suggesting autoregulatory control (see below). Neither DEF A nor GLO appears to be able to bind these motifs alone. It has been demonstrated that neither of the proteins encoded by the two previously mentioned *DEFICIENS* alleles, *def*-nic (an amino acid exchange at a conserved region of the MADS box) and *def*-101 (a single amino acid deletion at the N-terminus of the K-box) are able to substitute for wild-type DEF A in such in vitro translation/DNA-binding assays, presumably due to the mutants' inability to bind DNA or form dimers (Schwarz-Sommer et al. 1992). Various other combinations of plant MADS-box proteins (such as PLE, PLE/DEF A, PLE/GLO) have also been tested for their ability to bind DNA, although no binding has been observed (B. Davies unpubl.). This may reflect the fact that the appropriate heterodimerization partners have not yet been isolated or that unsuitable binding sites were used. The generation of antibodies against the different plant MADS-box proteins should allow dimerization to be studied without known binding sites and this may, in turn, lead to the determination of the optimal binding sites.

Some of the plant MADS-box proteins, such as GLO (Tröbner et al. 1992), fbp1 (Angenent et al. 1992) and the AGL (Ma et al. 1991) family, have C-termini which are rich in glutamine residues; a characteristic of certain transcription factors. However, others, such as DEF A, do not have an obvious transcription activation domain of a known type. It may be that for activation (or repression) of transcription, such MADS-box proteins rely on interactions with accessory proteins in the same way as the yeast and mammalian MADS-box proteins. Such accessory proteins could act in a cell-type or promoter-specific manner to influence transcription of target genes. The modification of such proteins would provide another point at which extracellular signals, known to affect flowering, could modulate the activity of the MADS-box genes. As described above, precedents for such controls exist in yeast and vertebrates.

4 MADS-Box Genes During Floral Organ Development

4.1 Genetic Models

Several models have been proposed to explain how the various homeotic gene products interact with each other to direct the formation of a flower (Haughn and Somerville 1988; Schwarz-Sommer et al. 1990; Coen and Meyerowitz 1991; Lord 1991). The models are broadly similar to each other; all involve different classes of homeotic genes being expressed in two adjacent whorls of the flower. In the *Arabidopsis* model (Haughn and Somerville 1988; Bowman et al. 1991; Coen and Meyerowitz 1991), three classes of homeotic gene activities are envisaged (A, B and C in Fig. 4). Within any one whorl, expression of A alone would specify sepal formation. The combinations AB and BC would specify the formation of petals and stamens, respectively. Expression of the C-function alone would direct the formation of carpels. Additionally, it is proposed that the A- and C-functions negatively regulate each other (Drews et al. 1991) and that the B-function is restricted to the second and third whorls independently of the A- and C-functions (Coen and Meyerowitz 1991). *Arabidopsis* mutants for all three homeotic activities are known. The A-function is represented by *apetala2*,

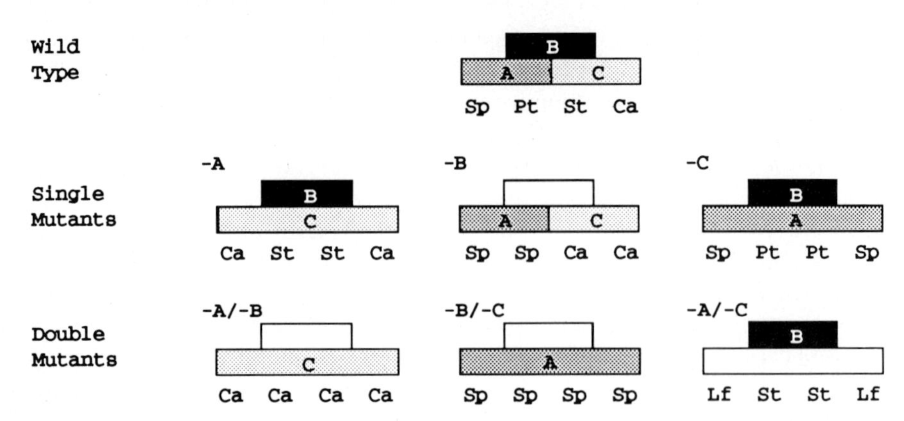

Fig. 4. The *A*, *B*, *C* model of flower development in *Arabidopsis* (see Coen and Meyerowitz 1991). The four whorls of the flower are described by *letters* denoting the organ present there. Thus, the wild-type phenotype is described as sepal (*Sp*), petal (*Pt*), stamen (*St*) and carpel (*Ca*). Three functions *A*, *B*, and *C* are represented by different degrees of *shading* within boxes the width of which indicates the expression patterns of the relevant functions; overlapping boxes denote overlapping expression patterns. Hence, the A-function in wild type is expressed in the first and second whorls and the B-function in the second and third whorls. The single and double mutants are illustrated beneath the wild type and their phenotypes and the functions expressed in their different whorls are shown. *White boxes* indicate the lack of certain functions. Mutants are described by a "-" symbol before the name of the function such that the double mutant -A/-B lacks both A- and B-functions. *Lf* in the first and fourth whorls of the -A/-C double mutant signifies the production of leaf-like organs at these positions. Since the A- and C-functions are mutually repressive (see discussion in text), a deficiency of one causes the expansion of the expression domain of the other (e.g., the C-function expands into the first and second whorls in the -A mutant)

the B-function by both *apetala3* and *pistillata* and the C-function by *agamous* (see Bowman et al. 1991 and refs. therein). The phenotypes of the single and double mutants of these *Arabidopsis* genes are as predicted in the model, as are the phenotypes produced by the ectopic expression of the *AGAMOUS* (function C) gene (Mandel et al. 1992a; Mizukami and Ma 1992).

The expression of the B-function is prevented from extending into the fourth whorl by the action of another homeotic gene, *FLO10* (Schultz et al. 1991) also named *SUPERMAN* (Bowman et al. 1992). *FLO10* is expressed only in the fourth whorl and does not control organ identity since all different organ types can be formed within the fourth whorl of *flo10* mutants (Schultz et al. 1991). However, *FLO10* negatively regulates the expression of the B-function genes *AP3* and *PI* in the fourth whorl, thus preventing the B-function from extending into the fourth whorl. In *flo10* mutants the expression of the B-function genes is not prevented in the fourth whorl and stamens are produced, at the expense of carpels, as would be predicted by the model (Schultz et al. 1991; Bowman et al. 1992).

In *Antirrhinum* there is as yet no reported A-function gene. Mutants with similar phenotypes to the *Arabidopsis* A-function mutants have been described (Carpenter and Coen 1990; Schwarz-Sommer et al. 1990) but, in contrast to the other homeotic mutants, these are semidominant. Recent results indicate that the *Antirrhinum* A-function mutant phenotype actually results from ectopic expression of the C-function gene *PLENA* (Bradley et al. 1993). It remains possible that a gene strictly analogous to *AP2* of *Arabidopsis* does not exist in *Antirrhinum*, thus representing a basic difference between the two plants.

There are also problems in fitting the homeotic mutants of petunia into the A, B, C model (van Tunen and Angenent 1991; Angenent et al. 1992). Two homeotic mutants of petunia are known, *blind* and *green petals*, each of which affects only one whorl of the flower. Recently, two MADS-box genes, fbp1 and fbp2, have been isolated from petunia (Angenent et al. 1992; see Table 1). The predicted protein product of fbp1 is 80% homologous to that of the *Antirrhinum* gene *GLOBOSA* (Angenent et al. 1992; Tröbner et al. 1992). In common with the *GLOBOSA* gene, fbp1 is transcribed in the second and third whorls of the flower. However, due to some form of posttranscriptional control, the fbp1 protein product is restricted to the second whorl (Angenent et al. 1992). Circumstantial evidence that the GLO protein is produced in the third whorl of *Antirrhinum* is provided by the partial or complete reversion to third whorl stamens arising from somatic excision of a transposon from the *GLOBOSA* gene (Tröbner et al. 1992). Since reactivation of the *GLOBOSA* gene leads to a concomitant increase in transcription of the *DEF* gene (see Sect. 4.2 below), the translation of the *GLOBOSA* transcript is demonstrated by its function. Currently, the tissue distribution of other floral MADS-box genes has not been investigated at the protein level and it will be interesting to discover whether posttranscriptional regulatory controls are widespread.

Detailed investigation of the phenotypical and biochemical differences between similar mutants in different species reveals more slight differences. The protein product of the *PISTILLATA* gene of *Arabidopsis*, PI, is suggested to be the cognate homologue of GLO (Jack et al. 1992). They share extensive amino acid homology (see Figs. 2 and 3) and mutations in their genes lead to similar effects in equivalent whorls (see Fig. 1). However, morphological differences are observed between the mutant phenotypes, such as organ number and extent of homeotic transformation (Tröbner et al. 1992). There are also differences in the interactions of the two proteins with other homeotic gene products. As described below, *DEF* and *GLO* are mutually dependent to achieve full levels of expression in the second and third whorls of *Antirrhinum*. (Schwarz-Sommer et al. 1992; Tröbner et al. 1992). Whilst this appears to be true for *AP3* and *PISTILLATA* in the third whorl of *Arabidopsis*, it is apparently not the case in the second whorl where their expression appears to be mutually independent (Jack et al. 1992; Tröbner et al. 1992). Such results provide further evidence that whilst the basic principles of the involvement of MADS-box transcription factors appear to be applicable to a wide variety of flowering plants, the regulation and interactions of these genes may be subject to subtle variation.

4.2 Molecular Models

The *Antirrhinum* MADS-box-containing proteins DEF A and GLO are the only ones whose interactions have been at least partially characterized at the molecular level (Schwarz-Sommer et al. 1992; Tröbner et al. 1992). Studies of the temporal and spatial transcription patterns of these two genes in a variety of different mutants have indicated that, although the transcription of each gene is independently induced, the later organ-specific reinforcement of transcription is mutually interdependent (Tröbner et al. 1992). DEF A and GLO can form heterodimers which are capable of binding DNA at specific CArG-box-like sequences (Schwarz-Sommer et al. 1992) and such sequences are found in the promoters of both their genes: *DEF* (Schwarz-Sommer et al. 1992) and *GLO* (Tröbner et al. 1992). A model has been proposed to account for the interdependence of these genes for maximal organ-specific expression (Tröbner et al. 1992; see Fig. 5). The two-stage model consists of an induction phase and a later reinforcement of transcription. The induction phase relies on the promoters of *DEF* and *GLO* independently recognizing an unidentified signal or signals to initiate a low level of transcription and hence to produce small amounts of the DEF A and GLO proteins. Since both *DEF* and *GLO* transcription is first observed by in situ analysis before the petal and stamen primordia are morphologically visible (Schwarz-Sommer et al. 1992; Tröbner et al. 1992), this signal must be received early in the development of these organs. The DEF A and GLO proteins can then form heterodimers which can bind to CArG boxes in each of their own promoters and boost transcription.

The purpose of the autoregulation of *DEF* and *GLO* may be to commit responsive cells which receive a transient early signal to a petal or stamen developmental pathway. Autoregulation could serve to ensure a balanced synthesis of DEF A and GLO (Schwarz-Sommer et al. 1992; Tröbner et al. 1992). Since low levels of transcription of *DEF* and *GLO* are observed in the fourth whorl, where no autoregulatory reinforcement of transcription is observed, there must be other influences on this process. Possibly autoregulation in the fourth whorl is inhibited by a negative regulator of *DEF* and/or *GLO* transcription or DEF A and GLO heterodimerization. Alternatively, the autoregulatory boosting of *DEF* and *GLO* transcription in petals and stamens may require the presence of another factor in those organs. Evidence for the existence of such a factor is provided by the *chlorantha* allele of *DEF* (*def*-chl) (Schwarz-Sommer et al. 1992). This mutation affects only four bases, more than 1000 bases upstream of the transcriptional start point of the *DEFICIENS* gene (wild type: CCCCTG, *def*-chl CGG), yet transcription is greatly reduced in petals and stamens when compared to wild type. The *def*-chl mutation is sited 32 bases up-

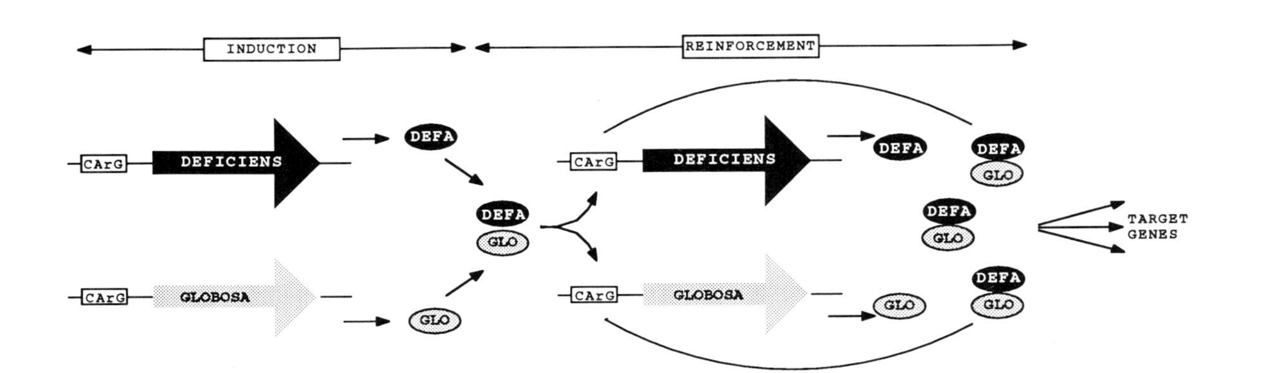

Fig. 5. The autoregulatory model for *DEFICIENS* and *GLOBOSA* gene expression. The *DEFICIENS* and *GLOBOSA* genes are depicted by *large*, *shaded arrows*. Expression of the *DEFICIENS* and *GLOBOSA* genes is a two-step process. The DEF A and GLO independent induction phase produces a small amount of each protein (represented as *shaded ovals*) which combine to form heterodimers. During the reinforcement phase, these heterodimers bind to the CArG motifs upstream of each gene to boost transcription, thus increasing the production of the proteins and hence the heterodimers. The amplified production of the heterodimers thus commits the cell to the developmental pathway and the specific target genes become activated

stream of a CArG box in the promoter of *DEF* and the wild-type sequence shows some homology to the consensus binding sequence of a transcription factor MyoD (Schwarz-Sommer et al. 1992). It has been proposed that a protein binding to the "chlorantha site" could interact with a DEF A/GLO complex binding to the nearby CArG box, resulting in tissue-specific upregulation of *DEF* (Schwarz-Sommer et al. 1992). Interestingly, the same sequence, CCCCTG, is also found 40 bases downstream of the strongest DEF A/GLO binding CArG box in the promoter of *GLOBOSA* (Tröbner et al. 1992).

As described above, both MCM1 and SRF are capable of forming ternary complexes with proteins which bind DNA at sites adjacent to their own binding sites, thus altering their regulatory activities. Further investigation is required to determine whether the same is true for plant MADS-box genes and the *def*-chl allele should prove to be useful for this.

4.3 Activation of the Floral MADS-Box Genes

Mutants of *Antirrhinum* and *Arabidopsis* exist in which flowering is impaired and some of these genes have been cloned (Coen et al. 1990; Huijser et al. 1992; Mandel et al. 1992b; Weigel et al. 1992). SQUA (*Antirrhinum*) and AP1 (*Arabidopsis*) show extensive homology and probably represent homologues; both encode MADS-box proteins and their expression patterns are very similar (Fig. 2). FLO (*Antirrhinum*) and LEAFY (*Arabidopsis*) are also homologous and are probably equivalents, although they are not MADS-box proteins and show no significant homology to any other genes. Expression of all of these genes is detected at a very early stage and thus they may activate the later MADS-box genes involved in organogenesis. In this respect the later MADS-box genes may represent potential targets of the early genes. The expression of the early genes is also observed in some of the floral organ primordia and thus their functions may not be confined to controlling the transition of the inflorescence meristem to the floral meristems. For example, it has been proposed that in *plena* mutants it is the transient presence of FLO in the fourth whorl that causes reiteration of the flowering process, leading to a loss of determinacy (see below). This does not happen in the wild type when PLE is established in the third whorl (where *FLO* is not expressed) and thus preexists in the fourth whorl when *FLO* is reactivated (Coen et al. 1990). The early genes also represent likely targets for more general, possibly environmental, factors which promote flowering. More work needs to be done to establish the nature of these signals and the interactions between the early genes and their target genes.

4.4 Determinacy

In the phenotypes of some of the homeotic mutants (such as *plena* in *Antirrhinum* and *agamous* in *Arabidopsis*), a loss of determinacy is observed. Thus, within the fourth whorl many further whorls are produced (Yanofsky et al. 1990; Bradley et al. 1993). In contrast, in the absence of DEF A or GLO, development of the fourth whorl is inhibited (Sommer et al. 1990; Schwarz-Sommer et al. 1992; Tröbner et al. 1992). Hence, in *Antirrhinum*, DEF A and GLO may act in the third whorl to prevent termination of flower development, whilst PLE in the fourth whorl might act to promote termination. Since mutants which ectopically express *PLE* form four whorls, the first of which consists of carpeloid sepals (Bradley et al. 1993), it appears that the presence of PLE in the absence of DEF A or GLO is not sufficient to achieve termination. The relation between DEF A and PLE appears to be different in the third and fourth whorls. In the third whorl both DEF A and PLE are required for the formation of stamens and so their functions could be viewed as cooperative for proper third whorl development. In the fourth whorl DEF A and PLE exert antagonistic effects with respect to carpel formation and determinacy. Future experiments in which the various combinations of homeotic genes are ectopically expressed in different combinations may lead to a greater understanding of the termination of flowering.

4.5 Target Genes

The downstream target genes for the plant MADS-box transcription factors are presently unknown. Since many more flower-specific MADS-box genes have been isolated than are required by the simple model, it is possible that other sets of MADS-box genes, with increased organ or tissue specificity, exist under the control of the hierarchically higher homeotic genes. These MADS-box genes could interact with each other or with the "upper level" genes, possibly generating a complex series of heterodimers with subtly altered DNA-binding specificities and affinities. Such a differential DNA sequence specificity has been implicated in determining the specificity of action of homeobox genes (Ekker et al. 1992). Alternatively, the homeotic MADS-box genes might control the activation or repression of both regulatory and structural target genes. It is apparent from an analysis of the mutant alleles of *DEFICIENS* that DEF A controls different sets of genes in stamens than in petals (Schwarz-Sommer et al. 1992). Attempts are currently being made to isolate the genes which are regulated by the homeotic MADS-box transcription factors. Several genes have been isolated, the transcription of which is reduced in *deficiens* mutants (Sommer et al. 1990). The promoters of some of these genes contain putative binding sites for DEF A/GLO (Nacken et al. 1991) and future work will establish whether these genes are under the direct or indirect control of DEF A and GLO and will also attempt to elucidate their functions.

5 Outlook

Almost all the floral homeotic mutants which have been characterized are produced by alterations to the structure or expression of a single type of transcription factor – the MADS-box-containing proteins. There are striking and extensive interspecies homologies between the sequences of MADS-box proteins which share a common spatial and temporal expression pattern. The existence of homologous genes with similar expression patterns and functions, across a wide range of species, suggests that the fundamental mechanisms of flower development are relatively ancient and that perhaps later modifications are responsible for the variety of forms apparent today. However, within the broad framework of MADS-box gene regulation there appear to be some differences between different species.

We are currently at an exciting stage in the study of flower development and the MADS-box gene family will be useful tools for future investigations. Nevertheless, an understanding of the processes involved in floral organogenesis will be achieved only after the identification and study of the factors which activate the floral MADS-box genes, those with which they interact and the targets of those interactions.

Acknowledgements. We thank Hans Sommer and Candice Sheldon for critical reading and helpful comments on this manuscript. B.D. is supported by a fellowship from the Human Frontier Science Program (HFSP).

References

Ammerer G (1990) Identification, purification and cloning of a polypeptide (PRTF/GRM) that binds to mating-specific promoter elements in yeast. Genes Dev 4:299–312

Angenent GC, Busscher M, Franken J, Mol JNM, van Tunen AJ (1992) Differential expression of two MADS box genes in wild-type and mutant petunia flowers. Plant Cell 4:983–993

Bercy J, Dubois E, Messenguy F (1987) Regulation of arginine metabolism in *Saccharomyces cerevisiae*: expression of the three ARGR regulatory genes and cellular localization of their products. Gene 55:277–285

Bowman JL, Smyth DR, Meyerowitz EM (1991) Genetic interactions among floral homeotic genes of *Arabidopsis*. Development 112:1–20

Bowman JL, Sakai H, Jack T, Weigel D, Mayer U, Meyerowitz EM (1992) *SUPERMAN*, a regulator of floral homeotic genes in *Arabidopsis*. Development 114:599–615

Bradley D, Carpenter R, Sommer H, Hartley N, Coen E (1993) Complementry floral homeotic phenotypes result from opposite orientations of a transposon at the *plena* locus of *Antirrhinum*. Cell 72:85–95

Carpenter R, Coen ES (1990) Floral homeotic mutations produced by transposon-mutagenesis in *Antirrhinum majus*. Genes Dev 4:1483–93

Coen ES, Meyerowitz EM (1991) The war of the whorls: genetic interactions controlling flower development. Nature 353:31–37

Coen ES, Romero JM, Doyle S, Elliott R, Murphy G, Carpenter R (1990) *floricaula*: A homeotic gene required for flower development in *Antirrhinum majus*. Cell 63:1311–1322

Dalton S, Treisman R (1992) Characterization of SAP-1, a protein recruited by serum response factor to the c-*fos* serum response element. Cell 68:597–612

Dolan JW, Fields S (1991) Cell-type-specific transcription in yeast. Biochim Biophys Acta 1088:155–169

Drews GN, Bowman JL, Meyerowitz EM (1991) Negative regulation of the *Arabidopsis* homeotic gene *Agamous* by the *Apetala2* product. Cell 65:991–1002

Dubois E, Messenguy F (1991) In vitro studies of the binding of the ARGR proteins to the *ARG5,6* promoter. Mol Cell Biol 11:2162–2168

Ekker SC, von Kessler DP, Beachy PA (1992) Differential DNA sequence recognition is a determinant of specificity in homeobox gene action. EMBO J 11:4059–4072

Errede B, Ammerer G (1989) STE12, a protein involved in cell-type-specific transcription and signal transduction in yeast, is part of protein-DNA complexes. Genes Dev 3:1349–1361

Hansen G, Estruch JJ, Sommer H, Spena A (1993) NTGLO: tobacco homologue of the *GLOBOSA* floral homeotic gene of *Antirrhinum majus*, cDNA sequence and expression pattern. Mol Gen Genet 239:310–312

Haughn GW, Somerville CR (1988) Genetic control of morphogenesis in *Arabidopsis*. Dev Genet 9:73–89

Herskowitz I (1989) A regulatory hierarchy for cell specialization in yeast. Nature 342:749–757

Huijser P, Klein J, Lönnig W-E, Meijer H, Saedler H, Sommer H (1992) Bractomania, an inflorescence anomaly, is caused by the loss of function of the MADS-box gene *squamosa* in *Antirrhinum majus*. EMBO J 11:1239–1249

Jack T, Brockman LL, Meyerowitz EM (1992) The homeotic gene *APETALA3* of *Arabidopsis thaliana* encodes a MADS box and is expressed in petals and stamens. Cell 68:683–697

Landschulz WH, Johnson PF, McKnight SL (1988) The leucine zipper: a hypothetical structure common to a new class of DNA binding proteins. Science 240:1759–1764

Lord EM (1991) The concepts of heterochrony and homeosis in the study of floral morphogenesis. Flowering Newslett 11:4–13

Ma H, Yanofsky MF, Meyerowitz EM (1991) AGL1–AGL6, an *Arabidopsis* gene family with similarity to floral homeotic and transcription factor genes. Genes Dev 5:484–495

Mandel MA, Bowman JL, Kempin SA, Ma H, Meyerowitz EM, Yanofsky MF (1992a) Manipulation of flower structure in transgenic tobacco. Cell 71:133–134

Mandel MA, Gustafson-Brown C, Savidge B, Yanofsky MF (1992b) Molecular characterization of the *Arabidopsis* floral homeotic gene *APETALA1*. Nature 360:273–277

Marais RM, Hsuan JJ, McGuigan C, Wynne J, Treisman R (1992) Casein kinase II phosphorylation increases the rate of serum response factor-binding site exchange. EMBO J 11:97–105

Meyerowitz EM, Smyth DR, Bowman JL (1989) Abnormal flowers and pattern formation in floral development. Development 106:209–217

Mizukami Y, Ma H (1992) Ectopic expression of the floral homeotic gene *AGAMOUS* in transgenic arabidopsis plants alters floral organ identity. Cell 71:119–131

Mueller CGF, Nordheim A (1991) A protein domain conserved between yeast MCM1 and human SRF directs ternary complex formation. EMBO J 10:4219–4229

Nacken WKF, Huijser P, Beltran J-P, Saedler H, Sommer H (1991) Molecular characterization of two stamen-specific genes, *tap1* and *fil1*, that are expressed in the wild type, but not in the *deficiens* mutant of *Antirrhinum majus*. Mol Gen Genet 229:129–136

Norman C, Runswick M, Pollock R, Treisman R (1988) Isolation and properties of cDNA clones encoding SRF, a transcription factor that binds to the c-*fos* serum response element. Cell 55:989–1003

Pnueli L, Abu-Abeid M, Zamir D, Nacken W, Schwarz-Sommer Zs, Lifschitz E (1991) The MADS box gene family in tomato: temporal expression during floral development, conserved secondary structures and homology with homeotic genes from *Antirrhinum* and *Arabidopsis*. Plant J 1:255–266

Pollock R, Treisman R (1991) Human SRF-related proteins: DNA-binding properties and potential regulatory targets. Genes Dev 5:2327–2341

Rosenthal N (1989) Muscle cell differentiation. Curr Opinion Cell Biol 1:1094–1101

Schultz EA, Pickett FB, Haughn GW (1991) The *FLO10* gene product regulates the expression domain of homeotic genes *AP3* and *PI* in *Arabidopsis* flowers. Plant Cell 3:1221–1237

Schwarz-Sommer Zs, Huijser P, Nacken W, Saedler H, Sommer H (1990) Genetic control of flower development by homeotic genes in *Antirrhinum majus*. Science 250:931–936

Schwarz-Sommer Zs, Hue I, Huijser P, Flor PJ, Hansen R, Tetens F, Lönnig WE, Saedler H, Sommer H (1992) Characterization of the *Antirrhinum* floral homeotic MADS-box gene *deficiens*: evidence for DNA binding and autoregulation of its persistent expression throughout flower development. EMBO J 11:251–263

Sommer H, Beltran JP, Huijser P, Pape H, Lönnig WE, Saedler H, Schwarz Zs (1990) *Deficiens*, a homeotic gene involved in the control of flower morphogenesis in *Antirrhinum majus*: the protein shows homology to transcription factors. EMBO J 9:605–613

Treisman R (1990) The SRE: a growth factor responsive transcriptional regulator. Semin Cancer Biol 1:47–58

Tröbner W, Ramirez L, Motte P, Hue I, Huijser P, Lönnig W-E, Saedler H, Sommer H, Schwarz-Sommer Zs (1992) *GLOBOSA*: a homeotic gene which interacts with *DEFICIENS* in the control of *ANTIRRHINUM* floral organogenesis. EMBO J 11:4693–4704

van Tunen AJ, Angenent GC (1991) How general are the models describing floral morphogenesis? Flowering Newslett 12:34–37

Weigel D, Alvarez J, Smyth DR, Yanofsky MF, Meyerowitz EM (1992) *LEAFY* controls floral meristem identity in *Arabidopsis*. Cell 69:843–859

Yanofsky MF, Ma H, Bowman JL, Drews GN, Feldmann KA, Meyerowitz EM (1990) The protein encoded by the *Arabidopsis* homeotic gene *agamous* resembles transcription factors. Nature 346:35–39

Yu Y-T, Breitbart RE, Smoot LB, Youngsook L, Mahdavi V, Nadel-Ginard B (1992) Human myocyte-specific enhancer factor 2 comprises a group of tissue-restricted MADS box transcription factors. Genes Dev 6:1783–1798

12 The *GL1* Gene and the Trichome Developmental Pathway in *Arabidopsis thaliana*

John C. Larkin, David G. Oppenheimer, and M. David Marks

1 Introduction

Developmental decisions in organisms as diverse as bacteriophage λ, *Saccaromyces cerevisia*, *Drosophila melanogaster*, *Caenorabditis elegans*, and *Arabidopsis thaliana* are regulated by transcription factors. In multicellular organisms, these transcription factors ultimately control the expression of genes responsible for the differentiation of various cell types. During the course of evolution, a relatively limited group of DNA-binding domains have diversified to take on a wide range of transcriptional regulation functions. One such conserved DNA-binding motif is present in the myb family of transcriptional activators, which has been identified in a wide variety of organisms (reviewed in Lüscher and Eisenman 1990; Graf 1992).

Our laboratory is studying trichome (hair) differentiation in the epidermis of *Arabidopsis thaliana* as a model for cell differentiation and elongation in plants (Marks et al. 1991; Marks and Esch 1992; Oppenheimer et al. 1992). *Arabidopsis* trichomes are single cells that are the result of an elaborate and highly specialized cell elongation process (Fig. 1). Because trichomes are nonessential and exposed on the surface of the plant, the trichome differentiation pathway is readily accessible to genetic analysis. Many mutations affecting trichome development have been isolated (Feenstra 1978; Koornneef et al. 1982, 1983; Haughn and Somerville 1988; Marks et al. 1991; Oppenheimer et al. 1992). Recessive mutations in two genes, *GL1* and *TTG*, fail to initiate trichomes on most shoot surfaces. Recessive mutations in two additional genes, *GL2* and *GL3*, result in the production of abortive trichomes. The phenotypes of these four genes affecting early stages in trichome development are summarized in Table 1. Mutations in several other genes have been identified that affect later stages of trichome development. Recently, our laboratory has cloned the *GL1* gene and shown that it encodes a member of the myb transcription factor family (Herman and Marks 1989; Marks and Feldmann 1989; Oppenheimer et al. 1991). Since trichome development as a whole has recently been reviewed (Marks et al. 1991; Marks and Esch 1992; Oppenheimer et al. 1992), this chapter will focus on the role of myb genes in plant development, with particular

School of Biological Sciences, University of Nebraska-Lincoln, Lincoln, Nebraska 68588-0118, USA

Results and Problems in Cell Differentiation 20
L. Nover (Ed.)
Plant Promoters and Transcription Factors

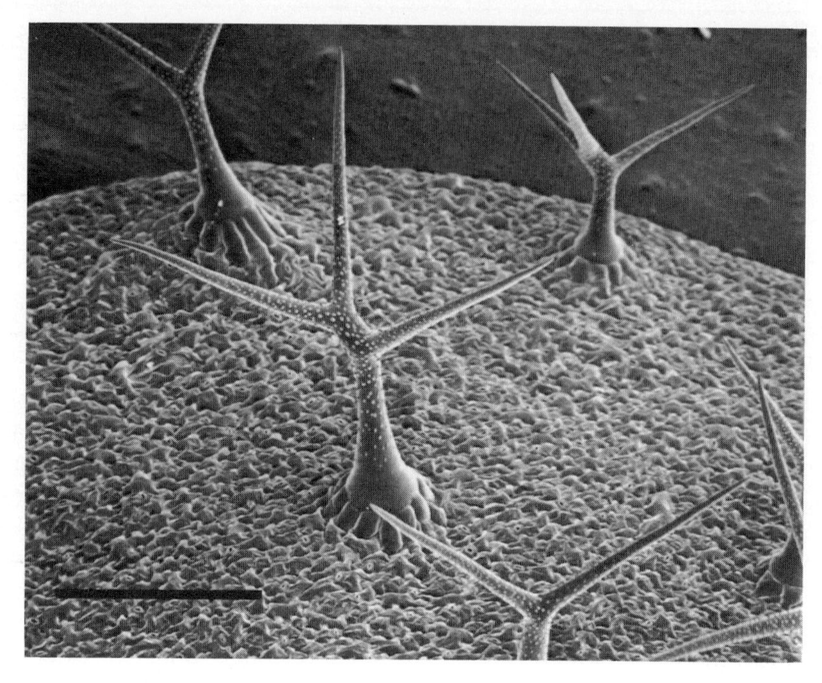

Fig. 1. Trichomes on the adaxial surface of a leaf from a wild-type plant of *Arabidopsis* (Columbia). *Bar* = 200 μm

Table 1. Genes involved in early stages of *Arabidopsis* trichome development

Mutation	Trichome phenotype	Other phenotypes	References[a]
gl1	No trichomes, except on leaf margins	None	1
ttg	No trichomes, except on leaf margins. Clusters of incompletely developed trichomes on some plants homozygous for some weak alleles	Lack of anthocyanin pigment. No seed coat mucileage	1,2,3
gl2	Few fully developed trichomes. Many trichomes aborted at very early stages	No seed coat mucileage	1
gl3	Few fully developed trichomes. Many trichomes completely absent. Some trichomes aborted at very early states	None	1

[a] 1, Koornneef et al. (1982); 2, Korrnneef (1981); 3, Haughn and Somerville (1988).

attention to the *GL1* gene and its regulation. Recent evidence for genetic interactions during trichome initiation, as well as a genetic model for the trichome initiation pathway, will also be presented.

2 Myb Genes in Animals and Fungi

The first myb gene identified was the v-*myb* oncogene of the avian myeloblastosis virus (Roussel et al. 1979). The cellular homologue of this gene in vertebrates is expressed primarily in developing hematopoietic cells, suggesting that it plays a role in hematopoiesis (see Lüscher and Eisenman 1990; Graf 1992). Recently, it has been shown that mice containing a mutant c-*myb* gene introduced by homologous recombination are defective for fetal hepatic hematopoiesis (Mucenski et al. 1991). The results to date indicate that c-*myb* is involved in the proliferation and differentiation of the hematopoietic precursor cells (Graf 1992).

Three functional domains have been identified in the c-*myb* protein: the N-terminal DNA-binding domain, an acidic transcriptional activation domain located downstream of the DNA-binding domain, and a negative regulatory domain in the carboxy-terminal half of the protein (Sakura et al. 1989). The DNA-binding domain of animal mybs consists of three imperfect repeats of a sequence with similarity to the helix-turn-helix motif found in bacterial repressors and homeodomain proteins (Frampton et al. 1989). Only the second and third repeats are necessary for DNA binding (Lüscher and Eisenman 1990). Mutations in repeat 2 of the v-*myb* DNA-binding domain that affect transformed cell phenotypes may represent a protein-protein interaction site (Graf 1992). Deletion of the negative regulatory domain increases transcriptional activation by c-*myb* (Sakura et al. 1989). This domain is also a candidate for interactions between c-*myb* and other proteins. A major phosphorylation site has been mapped just upstream of the DNA-binding domain (Lüscher et al. 1990). Phosphorylation of this site reduces the ability of c-*myb* to bind DNA. The amino-terminal region containing this site is often lost in oncogenic v-*mybs*, consistent with a regulatory role for phosphorylation of this site. Recent evidence suggests that the level of c-*myb* phosphorylation is increased during mitosis (Lüscher and Eisenman 1992). The vertebrate myb gene family appears to be relatively small, although two myb genes in addition to c-*myb* have been identified in humans (Nomura et al. 1988). One of these human myb genes, B-*myb*, lacks the transcriptional activation domain, and is capable of repressing c-*myb* and B-*myb* activation of a target gene (Foos et al. 1992).

Myb genes have also been identified in *Drosophila* (Katzen et al. 1985; Peters et al. 1987), *Dictyostelium* (Stober-Grasser et al. 1992), and yeast (Tice-Baldwin et al. 1989), in addition to vertebrates and plants. Evidence has also been presented indicating that myb genes in sponges are regulated by retinoic acid (Biesalski et al. 1992). The yeast *BAS1* gene is of particular interest. This gene activates basal expression of *HIS4* in conjunction with *BAS2*, a homeobox protein (Tice-Baldwin et al. 1989). Although *BAS1* and *BAS2* bind to adjacent sequences in the *HIS4* promoter, no evidence for a cooperative interaction between *BAS1* and *BAS2* was detected.

3 Plant Myb Genes

Myb genes have been isolated from several plant species (Paz-Ares et al. 1987; Marocco et al. 1989; Grotewold et al. 1991; Jackson et al. 1991; Oppenheimer et al. 1991; Shinozaki et al. 1992). In all cases, the plant myb genes contain only two repeats of the myb DNA-binding region, corresponding to the second and third repeats found in animal mybs. A compilation of plant myb DNA-binding domains is shown in Fig. 2. Although no sequence similarity between plant mybs and other mybs has been detected outside of the DNA binding domain, all plant myb genes encode an acidic region which may be equivalent to the c-*myb* transcriptional activation domain. Six myb genes expressed in flowers have been cloned from *Antirrhinum* (Jackson et al. 1991). In situ hybridization to RNA in tissue sections revealed that one of these genes was expressed primarily in floral nectary tissue and in the transmitting tract of the style. Both of these tissues secrete carbohydrates, and Jackson et al. (1991) have suggested that this myb gene might regulate secretion in these tissues.

The best characterized plant myb gene is the *C1* gene of maize, which regulates the transcription of the anthocyanin biosynthetic enzyme genes (Coe et al. 1988, Klein et al. 1989). In addition to *C1*, transcription of the anthocyanin biosynthetic pathway genes requires the gene product of either the *R* or the *B* locus (Klein et al. 1989). *R* and *B* are duplicate genetic factors (Styles 1970) encoding homologous proteins that contain a basic helix-loop-helix domain similar to the product of the c-*myc* oncogene (Chandler et al. 1989; Ludwig et al. 1989). The C1 and R proteins appear to bind to adjacent sites in the promoter of the *Bz1* gene (Roth et al. 1991). Domain-swapping experiments indicate that the acidic domain of C1 is functionally equivalent to the acidic transcriptional activation domain of the yeast *GAL4* gene (Goff et al. 1991). Recent evidence indicates that activation of the anthocyanin biosynthetic genes requires a direct interaction between B and C1 (Goff et al. 1992). This interaction appears to be mediated by the C1 myb DNA-binding region and the amino terminus of the B-protein. In the maize aleurone, *C1* expression is regulated by the product of the *Vp1* gene, discussed elsewhere in this volume (McCarty et al. 1989).

Repeat	**1**					
*At*Gl1	YKKGLWTVEE	DNILMDYVLN	HGTGQWNRIV	RKTGLKRCGK	SCRLRWMNYL	SPN
*At*CD2	....	.K......KA	..K.H....A	K.........	......V...	..H
*At*myb1	RV..P.SK..	.DV.SEL.KR	L.ARN.SF.A	.SIP-G.S..	......C.Q.	N..
*Am*305	VR..P..M..	.L..IN.IA.	..E.V..SLA	.SA....T..	......L...	R.D
*Am*340	VR..P..M..	.L..INFIS.	..E.V..T.A	.SA....T..	......L...	R.D
*Am*330	TN..A..K..	.QR.IN.IRA	..E.C.RSLP	KAA..L....	......I...	R.D
*Am*315	LKR.P..E..	.QK.TS...K	N.IQG.RV.P	KLA..S....	......M...	R.D
*Am*308	TN..A..K..	..R.VA.IRA	..E.C.RSLP	KAA..L....	......I...	R.D
*Am*306	V...P..P..	.I..VS.IQE	..P.N.RA.P	SN...L..S.	......T...	R.G
*Zm*C1	V.R.A..SK.	.DA.AA..KA	..E.K.REVP	Q.A..R....	......L...	R..
*Zm*P1	L.R.R..A..	.QL.AN.IAE	..E.S.RSLP	KNA..L....	......I...	RAD
*Zm*1	LNR.S..PQ.	.MR.IA.IQK	..HTN.RALP	KQA..L....	......I...	R.D
*Zm*38	TNR.A..K..	.ER.VA.IRA	..E.C.RSLP	KAA..L....	......I...	R.D
*Hv*1	TN..A..K..	.DR.TA.IKA	..E.C.RSLP	KAA..L....	......I...	R.D
*Hv*33			.V.C.SSVP	.LAA.N....	......I...	R.D
*Dm*myb	LI..P..RD.	.DMVIKL.R.	F.PKK.T-LI	ARYLNG.I..	Q..E..H.H.	N..
*Hs*myb	LI..P..K..	.QRVIEL.QK	Y.PKR.S-VI	A.HLKG.I..	Q..E..H.H.	N.E

Repeat	**2**					
*At*GL1	VNKGNFTEQE	EDLIIRLHKL	LGNRWSLIAK	RVPGRTDNQV	KNYWNTHLSK	K
*At*CD2	.KR.......	....	.....*			
*At*myb1	LIRNS...V.	DQA..AA.AI	H..K.AV...	LL......AI	..H..SA.RR	R
*Am*305	.RR..I.PE.	QL..ME..AK	W.....K...	TL......EI	....R.RIQ.	H
*Am*340	.RR..I.PE.	QL..ME..AK	W.....K...	HL......EI	....R.RIQ.	H
*Am*330	LKR.....E.	DEI..K..S.	...K.....A	.L......EI	.......IKR	.
*Am*315	LK..PL..M.	.NQ..E..AH	......K..L	HI......EI	.......IK.	.
*Am*308	LKR.....E.	DE...K..S.	...K.....G	.L......EI	.......IRR	.
*Am*306	IKR.D...H.	.KM..H.QA.	.....AA..S	YL.H....DI	........K.	.
*Zm*C1	IRR..ISYD.	........R.	.........G	.L......EI	.....ST.GR	R
*Zm*P1	.KR..ISKE.	..I..K..AT	.........S	HL......EI	.....SH.SR	Q
*Zm*1	LKR....DE.	.EA.....G.	...K..K..A	CL......EI	..V.....K.	.
*Zm*38	LKR....AD.	D.L.VK..S.	...K.....A	.L......EI	.......VRR	.
*Hv*1	LKR...SHE.	.EL..K..S.	...K.....G	.L......EI	.......IRR	.
*Hv*33	LKR.C.SQQ.	..H.VA..QI	......Q..S	HL......EI	..F..SCIK.	.
*Dm*myb	IK.TAW..K.	DEI.YQA.LE	...Q.AK...	.L......AI	..H..STMRR	.
*Hs*myb	.K.TSW..E.	DRI.YQA..R	.....AE...	LL......AI	..H..STMRR	.

Fig. 2. A comparison of the myb repeats of plant myb proteins. The second and third myb repeats of the *Drosophila* and human c-mybs are shown for comparison. *At* = *Arabidopsis thaliana*, *Am* = *Antihrrhinum majus*, *Zm* = *Zea mays*, *Hv* = *Hordeum vulgare*, *Dm* = *Drosophila melanogaster*, *Hs* = *Homo sapiens*. References for the sources of the sequences are given in the text

4 *GL1* Is an Myb Gene Regulating Trichome Development in *Arabidopsis*

Mutations in the *GL1* gene prevent the initiation of most trichomes. No other phenotypes have been detected in *gl1-1* mutant plants. The *GL1* gene has been cloned by T-DNA tagging, and the *gl1-1* allele was shown to be a deletion of the entire *GL1* coding region (Oppenheimer et al. 1991). Thus, *GL1* encodes a function specifically required for trichome initiation.

DNA sequence analysis revealed that the *GL1* gene encodes a typical plant myb protein containing two repeats of the myb DNA-binding consensus and an acidic C-terminal domain (Oppenheimer et al. 1991). No sequence similarity was found outside of the DNA-binding region between *GL1* and the myb genes of any other organism. Two introns are present, located in the same positions as the introns of the maize *C1* gene (Fig. 3; Oppenheimer et al. 1991).

The observation that *GL1* encodes an myb protein demonstrates that, in plants as well as in animals, myb genes play a role in controlling cellular differentiation. Do additional myb genes regulate other differentiation events in *Arabidopsis*? Under low stringency hybridization conditions, numerous *Arabidopsis* genomic DNA fragments hybridize to a *GL1* myb domain probe (Marks et al. 1991). We have isolated and partially sequenced genomic clones of five *Arabidopsis* myb genes in addition to *GL1* (Oppenheimer et al. 1991; D.G. Oppenheimer unpubl. data). We have studied one of these genes, CD2, in more detail. The myb domain of CD2 exhibits greater similarity to the *GL1* myb domain than is seen for any of the other *Arabidopsis* myb genes examined. When presumptive regulatory sequences from the 5′ end of the CD2 gene are used to drive a β-glucuronidase (GUS) reporter gene, expression is primarily detected in root apical meristems (D.G. Oppenheimer unpubl. data). GUS expression is still detectable as far

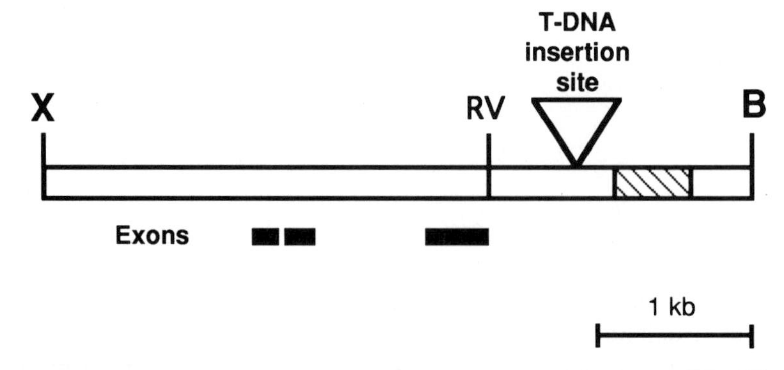

Fig. 3. Structure of the *GL1* gene (Oppenheimer et al. 1991; Larkin et al. in press). The exons are indicated by the *black boxes*. The location of the 492-bp restriction fragment containing the downstream enhancer described in the text is indicated by *diagonal cross-hatching*. *X* XhoI site; *RV* EcoRV site; *B* BamHI site

back as the zone of root hair initiation. Preliminary in situ hybridization results using a CD2-specific probe are consistent with this expression pattern. These results suggest that CD2 function is required in or near the root meristem. We are currently attempting to determine the function of the CD2 gene through the introduction of antisense and overexpression constructs into *Arabidopsis* plants.

5 A Functionally Required Enhancer Directing *GL1* Expression in Leaf Primordia is Located Downstream of the *GL1* Polyadenylation Site

The level of the *GL1* transcript is extremely low. To date, the *GL1* transcript has only been detected by PCR amplification of cDNA. Although attempts to localize the *GL1* message by in situ hybridization are in progress, it has so far been necessary to use indirect methods to deduce the *GL1* expression pattern. In preliminary studies reported previously (Oppenheimer et al. 1991), 5′ sequences from the putative *GL1* promoter were fused to a GUS reporter gene. Transgenic plants containing this construct expressed the reporter gene in the minute paired stipules present at the base of each leaf (Oppenheimer et al. 1991). This unexpected result was not consistent with the simplest models of trichome development, in which the GL1 protein would function directly in the protodermal cells fated to become trichomes.

However, one piece of evidence already suggested that the regulation of *GL1* might be complex. The T-DNA insertion allele *gl1-43*, used in the cloning of *GL1*, has trichomes on the leaves, but no trichomes on the stem. This allele is the result of an insertion of an estimated 65 kb of T-DNA sequences into a position 581 bp downstream of the major *GL1* polyadenylation site (Fig. 2; Oppenheimer et al. 1991; M.D. Marks unpubl. data). To examine the basis for the *gl1-43* phenotype, a series of deletions of the region downstream of the *GL1* gene were constructed (Larkin et al. in press). The starting point for these deletions was a 4.5-kb XhoI-BglII restriction fragment that was known to contain a functional *GL1* gene. The 3′ deletion constructs, which still included the entire *GL1* coding region, were introduced into *gl1-1* mutant plants and assayed for the production of trichomes on leaves and stems.

Constructs in which sequences were deleted beyond a point 1098 bp downstream of the polyadenylation site completely failed to complement the trichomeless phenotype of the *gl1-1* mutation (Larkin et al. in press). Additional deletion constructs indicated that at least a portion of the essential downstream sequences must be located between 946 and 1098 bp downstream of the polyadenylation site. A 492-bp fragment spanning this region restores *GL1* function when present in either orientation in a construct from which virtually all other downstream sequences have been eliminated. These results indicate that an enhancer sequence located approximately 1 kb downstream of the *GL1* coding region is required for *GL1* function (Fig. 2). All constructs that are not fully wild type affect both leaf and stem trichomes (Larkin et al. in press). These results provide no evidence for a stem-specific regulatory element. Apparently, the *gl1-43* stem-specific mutant phenotype results from the novel sequences introduced by the T-DNA insert.

When the same 492-bp fragment that is capable of restoring *GL1* function is inserted downstream of the *GL1* 5′/GUS gene, the expression pattern of the GUS reporter gene is altered. In constructs containing the downstream enhancer sequence, GUS expression is detected in leaf primordia and developing trichomes in addition to stipules. Since the same downstream sequences that are required for *GL1* function contain an enhancer directing reporter gene expression in leaf primordia, it seems likely that *GL1* expression in leaf primordia is necessary for trichome initiation (Larkin et al. in press).

6 The Overexpression Phenotype of *GL1*

A construct in which expression of the *GL1* coding region is driven by the Cauliflower Mosaic Virus (CaMV) 35S promoter (35S*GL1*) is capable of complementing the *gl1-1* mutant phenotype, although the number of trichomes produced in these plants is less than that of wild type. This result indicates that the 35S*GL1* construct makes a functional GL1 product. Since the CaMV 35S promoter is expressed in a wide variety of tissues (Benfey and Chua 1990), yet all epidermal cells in these plants do not differentiate as trichomes, it appears that the expression of *GL1* in an epidermal cell is not sufficient to cause the cell to differentiate as a trichome.

Unexpectedly, when the 35S*GL1* construct is introduced into wild-type *Arabidopsis* plants, the number of trichomes is reduced relative to that of wild-type plants (Larkin et al. in preparation). Wild-type plants (ecotype Columbia) have approximately 30 ± 5 trichomes on each of the first two leaves; while 6 of the 7 independent familes of 35S*GL1* plants examined had between 6.2 ± 1.6 and 18.7 ± 5 trichomes on each of the first two leaves. One family had an essentially wild-type number of trichomes per leaf. Many of the plants homozygous for 35S*GL1* also lacked stem trichomes, although this phenotype was not fully penetrant. In situ hybridization with a *GL1*-specific probe indicates that the 35S*GL1* transcript is expressed strongly in all regions of the shoot apex, including the stipules and young leaf primordia. This result indicates that the 35S*GL1* transgene is being expressed, at least at the RNA level, in the tissues where we believe *GL1* function is required (see above). Thus, it appears that the reduction in trichome number observed in 35S*GL1* plants cannot be explained by "cosuppression" of transcript levels of the endogenous *GL1* gene and the

35S*GL1* transgene, as has been observed in several instances (Matzke and Matzke 1990; Napoli et al. 1990).

The in situ hybridization results indicate that the level of the 35S*GL1* transcript is much higher than the wild-type levels of *GL1* transcript in stipules and developing leaf primordia. If levels of the GL1 protein are correspondingly increased, then overexpression of GL1 apparently results in a decrease in the number of trichomes initiated. One possible explanation for this observation is that the GL1 protein interacts with another factor(s) to form a heteromeric complex regulating trichome initiation. Uncomplexed GL1 might be able to bind to the target promoters, but might not be able to activate transcription except as part of this complex. If the concentration of the other factor(s) needed to form the complex is limiting, then in the presence of a vast excess of free GL1 most promoters would bind uncomplexed GL1, and transcription from the target promoter(s) would be "squelched" (Ptashne and Gann 1990). Plants containing the 35S*GL1* transgene do not exhibit any obvious phenotypes unrelated to trichome development. Thus, any factor(s) with which GL1 interacts must be fairly specific for trichome development.

One other important aspect of the 35S*GL1* phenotype remains to be mentioned. A few trichomes are produced on the cotyledons and on the abaxial surface of the first leaf pair in 35S*GL1* plants, locations which almost never produce trichomes in wild-type plants. However, only a few epidermal cells in these locations become trichomes. Thus, it appears that a few cells in the adaxial epidermis of the cotyledons and on the abaxial epidermis of the first two leaves are competent to respond to the *GL1* gene product by differentiating as trichomes. The factors limiting the competence of other epidermal cells in these locations to develop as trichomes in response to ectopic *GL1* expression remain unknown, although the *TTG* gene product is a good candidate (see below).

7 New Light on the Complex Phenotype of *ttg* Mutants

The issue of competence for trichome development is closely related to the question of how trichome precursor cells are selected. Less than 1 in 200 cells in the mature adaxial epidermis of an *Arabidopsis* leaf is a trichome (J. Larkin unpubl. observ.). Why do only a small subset of epidermal cells become trichomes? Recent results from our laboratory, as well as a recent report by Lloyd et al. (1992), suggest that the *TTG* gene plays a key role in both selecting the trichome precursor cells and in determining the competence of epidermal cells in various tissues to initiate trichome development.

Like *GL1*, *TTG* has been implicated in trichome initiation by virtue of the failure of *ttg* mutant plants to produce trichomes in most locations (Table 1). Eight *ttg* alleles were reported (Koorneef 1981; Koorneef et al. 1982). The most generally available of these alleles, *ttg-1*, produces essentially no trichomes on the surfaces of leaves. However, *ttg* mutants differ from *gl1* mutants in that the *ttg* mutant phenotype is pleiotropic; all eight of the *ttg* alleles discussed by Koornneeff are deficient in anthocyanin biosynthesis and fail to produce seed coat mucilage, in addition to the defect in trichome development (Koornneef 1981). Given the frequency of isolation of alleles with this complex phenotype, these alleles are probably representative of the phenotype due to a complete loss of the *TTG* function. This pleiotropic phenotype suggests that *TTG* is a regulatory gene that plays a major role in several rather different pathways.

What is the relative order of function of the *GL1* and *TTG* genes in the trichome initiation pathway? The pleiotropic phenotype of *TTG* might indicate that *TTG*, having a more general function than *GL1*, acts upstream of *GL1* in a linear pathway. Thus, *TTG* expression might be necessary for the expression of *GL1* in leaves, for the expression of some other transcription factor in anthocyanin-producing tissues, and for the expression of yet another transcription factor in the developing seed coat. Alternatively, the complex phenotype of *ttg* mutants could be the result of a combinatorial interaction of the *TTG* gene product with different transcription factors expressed in the different tissues, each combination triggering the transcription of a different set of target genes. Thus, the TTG and GL1 proteins might bind to the same promoters, with both proteins being required to activate transcription of genes required for trichome differentiation. An example of this type of interaction is provided by the combined action of C1 and R to induce transcription from the promoter of the maize *Bz1* gene (Klein et al. 1989; Roth et al. 1991). It is even possible, using more elaborate versions of the combinatorial interaction model, to imagine a pathway in which *TTG* acts downstream of *GL1*.

The 35S*GL1* construct, which expresses the *GL1* coding region ectopically under the control of the CaMV 35S promoter, provides a tool to distinguish the first of these models from the latter two. In the first model, *TTG* expression is required for *GL1* expression. This model predicts that *TTG* function would not be required in plants ectopically expressing *GL1*. In contrast, the other two models predict that *TTG* function would be required for trichome initiation even in plants with ectopic *GL1* expression. We have found that plants of the genotypes *ttg*/*ttg*; 35S*GL1*/35S*GL1* and *ttg*/*ttg*; 35S*GL1*/+ fail to produce trichomes, demonstrating that *TTG* does not act upstream of *GL1* in the trichome initiation pathway (Larkin et al. in preparation). Thus, *TTG* must act either at the same point in the trichome initiation pathway as *GL1*, or at some point downstream of *GL1*.

In the course of these experiments involving 35S*GL1* and *ttg*, an unexpected observation was made. Plants heterozygous for *ttg* and containing at least one copy of 35S*GL1* were found to produce clusters of adjacent trichomes on their leaves at a high frequency (Larkin et al. in preparation). Such clusters are quite rare on wild-type leaves; approximately 0.6% of the trichomes on wild-type leaves is adjacent to another trichome. However, in plants of the genotype *ttg*/*TTG*; 35S*GL1*/+, 28% of the trichomes on the first two leaves is part of a cluster of two or more adjacent trichomes. The total number of trichomes on each of the first two leaves of *ttg/TTG*; 35S*GL1*/+ plants is also greater than the number of trichomes found on each of the first two leaves of 35S*GL1* plants. Genetic analysis of the segregation of these phenotypes in backcrosses and in F3 families demonstrated that the clustered trichome phenotype was due to the *ttg* mutation or a mutation very tightly linked to *ttg*. Scanning electron microscopy revealed that these clusters of trichomes were due to the simultaneous initiation of trichome development in two or more adjacent cells (Larkin et al. in preparation).

We have isolated a new EMS-induced *ttg* allele (*ttg-10*) that produces large numbers of abortive trichomes on the early leaves. A large proportion of these trichomes occur as clusters of trichomes in various stages of development, ranging from branched trichomes to barely expanded aborted trichomes resembling those found on *gl2* plants (Larkin et al. in preparation). Another EMS-induced *ttg* allele, *ttg-9*, was obtained from Dr. George Haughn (Haughn and Somerville 1988). Plants homozygous for this allele produce a few abortive trichomes that occasionally occur in clusters (Fig. 4). These results suggest that *TTG* is involved in selecting the trichome precursor cells, in addition to its overt role in trichome initiation (see below). The production of trichomes aborted at various developmental stages suggests

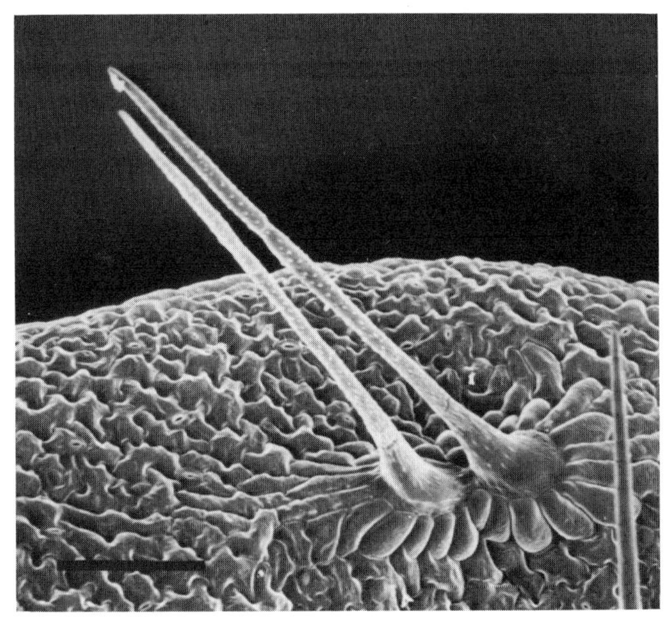

Fig. 4. A cluster of two unbranched trichomes on the adaxial leaf surface of a *ttg-9* plant (Haughn and Somerville 1988). Unbranched trichomes are very rare on wild-type leaves. *Bar* = 100 μm

that *TTG* is required for the maintenance of trichome development, as well as for initiation.

Another recent result illuminates the role of *TTG* in both trichome development and anthocyanin pigment formation. Lloyd et al. (1992) have expressed the maize *R* gene in *Arabidopsis* using the CaMV 35S promoter (35S*R*). They have found that *ttg* mutant plants containing the 35S*R* transgene produce both trichomes and anthocyanins. Trichomes were produced on many tissues where trichomes are not normally found, including petals, stamens, and pistils, in addition to producing trichomes in all normal locations (Lloyd et al. 1992). When 35S*R* was expressed in *TTG* plants, the leaves of these plants produced two- to fivefold more leaf trichomes than in wild-type plants. As described above, the maize *R* gene is an myc, helix-loop-helix homologue that induces the genes encoding anthocyanin biosynthetic enzymes in conjunction with *C1*, an myb homologue. The results of Lloyd et al. (1992) suggest that the *TTG* gene product may be an *R* homologue, although it is also possible that *TTG* activates an *R* homologue. The observation that expression of the maize *R* gene, a potential *TTG* homologue, in *Arabidopsis* results in the production of trichomes in locations where they are not normally found suggests that *TTG* may be involved in controlling the competence of protodermal cells to form trichomes.

8 A Model for the Trichome Initiation Pathway

A genetic model for the trichome initiation pathway is presented in Fig. 5. In this model, *GL1* and *TTG* are required together to initiate trichome development. In addition, *TTG* is envisioned to play a *GL1*-independent role in inhibiting trichome development in adjacent protodermal cells ("lateral inhibition" in Fig. 5). Complete loss of *TTG* function would pre-

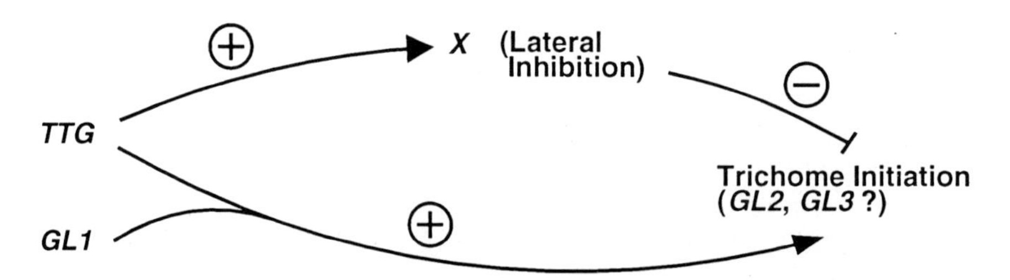

Fig. 5. Genetic model for the trichome initiation pathway. *TTG* and *GL1* are assumed to cooperate to promote trichome initiation, possibly by activating the *GL2* and *GL3* genes. *TTG* also is envisioned to suppress trichome initiation in adjacent cells via an independent pathway (lateral inhibition). *X* represents an unknown gene or genes on the lateral inhibition pathway. See text for details

vent trichome initiation by blocking the *GL1*-dependent initiation pathway. A moderate reduction in *TTG* function would allow trichome development to proceed, but in the presence of excess GL1 from a 35S*GL1* transgene, some adjacent cells would escape lateral inhibition and form trichomes. A more severe reduction in the level or activity of the *TTG* gene product might allow the formation of clusters of abortive trichomes without the presence of excess *GL1*, as are seen in the *ttg-9* and *ttg-10* alleles. Our model predicts that mutations in components of the lateral inhibition pathway downstream of *TTG* ("gene X" in Fig. 5) should also produce clusters of trichomes. We are currently screening for such mutations.

Precedent for the lateral inhibition pathway is found in bristle development in *Drosophila* (Held 1991; Campuzano and Modolell 1992) and in some cell fate decisions in *Caenorabditis* (Sternberg and Horvitz 1984). In these cell fate decisions, several cells have been shown to have equivalent development potential, even though only a single cell will ultimately be selected for differentiation. This cell is selected by negotiations between the cells of the "equivalence group" (Stern 1954; Sternberg and Horvitz 1984; Held 1991). The protoderm of the *Arabidopsis* leaf may contain trichome equivalence groups, any member of which could become a trichome. The first cell in the equivalence group to become committed to trichome development would then inhibit its neighbors from becoming trichomes via the *TTG*-dependent lateral inhibition pathway. Precedent also exists in both *Drosophila* (Bang and Posakony 1992) and *Caenorabditis* (Sternberg and Horvitz 1984) for changes in gene dosage as small as twofold to have a dramatic effect on cell fate, as we have observed for *ttg* heterozygotes in a 35S*GL1* background.

Other explanations for the occurrence of trichome clusters are also possible. For example, *TTG* could be required to cause the trichome precursor cell to stop dividing, in conjunction with the commitment of the precursor cell to differentiate as a trichome. Reducing the activity of the *TTG* gene product might allow the precursor cell to divide one or two extra times after becoming committed to trichome differentiation, resulting in a cluster of trichomes. This model requires the cells of the trichome cluster to be clonally related. Since the cell-cell communication model outlined above does not require a clonal relationship between the cells in a cluster, genetic mosaic analysis could provide a test to distinguish between the two models. Note that the production of trichomes aborted at a wide range of stages in *ttg-9* and *ttg-10* mutants shows that TTG function is required even after cell division ceases in the trichome precursor cell. Thus, even this alternative model predicts that *TTG* has more than one function in the trichome developmental pathway.

Do the *GL1* and *TTG* gene products interact directly to promote trichome initiation? This hypothesis is particularly attractive because *GL1* is a myb gene homologous to the *C1* gene of maize, and Lloyd et al. (1992) have presented evidence that *TTG* is functionally equivalent to the *R* and *B* genes of maize, the products of which appear to interact directly with the *C1* product (Goff et al. 1992). Furthermore, we have presented evidence, based on the phenotype of plants containing the 35S*GL1* transgene, suggesting that the *GL1* gene product interacts directly with another factor involved in trichome initiation. The *TTG* gene product is a likely candidate for this factor. However, it should be noted that while 35S*GL1* plants have a reduced number of trichomes in comparison with wild type, they have normal anthocyanin pigmentation and seed coat mucilage. Perhaps the TTG protein is not limiting in the tissues where anthocyanins and mucilage are produced. Alternatively, perhaps some other factor involved in trichome initiation has been overlooked. We are currently searching for proteins that directly interact with GL1.

What are the genes directly downstream from *GL1* and *TTG* in the trichome developmental pathway? The mutations with the next earliest block in trichome development are mutations in the *GL1* and *GL3* genes (Table 1). The trichome phenotypes of *gl2* and *gl3* mutants are quite similar. Mutants in both genes have fewer than normal trichomes, and the trichomes that are produced are less branched than normal. The most severe alleles of *gl2* produce large numbers of abortive trichomes blocked very early in cell expansion. *gl3* mutants also produce a few abortive trichomes. The abortive trichomes on the leaves of *gl2* mutants are of particular interest. Many of these abortive trichomes produce an elongated "branch" in the plane of the leaf (Marks et al. 1991). Similar abortive trichomes are found in both apparently hypomorphic (partial loss of function) alleles of *TTG*, *ttg-9* and *ttg-10*, although the abortive trichomes on *gl2* mutants never occur in clusters. The *gl2* phenotype is also pleiotropic. All *gl2* alleles in our possession lack seed coat mucilage, indicating that *GL2* participates with *TTG* in at least one other pathway. These observations suggest that *GL2* may be just downstream of *TTG* in the trichome initiation branch of the pathway. William Rerie in our laboratory has recently cloned the *GL2* gene by T-DNA tagging (unpubl. observ.). The cloning of *GL2* should afford us further insights into the trichome development pathway.

Have the genes regulating trichome development in *Arabidopsis* played a role in the evolution of more complex trichome types? A great diversity of trichome types are found among the angiosperms, including all combinations of single-celled and multicellular, glandular and nonglandular types (Rollins and Banerjee 1976; Theobald et al. 1979). These widely varied structures play a number of different roles in defense against herbivores and protection from the environment (Uphof 1962; Johnson 1975). The evolutionary relationship among these structures is unclear at present. However, it is interesting to note that some genera of plants, such as *Quercus* (oaks) (Harden 1979; Thomson and Mohlenbrock 1979) and *Polygonum* (Lersten and Curtis 1992) have some trichomes consisting of clusters of single-celled hairs joined at the base. It is possible that these trichome types originated by genetic changes similar to those we have studied in *Arabidopsis*.

9 Conclusion

Although we are still at an early stage in our understanding of the control of trichome development, several themes are already apparent. First, the role of *GL1* in trichome initiation extends the already wide range of functions of the plant myb genes to include the control of cell differentiation. It seems clear that the myb gene family in plants has undergone a diversification of function unprecedented in animals. Secondly, our genetic analysis suggests that trichome development does not proceed by a simple linear pathway. The multiple roles of *TTG* in trichome development, the evidence for factors interacting with the *GL1* gene product, and the incomplete overlap between the pleiotropic phenotypes of *TTG* and *GL2* all suggest that combinatorial interactions of regulatory factors play an important part in this pathway. Finally, our model for the trichome developmental pathway suggests that cell-cell communication may be important in selecting the trichome precursor cells. In any event, it is clear that the study of the development pathway for this single cell type has revealed unexpected complexities.

Acknowledgments. This work was supported by NSF (DCB-9118306), USDA (91-37304-6470) and the University of Nebraska Center for Biotechnology.

References

Bang AG, Posakony JW (1992) The *Drosophila* gene *Hairless* encodes a novel basic protein that controls alternative cell fates in adult sensory organ development. Genes Dev 6:1752–1769

Denfey PN, Chua N-H (1990) The cauliflower mosaic virus 35S promoter: combinatorial regulation of transcription in plants. Science 250:959–966

Biesalski HK, Doepner G, Tzimas G, Gamulin V, Schroder HC, Batel R, Nau H, Muller WEG (1992) Modulation of *myb* gene expression in sponges by retinoic acid. Oncogene 7:1765–1774

Campuzano S, Modolell J (1992) Patterning of the *Drosophila* nervous system: the *achaete-scute* gene complex. Trends Genet 8:202–208

Chandler VL, Radicella JP, Robbins J, Chen J, Turks D (1989) Two regulatory genes of the maize anthocyanin pathway are homologous: isolation of *B* using *R* genomic sequences. Plant Cell 1:1175–1183

Coe EH, Neuffer MG, Hoisington DA (1988) The genetics of corn. In: Sprague GF, Dudley JW (eds) Corn and corn improvement. Agronomy Monograph 18, 3rd edn. American Society of Agronomy, Madison, pp 81–258

Feenstra WJ (1978) Contiguity of linkage groups I and IV as revealed by linkage relationship of two newly isolated markers *dis-1* and *dis-2*. Arabidopsis Inf Serv 15:35–38

Foos G, Grimm S, Klempnauer K-H (1992) Functional antagonism between members of the myb family: B-*myb* inhibits v-*myb*-induced gene activation. EMBO J 11:4619–4629

Frampton J, Leutz A, Gibson T, Graf T (1989) DNA-binding domain ancestry. Nature 342:134

Goff SA, Cone KC, Fromm ME (1991) Identification of functional domains in the maize transcriptional activator *C1*: comparison of wild-type and dominant inhibitor proteins. Genes Dev 5:298–309

Goff SA, Cone KC, Chandler VL (1992) Functional analysis of the transcriptional activator encoded by the maize *B* gene: evidence for direct functional interaction between two classes of regulatory proteins. Genes Dev 6:864–875
Graf T (1992) Myb: a transcriptional activator linking proliferation and differentiation in hematopoietic cells. Curr Opinion Genet Dev 2:249–255
Grotewold E, Athma P, Peterson T (1991) Alternatively spliced products of the maize P gene encode proteins with homology to the DNA binding domain of myb-like transcription factors. Proc Natl Acad Sci USA 88:4587–4591
Harden JW (1979) Patterns of variation in foliar trichomes of eastern North American *Quercus*. Am J Bot 66:576–585
Haughn GW, Somerville CR (1988) Genetic control of morphogenesis in *Arabidopsis*. Dev Genet 9:73–89
Held LI (1991) Bristle patterning in *Drosophila*. BioEssays 13:633–640
Herman PL, Marks MD (1989) Trichome development in *Arabidopsis thaliana*. II. Isolation and complementation of the *GLABROUSl* gene. Plant Cell 1:1051–1055
Jackson D, Culianez-Macia F, Prescott AG, Roberts K, Martin C (1991) Expression patterns of *myb* genes from *Antirrhinum* flowers. Plant Cell 3:115–125
Johnson HB (1975) Plant pubescence: an ecological perspective. Bot Rev 41(3): 233–258
Katzen AL, Kornberg TB, Bishop JM (1985) Isolation of the proto-oncogene c-*myb* from *D. melanogaster*. Cell 41:449–456
Klein TM, Roth BA, Fromm ME (1989) Regulation of anthocyanin biosynthetic genes introduced into intact maize tissues by microprojectiles. Proc Natl Acad Sic USA 86:6681–6685
Koornneef M (1981) The complex syndrome of *ttg* mutants. Arabidopsis Inf Serv 18:45–51
Koornneef M, Dellaert SWM, van der Veen JH (1982) EMS- and radiation-induced mutation frequencies at individual loci in *Arabidopsis thaliana* (L.) Heynh. Mutat Res 93:109–123
Koornneef M, van Eden J, Hanhart CJ, Stam P, Braaksma FJ, Feenstra WJ (1983) Linkage map of *Arabidopsis thaliana*. J Hered 74:265–272
Lersten NR, Curtis JD (1992) Foliar anatomy of *Polygonum* (Polygonaceae): survey of epidermal and selected internal structures. Plant Syst Evol 182:71–106
Lloyd AM, Walbot V, Davis RW (1992) Anthocyanin production in dicots activated by maize anthocyanin-specific regulators, *R* and *Cl*. Science 258:1773–1775
Ludwig SR, Habera LF, Dellaporta SL, Wessler SR (1989) *Lc*, a member of the maize *R* gene family responsible for tissue-specific anthocyanin production, encodes a protein similar to transcription factors and contains the myc homology region. Proc Natl Acad Sci USA 86:7092–7096
Lüscher B, Eisenman RN (1990) New light on Myc and Myb. Part II. Myb. Genes Dev 4:2235–2241
Lüscher B, Eisenman RN (1992) Mitosis-specific phosphorylation of the nuclear oncoprotein myc and myb. Cell Biol 118:775–784
Lüscher B, Christenson E, Litchfield DW, Krebs EG, Eisenman RN (1990) Myb DNA binding inhibited by phosphorylation at a site deleted during oncogenic activation. Nature 344:517–522
Marks MD, Esch JJ (1992) Trichome formation in *Arabidopsis* as a genetic model for studying cell expansison. Current Top Plant Biochem Physiol 11:131–142
Marks MD, Feldmann KA (1989) Trichome development in *Arabidopsis thaliana*. I. T-DNA tagging of the *GLABROUSl* gene. Plant Cell 1:1043–1050
Marks MD, Esch J, Herman P, Sivakumaran S, Oppenheimer D (1991) A model for cell-type determination and differentiation in plants. In: Jenkins GI, Schuch W (eds) Molecular biology of plant development. The Company of Biologists Limited, Cambridge, pp 77–87
Marocco A, Wissenbach M, Becker D, Paz-Ares J, Saedler H, Salamini F (1989) Multiple genes are transcribed in *Hordeum vulgare* and *Zea mays* that carry the DNA binding domain of the myb oncoproteins. Mol Gen Genet 210:183–187

Matzke MA, Matzke AJM (1990) Gene interactions and epigenetic variation in transgenic plants. Dev Genet 11:214–223

McCarty DR, Carson CB, Stinard PS, Robertson DS (1989) Molecular analysis of *viviparous-1*: an abscisic acid insensitive mutant of maize. Plant Cell 1:523–532

Mucenski ML, McLain K, Kier AB, Swerdlow SH, Schreiner CM, Miller TA, Pietryga DW, Scott WJJ, Potter SS (1991) A functional c-*myb* gene is required for normal nurine fetal hepatic hematopoiesis. Cell 65:677–689

Napoli C, Lemieux C, Jorgensen R (1990) Introduction of a chimeric chalcone synthase gene into petunia results in reversible co-suppression of homologous genes in *trans*. Plant Cell 2:279–289

Nomura N, Takahashi M, Matsui M, Ishii S, Date T, Sasamoto S, Ishizaki R (1988) Isolation of human c-DNA clones of myb-related genes, A-myb and B-myb. Nucleic Acids Res 16:11075–11089

Oppenheimer DG, Herman PL, Esch J, Sivakumaran S, Marks MD (1991) A *myb*-related gene required for leaf trichome differentiation in *Arabidopsis* is expressed in stipules. Cell 67:483–493

Oppenheimer DG, Esch J, Marks MD (1992) Molecular genetics of *Arabidopsis* trichome development. In: Verma DPS (ed) Control of plant gene expression. CRC Press, Boca Raton, pp 275–286

Paz-Ares J, Ghosal D, Wienand U, Peterson PA, Saedler H (1987) The regulatory *c1* locus of *Zea mays* encodes a protein with homology to *myb* proto-oncogene products and with structural similarities to transcriptional activators. EMBO J 6(12):3553–3558

Peters CWB, Sipel AE, Vingron M, Klempnauer KH (1987) *Drosophila* and vertebrate *myb* proteins share two conserved regions, one of which functions as a DNA-binding domain. EMBO J 6:3085–3090

Ptashne M, Gann AF (1990) Activators and targets. Nature 346:329–331

Rollins RC, Banerjee UC (1976) Trichomes in studies of the Crucifereae. In: Vaughan JG, Macleod AJ, Jones BMG (eds) The biology and chemistry of the Cruciferae. Academic Press, New York, pp 145–166

Roth BA, Goff SA, Klein TM, Fromm ME (1991) *C1*- and *R*-dependent expression of the maize *Bz1* gene requires sequences with homology to mammalian myb and myc binding sites. Plant Cell 3:317–325

Roussel M, Saule S, Lagrou C, Rommens C, Beug H, Graf T, Stehelin D (1979) Three new types of viral oncogene of cellular origin specific for haematopoietic cell transformation. Nature 281:452–455

Sakura H, Kanei-Ishii C, Nagase T, Nakagoshi H, Gonda TJ, Ishii S (1989) Delineation of three functional domains of the transcriptional activator encoded by the c-myb protooncogene. Proc Natl Acad Sci USA 86:5758–5762

Shinozaki K, Yamaguchi-Shinozaki K, Urao T, Koizumi M (1992) Nucleotide sequence of a gene from *Arabidopsis thaliana* encoding a *myb* homologue. Plant Mol Biol 19:493–499

Stern C (1954) Two or three bristles. Am Sci 42:213–247

Sternberg PW, Horvitz HR (1984) The genetic control of cell lineage during nematode development. Annu Rev Genet 18:489–524

Stober-Grasser U, Brydolf B, Bin X, Grasser F, Firtel RA, Lipsick JS (1992) *Dictyostelium* MYB: evolution of a DNA-binding domain. Oncogene 7:589–596

Styles ED (1970) Functionally duplicate genes conditioning anthocyanin formation in maize. Can J Genet Cytol 12:397

Theobald WL, Krahulik JL, Rollins RC (1979) Trichome description and classification. In: Metcalfe CR, Chalk L (eds) Anatomy of the dicotyledons, vol I. Clarendon Press, Oxford, pp 40–53

Thomson PM, Mohlenbrock RH (1979) Foliar trichomes of *Quercus* subgenus *Quercus* in the eastern United States. J Arnold Arb Harv Univ 60:350–366

Tice-Baldwin K, Fink GR, Arndt KT (1989) *BAS1* has an Myb motif and activates *HIS4* transcription only in combination with *BAS2*. Science 246:931–935

Uphof JCT (1962) Plant hairs. Encyclopedia of Plant Anatomy. Borntraeger, Berlin, pp 148–177

Printing: Saladruck, Berlin
Binding: Buchbinderei Lüderitz & Bauer, Berlin